Evangelos Pantazis
Designing with Multi-Agent Systems

Also of interest

Computational Physics.
With Worked Out Examples in FORTRAN® and MATLAB®
Michael Bestehorn, 2023
ISBN 9783110782363, e-ISBN (PDF) 9783110782523

Methods for Scientific Research.
A Guide for Engineers
Vinay Prasad, 2023
ISBN 9783110625295, e-ISBN (PDF) 9783110625301

Multicriteria Decision Making.
Systems Modeling, Risk Assessment, and Financial Analysis for Technical Projects
Timothy Havranek and Doug MacNair, 2023
ISBN 9783110765649, e-ISBN (PDF) 9783110765861

Data Management for Natural Scientists.
A Practical Guide to Data Extraction and Storage Using Python
Matthias Hofmann, 2023
ISBN 9783110788402, e-ISBN (PDF) 9783110788433

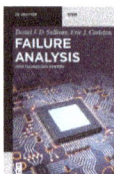

Failure Analysis.
High Technology Devices
Daniel J. D. Sullivan and Eric J. Carleton, 2022
ISBN 9781501524783, e-ISBN (PDF) 9781501524790

Evangelos Pantazis

Designing with Multi-Agent Systems

A Computational Methodology for Form-Finding
Using Behaviors

DE GRUYTER

Author
Evangelos Pantazis, Ph.d.
Topotheque Design Research
Uetlibergstrasse 123
8045 Zurich
Switzerland
vague@topotheque.com

This edition is supported by the Onassis foundation

ISBN 978-3-11-079704-6
e-ISBN (PDF) 978-3-11-079743-5
e-ISBN (EPUB) 978-3-11-079747-3

Library of Congress Control Number: 2023944095

Bibliographic information published by the Deutsche Nationalbibliothek
The Deutsche Nationalbibliothek lists this publication in the Deutsche Nationalbibliografie;
detailed bibliographic data are available on the Internet at http://dnb.dnb.de.

Abstract

The current architectural design models are in general not capable of integrating the design formation processes that are directly informed by building performance simulation and construction processes. Developing computational design methods, which consider building design holistically and help reduce the increasing complexity in the fields of architecture, engineering and construction (AEC), are considered crucial as we move rapidly towards the full digitalization of our built environment.

This book presents a computational design methodology for exploring architectural design alternatives in the early design stage based on the coupling of geometrical parameters with environmental and structural analysis, as well as autonomous construction processes. The methodology is implemented in a multi-agent systems framework, in which building elements are represented as agents. The framework considers the design-to-construction process holistically; and to facilitate an integrative design process, domain-specific data, which relates to different design phases (architectural, structural, environmental design), can be used to inform the agent's behaviors.

The proposed framework is applicable to the early design stage and especially design problems which traditionally require the close collaboration of architects and engineers, such as façade design and shell structures. The objective is to facilitate passing data between the different types of models (architectural, structural, environmental) and to enable architectural designers to explore larger solution spaces and gain deeper insight into complex building design decisions. To achieve this, a prototype design tool has been developed that allows the decomposition of design problems into agents and behaviors that automatically generate and evaluate design alternatives based on performance targets set by the designer in the form of heuristic functions. The agents operate within user-specified geometric and data environments, meaning their behavior and motion are informed by data sets related to geometric parameters and constraints or to environmental parameters and material properties. For the purposes of analysis and validation of the methodology, as well as to show how the developed tools and workflows can be used, a series of experimental designs are presented with cases that vary in complexity and scope. The work presented in this book was developed during my Ph.D. studies at the Viterbi School of Engineering at the University of Southern California between January 2014 and May 2019.

https://doi.org/10.1515/9783110797435-202

Preface

This book compiles research work on the use of computational methods for developing data-driven generative design strategies in architecture, whose express purpose is to extend the creative capacity of architects by integrating performance analysis in the early design stage. It brings together design, engineering and scientific perspectives and investigates how we might reconsider our design methods, seen through the lens of complexity and by utilizing the modular and distributed nature of multi agent systems.

It remains unapologetically transdisciplinary, both in its perspective and scope: it is a study on architectural theory, in conjunction with complexity theory, design computing theory and robotics. Its interests are inquisitive and projective, as well as analytical; it is about drafting things before their arrival, as much as correlating things as they are. The objective is not to produce a final solution, but rather to propose design workflows that focus on how engineering parameters and fabrication constraints can be represented abstractly and become design drivers that lead to complex, yet performative design outcomes that would be impossible to design and build using traditional methods.

The book starts out by looking into the creation and evolution of the term complexity in the 20th century and how it relates to computation and the fields of AEC, as well as their respective cultures – closely interrelated yet different. Computation is regarded as a logic of culture and, therefore, a logic of design and not just a means of developing new design tools. Computational logic is how different disciplines design but, by the same token, requires better design itself as it becomes more embedded in our lives. Consider complexity and its underlying principles as a special lens through which we can better understand how computational logic operates and how the buildings we design and construct, i.e., artificial complex systems, impact people across the world and affect our planet in the long term.

I emphasize the discrepancies of understanding complexity created by the multitude of definitions that have occurred in recent years across different disciplines. The book also includes a taxonomy of complexity that helps outline how certain concepts can be applied in the field of architecture. Based on this research, I propose a design methodology and apply it to a series of design experiments dealing with problems that typically require collaboration between architects and engineers. These experimental case studies are indicative of typical design problems, but by no means capture the full breadth of architectural design. The work concludes with a series of thoughts on future research directions. Like any other project that tries to achieve a certain holism, this thesis produces its own terminology (e.g., behavioral form finding) that becomes clearer the further along you progress. To help the reader on this journey, I have attempted to create a unifying visual language in the accompanying figures.

Since this work reflects on many disciplinary discourses, it is inevitable that some parts may seem opaque and others obvious, depending on the reader's background. What is important are the lines of connection between different concepts, design methods and their application. By stepping out of the architectural domain and the definitions of complexity offered by architects, such as R. Venturi, we discern an emerging picture of complexity that may be different to what has been predicted

https://doi.org/10.1515/9783110797435-203

within our field and has been transformative for other disciplines (i.e., tech industries, manufacturing etc.) in the last 30 years. We may then realize that this emerging picture is based upon models that have been formulated before our eyes and are more related to the mantra of "more is different" rather than "less is more." But what does this mean to us? What can we design with it? The answers lie in the theories and tools we develop, on the models and codes we use.

It is my firm belief that as architects and engineers zoom in on their process, upping the ante on precision and resolution, it becomes progressively harder to grasp the big picture. The price you must pay for analysis is the dissipation of synthesis. As such, design software may require the development of design theory as much as design theory requires the development of design software. In this work I assume that computation and digitalization in the AEC does not merely imply the use of code and machinery but a whole new way of thinking which, in turn, requires a new level of literacy among designers, including a new set of design abstractions. A review of the literature within the architectural realm in the last 20 years reveals that digital literacy has often been mistaken for the ability to operate digital design tools. Although being able to operate design tools is very important, it is not equal to understanding what a tool can and cannot do and how well it serves the theoretical model it was based upon.

Despite the acceleration of information technologies, the AEC industry has not picked up the pace since I started studying. It takes almost as long to build a complex project as it did 20 years ago. Although there are traces of lightning speed progress, we are still in the early stages of adopting sector-scale computation. How the algorithmic design tools within AEC will evolve and how our cultural systems will influence them and be influenced by them is still very open-ended. Theories and educational curricula will play an integral role in that. I hope that we will look back on this moment, when you could attend an architecture school without taking algorithm design or programming courses, as one curtailed by a bizarre intellectual paranoia. The same goes for obtaining an engineering degree without being fluent in any of the concepts central to the philosophy of technology or the underlying principles of complexity and contemporary art.

The next generation of design tools and buildings will require a different relationship to machines (computers, sensors and robots), as well as a more promiscuous figurative imagination. Adopting a holistic view becomes instrumental in navigating the information age we are living in. I consider the work presented here as a small step towards this direction.

Borrowing the words of D. Engelbart, a pioneer in the field of computing, "these days, the problem isn't how to innovate; it's how to get society to adopt the good ideas that already exist." My goal with this book is to take a step towards correlating ideas that exist within the domain of architecture, as well as outside of it, to develop a more holistic design approach towards architecture, one that allows designers to explore the full potential of multi-agent systems without introducing and compounding complexity, from the design to the construction process.

Contents

Part III: **Methodology**

Part IV: **Case studies**

Part V: **Evaluating alternatives**

Part VI: **Bibliography**

Part I: **Prologue**

Situating complexity within the AEC

1 Prologue

1.1 Finding simplicity among building complexity

It is arguable that complexity is one of the most challenging intellectual, scientific and technological topics of the twenty-first century [1]. Nowadays, the complexity of building design can be witnessed in mixed-use superstructures towering over modern cities in the United Arab Emirates, while geometric complexity has been celebrated in many sophisticated cultural and residential buildings around the world.

Undoubtedly, computational advancements have played a big role in the transformation of the built environment, from the materialization of drawings into their constructed forms – as introduced in the Renaissance – to the materialization of digital information [2]. The technological breakthroughs achieved throughout the different industrial revolutions have enabled the creation of different systems and have introduced multiple types of complexity (Figure 1.1). What is remarkable is the curtailing of the time span from the introduction to the full adoption of information technologies due to their ubiquitous nature. This time span has become even shorter with the wide adoption of sensor technologies, the Internet of things and robotics, among other technologies, that marked the onslaught of the Fourth Industrial Revolution.

The introduction of embedded building systems and sensor technologies is turning buildings from static structures into cyber-physical systems that can sense and respond to climatic or temporal changes of the environment [3, 4]. Architectural design thinking has not been greatly affected, even though information technologies have radically changed the way we design and construct our buildings. Unlike many scientific fields, such as biology and physics, where interest in computational methods and complexity theory arose due to the incapacity of existing methodologies to efficiently tackle complex problems, in architecture the first digital tools were not really used to solve design problems [5] but to automate the production of drawings of complex shapes.

With a few exceptions, in its early manifestations, digitally mediated architecture focused on geometry and approached complexity in a diagrammatic rather than a scientific manner. Digital design approaches remained rooted in representation-based design paradigms, instead of developing a deeper understanding of complexity and reconsidering the design process in the light of computation [6]. To date, each domain in architecture, engineering and construction (AEC) still provides largely independent solutions that are already defined outcomes before they are passed from one discipline to another. This lack of integration between disciplines and inefficiencies in the design process has resulted in a built environment that accounts for 40% of global energy consumption and up to 30% of global greenhouse gas emissions. Static building models that do not consider the parameter of time (a building's life cycle) are still primarily used. Therefore, up to 46% of the buildings' energy consumption is locked in for long periods of time due to the life span of these buildings [7].

As concerns about the environmental footprint of buildings have increased and the environmental performance standards have become more stringent, architectural de-

https://doi.org/10.1515/9783110797435-001

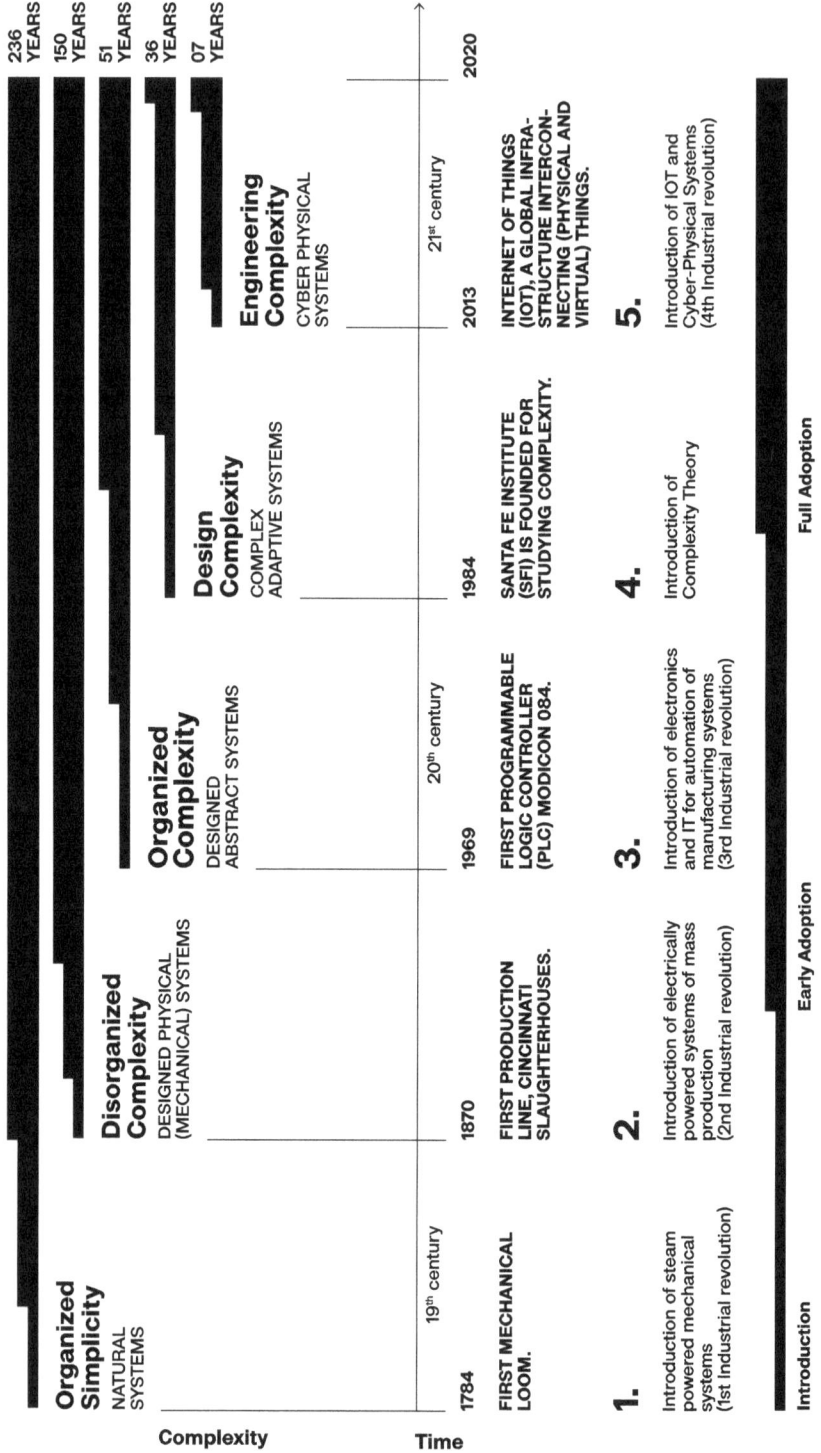

Figure 1.1: Timeline illustrating the introduction and full adoption of different types of complexity.

signers have gained the important responsibilities of employing designs and creativity to reduce energy use and to achieve more sustainable solutions. This is achievable by employing smart and creative design solutions to reduce building complexity. As A. Pottinger, lead structural engineer at Buro Happold Engineering, stated regarding the successful completion of the Abu Dhabi Louvre, a challenging project in terms of both architecture and engineering: "One of the absolutely overriding things we had to do was to find simplicity amongst the complexity. If we didn't do that the project wouldn't be buildable."

1.2 Are we reaching a new maturity in architectural design?

Architectural design thus far has been rooted in descriptive (perspective) modeling, which in turn is passed on to engineering disciplines so as to conduct further analysis prior to commencing construction. Advances in digital design and fabrication have presented an exciting opportunity to merge digital and physical tools and processes for constructing nonstandard building structures with better design performance. In Figure 1.2, a timeline of significant technologies that influenced the AEC shows the steadily growing pace of these changes from their first appearance in the 1950s until today. The wider access to technology and digital tools has also created a demand for architects to address the complex character of design problems in a more methodological and integrated way in order to provide sustainable solutions that can cater for environmental and structural parameters, as well as user preferences [4].

The ability to evaluate a larger amount of possible design alternatives (solution space) is essential in architectural design, as has been shown by Woodbury et al. [8] and Gero and Sosa [9]. To enable architects to explore larger solutions and generate designs that reduce the energy footprint of both the design-to-construction processes and the life cycle of buildings, computational design approaches are considered crucial to successfully manage the conflicting interdomain constraints of building design. Gero and Sosa argue that agent-based approaches can be used to increase the capacity of designers to explore a larger number of designs and also to enable them to develop complex designs by formulating design goals into agents' behaviors.

Early research efforts emphasized developing computer-aided design tools that reduced the complexities relating to drafting and the automation of drawing production, rather than developing new design methods and tools [10]. Subsequently, research in the field of design computing largely focused on bridging the gap between the physical and digital realms and the integration of fabrication and material constraints in the early design stage [11]. More recently, performance-based design emerged as an integrated approach, which allowed designers to consider environmental and structural parameters in the early design stage. This has brought a new maturity that promises to transcend the formal and geometric innovations, mainly driving the architects' interest in using digital technologies, and to transfer the focus on how computational techniques can be used to transcribe fundamental formative processes into architectural design [12].

CAD Technology Timeline

Time

Hardware	Time	Software
		MidJourney/ChatGPT
Large scale 3d printing	2020	
Sycamore Processor (54-qubit machine)		
Occulus Rift Released		
Digital Construction Platform (DCP) released		OpenAI Gym is released
		IFCv4 and aecXML, ISO 16739
	2010	Dynamo Released
Digital Construction Platform introduced		
		Grasshopper Released
Industrial Robots in Architectural Education		Unity Released
Cloud Computing introduced	2000	Building Information Modelling (Revit)
		Virtual Reality
Contour Crafting formed		NURBS Modelling
Word Wide Web formed		PC solid modelling
	1990	Form Z 1.0 Released
Construction robots introduced		Archicad
		Wavefront formed
		Space Invader- First Video Game
	1980	Autodesk formed
Apple III released		CATIA Parametric Modelling
IBM Personal Computer released		
Internet created (ARPANET)		First 3D Building Model
Color Raster Displays		First Solid Modeler
Storage fuse display		
	1970	
		M&S Computing (Intergraph) formed
Holographic displays		
		Computer Vision Formed
First VR Headset / I. Sutherland	1960	
		Sketchpad- Sutherland
First Mouse / Doug Englebert		
GM & IBM begin DAC (Deisgn Augmented By Computer)		
First Numerical Controlled Machine (MIT)		
Hardware	1950	**Software**

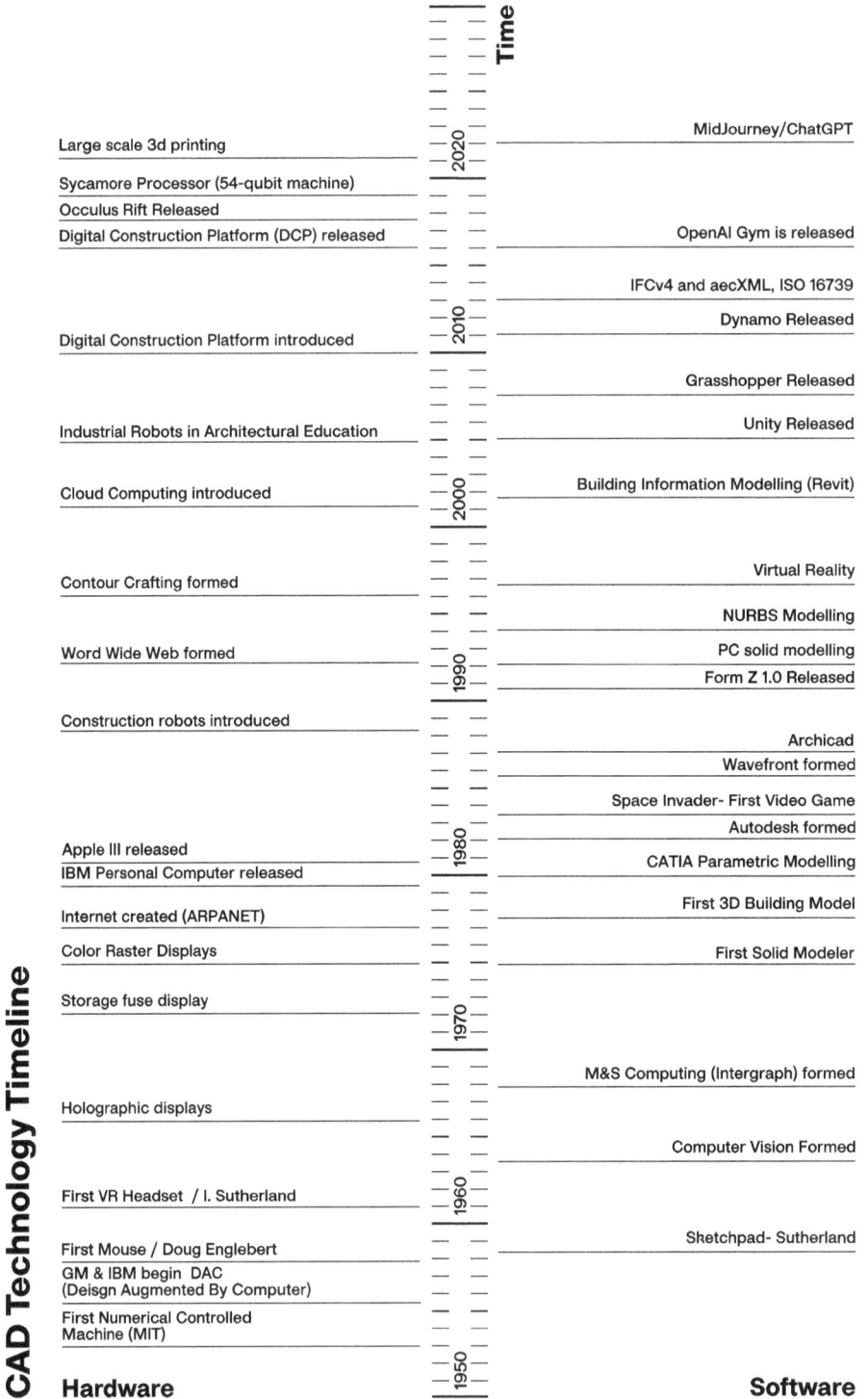

Figure 1.2: Timeline of technology (Hardware & Software) impacting the AEC.

The shift toward maturity in this digital era of architecture, also referred to as the third digital turn by architectural theorist Carpo [13] or Construction Industry 4.0 [14], requires architectural researchers to develop a deeper understanding of complexity and evolutionary processes via computational means if they wish to use them wisely for architectural purposes [12, 15]. In Figure 2.3, we provide a table where we correlate the different groups of tools that we mentioned above with a design approach and a model type in an attempt to help the reader connect concepts presented in the following pages.

This promise to use advanced digital technologies to design and construct structures that reduce environmental impact is yet to be fulfilled due to a number of reasons that range from the slow rate of adopting innovations within the AEC industry to the digital illiteracy of the workforce but also the lack of methodologies and tools that support computational design approaches. Most design tools widely used in the AEC can be classified under two main categories: (a) disconnected design tools that rely on proprietary file types and (b) centralized building information modeling tools. While both have advantages, they suffer from significant limitations.

Take, for example, state-of-the-art building performance software: there are numerous tools specializing in different building-related problems, that is, daylight/glare modeling, embodied carbon estimation, HVAC sizing, energy modeling, and none of which communicate with each other. As a result, designers frequently find themselves having to recreate models for each specific software, which is an overcomplicated and time-consuming process. On the contrary, BIM tools have been developed in a more centralized fashion assuming, for instance, that all design members will put their data into a single model which results in inflexible and hard to operate on models. What disconnected and centralized tools have in common is that they focus on the tools themselves, rather than the design workflow.

In the last decade, however, more emphasis has been placed on the importance of developing tools that support existing and novel design workflows, rather than the opposite which was the norm until the early 2000s. We have seen the emergence of toolkits and libraries from developers within the AEC that focuses on the development of cohesive toolkits that operate on top of standards, open frameworks and existing software and seek to enhance workflows instead of creating a "magic button." Unlike centralized tools, toolkits and libraries have the flexibility to address multiple design objects if they become relevant. For instance, one does not have to specify all properties of a building element when drawing a line but can do that incrementally as the project progresses and the use of the pertinent tool is deemed necessary. On the other hand, unlike disconnected, stand-alone tools, the development of integrated tools within a kit means that they are expected to work together, allowing for the seamless exchange of data. The increasing popularity of Visual Programming Languages (VPL) and tools like Grasshopper (McNeel Associates) and Dynamo (Autodesk Inc.) are examples of how the notion of a toolkit has been successfully manifested in the AEC. The open-ended character of these platforms has provided fertile ground for designers and engineers to develop their own libraries and toolkits based on specific workflows or design methods. The development efforts in both academia and industry are reflected in the evolution from passive design tools (i.e., 2D digital drafting) to generative design tools (i.e., parametric design) and, more recently, intuitive design tools (i.e., generative AI tools) (Figure 1.3).

Although most designers agree that new computational technologies should be used for the improvement of building designs and workflows, there has been a growing chasm between the "technologists" and the "designers" due to the disconnectedness of tools and restraints in the adoption of new methods. A lot of times, the disconnectedness of tools and lack of understanding of computational processes lead to labor-intensive and rather "manual" parametric design workflows, despite the integration of new technologies and the reliance on increased computational capacity. Despite the sophisticated tools used in the design phase and the automations assumed in the beginning of a lot of projects nowadays, the reality in the construction site is usually different as depending on the location, contractors tend to operate in old-fashioned and fragmented ways. This results in a lot of unnecessary remodeling and reworking that could be avoided. This is especially true for large-scale and complex building projects, where the use of digital design tools is more common due to the capacity to handle their size and complexity. Therefore, digital design practices can be characterized as computer-based yet not computational.

Despite the increasing use of VPLs and the release of many toolkits, there is still a lack of common open standards and integrated computational tools that would further extend the designers' creativity by integrating form generation with design performance feedback (i.e., structural or daylight analysis) and physical feedback (fabrication) in the early design stage.

All of the above suggest that the future direction of architectural design should be based upon the holistic consideration of the design process, as well as the comprehensive study of computational morphogenesis with regard to building complexity-inclusive design performance, dynamic occupant behavior and construction processes. A holistic design process considers a building as an interconnected artefact which is an outcome of the overlaying of multiple layers, namely: the social context within which it is being produced (Figure 1.1), the available design and building technology (Figure 1.2), the tools used by the architects and engineers to create and analyze it (Fig 1.3), as well as the underlying design principles and models followed to design it (Figure 2.1–2.5) , and lastly the prevailing building paradigms that were employed for its construction (Fig 2.6). In Figure 1.4 we classify the models, tools and technologies which will be investigated in this work. Holistic design processes have been employed in vernacular structures since antiquity when architecture developed from trial and error to successful replication. People and societies always strived to inscribe the cosmos in their structures and shape them based on the propitious points of the universe or represent them in their details that can be found in different structures across the seven continents. From the 1st AD with the first formal architectural theory from Vitruvius, to the Renaissance and the formalization of proportions and the reductionism of the 20th century, there has been a close connection between the underlying theories, the available technologies and tools and the architectural work. In the digital age that we are currently experiencing, architects and engineers instead of using current technologies (i.e. computers, robots) to serve

Digital Design Tools' Timeline

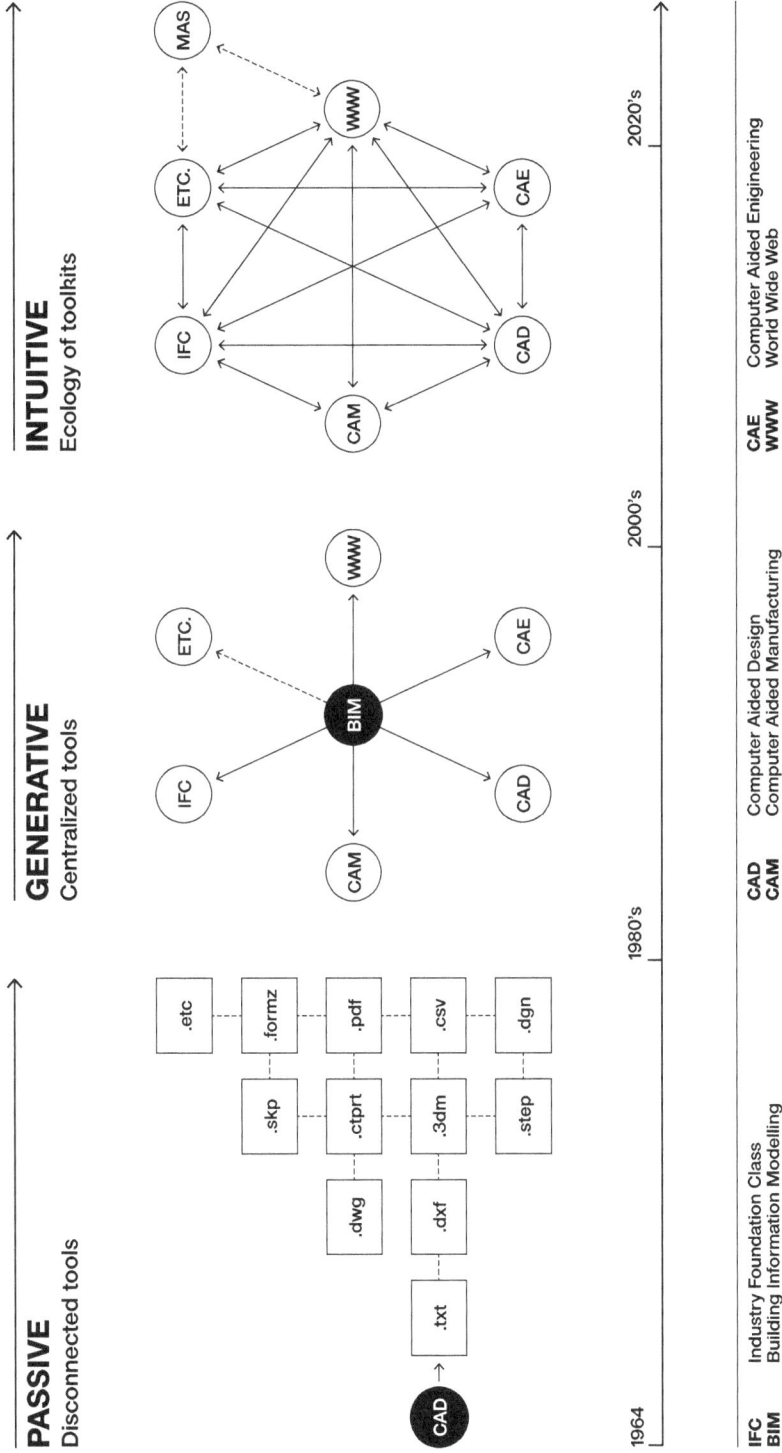

Figure 1.3: Diagram illustrating the evolution of digital design tools.

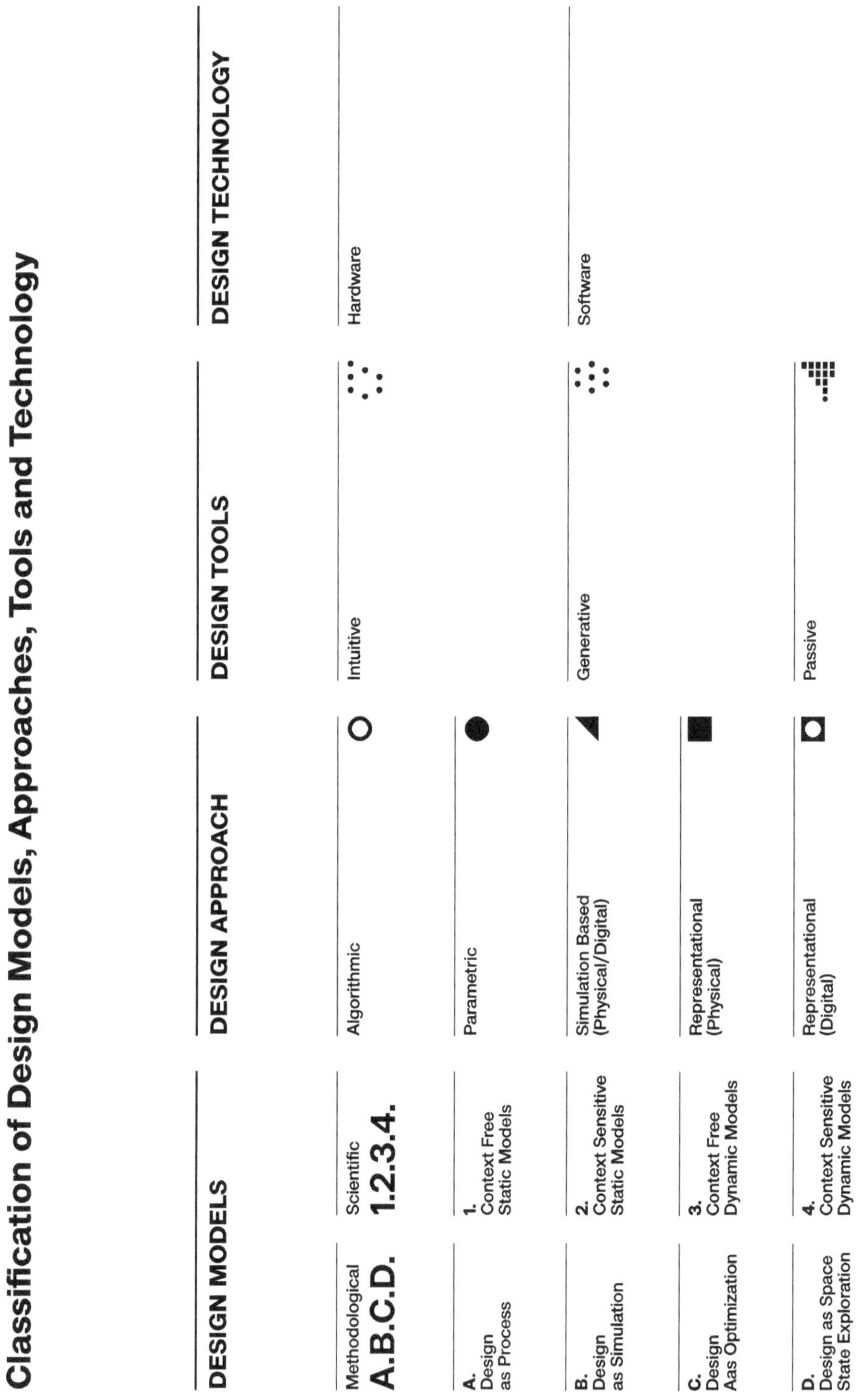

Classification of Design Models, Approaches, Tools and Technology

DESIGN MODELS		DESIGN APPROACH	DESIGN TOOLS	DESIGN TECHNOLOGY
Methodological **A.B.C.D.**	Scientific **1.2.3.4.**	Algorithmic	Intuitive	Hardware
A. Design as Process	**1.** Context Free Static Models	Parametric		
B. Design as Simulation	**2.** Context Sensitive Static Models	Simulation Based (Physical/Digital)	Generative	Software
C. Design Aas Optimization	**3.** Context Free Dynamic Models	Representational (Physical)		
D. Design as Space State Exploration	**4.** Context Sensitive Dynamic Models	Representational (Digital)	Passive	

Figure 1.4: Relationship map of design tools, design models and design approaches.

old design models, we need to reconsider the role of technology in the design process and how computational methods can be utilized to achieve a level of cosmogenesis in the cyber-physical realm.

Today there is an availability of sophisticated design tools to go beyond physical space,and visualize multiple models, but the design approaches have remained very traditional and compartmentalized in the silos of different disciplines and cannot always address the rising building complexity.

Despite the appearance of the first digital design tool as early as 1950's and an initial enthusiasm around computational methods and the reconsideration of design processes to consider things such as evolutionary concepts there was a hiatus of almost 50 years. Since the 2010's onwards there is an increasing number of researchers within the AEC who are focused on developing computational-based approaches and toolkits that enable the exploration of architectural forms based on the concepts of form-finding and optimization evolutionary computation emergent behaviors [17], digital fabrication [18] and rule-based models [19]. This work is a small step in this direction.

1.3 What problem are we trying to solve?

From the deliberation above, two main issues arise:
1. Due to the paradigm shift that information technologies have brought to the ways we conceive, design and construct buildings, and due to the introduction of sophisticated smart building systems (sensors, smart devices, etc.) that can capture dynamic changes in the environment (occupant behavior), the complexity of building design is rapidly increasing – and so is the quantity of data that architects need to take into consideration when designing a new building.
2. Although various sophisticated computational tools – really helpful for analytical purposes – have been made available to designers, they do not aid the creative aspect of design. So, what we are looking at is a lack of computational tools that can combine form generation and design performance in the early design stage to relate evolutionary processes with the principles of geometric forms, structure, material and constructability.

1.4 What is the motivation behind this work?

There are a lot of examples in nature, where complex, large-scale functional nests are built collectively by millions of social insects, without any global coordination or blueprint. Not only that, but they use the minimum resources possible. For instance, the routes termites choose when foraging or building their colonies may appear complex; but on closer inspection, they are a series of simple responses to environmental conditions. The complexity of a termite colony is due to environmental parameters, not the "program" that steers the ant [20]. As we continue to investigate the design of cyber-physical systems and buildings that respond to both the environment and the user, we ought to focus our research efforts on developing novel design methods and synthesizing design systems that can handle complexity, while remaining intuitive and robust [21].

Literature suggests that this kind of design approach is achievable by developing computational methods that integrate multiple domains. In fact, given a set of specifications, they can generate design alternatives that proactively balance the requirements of the primary domain (e.g., architectural design) with the impact on the energy domain (e.g., environmental and structural engineering), as well as the process domain (e.g., construction). Current advances in robotic engineering offer a taste of what's to come, suggesting construction robots will become the standard [22]. What becomes critical when we consider the design of a robotic construction system which involves multiple interacting robots in a dynamically changing construction site is robustness and scalability. In looking for inspiration on how to engineer and control collectives of robots aimed at achieving human-specific goals, the study of complex adaptive systems – the construction activities of ants and beavers, for instance – serves as a very good example [23, 24].

In this work, I zero in on a methodology based upon evolutionary, generative and modular design tools capable of aiding architects to better deal with complex design problems. The proposed methodology integrates the principles of design generation and performance with the help of behavioral modeling techniques and stochastic search methods. I map out a multi-agent systems (MAS) framework for architectural engineering design and apply it on architectural design problems, which typically require the close collaboration between architects and civil engineers (i.e., facade design and shell design).

1.5 What is the hypothesis?

The objective of this work is to propose and test a computational methodology that reduces design complexity by decomposing problems into elementary design agencies and formulating behaviors that incorporate geometric variables with numerical analysis, fabrication and material parameters. Next, it develops a MAS design tool that enables the automated generation of design alternatives, based on the coupling of bottom-up rules that relate to design intentions and top-down rules that relate to regulations and constraints.

The hypothesis of this work is that in order to manage the increasing complexity of building, the current analogy between designer (user) and machine (tool) in digital design should be reversed. In this analogy, the designer acts as apprentice using a specific language (Python, C++ and Java) or interface to pose questions to their master (computer) and respectfully awaits the answer. To promote the designers' creativity, computational design tools should be conceived as a designer's apprentice instead of drafting aids, capable of generating proposals (design alternatives), when given a set of specifications that the designer (master craftsman) can evaluate and critique. In this scenario, the interaction with the designer and the processing of multiple data sets will allow the tool to build up knowledge. The specific research questions this proposal considers are as follows:

1. How can architects handle complexity in building design via the formalization of design thinking and by means of computable functions and the holistic consideration of building design?
2. How can the coupling of generative design approaches with analytical solvers in the early design stage enable architects to search for larger solutions more efficiently?
3. How can designers explore emergent forms by applying stochastic search methods to architectural design problems via their methodological decomposition into autonomous agencies?
4. What type of stochastic search methods are best suited to architectural design problems?
5. How can autonomous and adaptive models that integrate geometric properties with performance simulations and robotic construction be used for architectural form exploration?

1.6 How does this work add to architectural research?

If the objectives stated are fulfilled, this book will contribute to fundamental knowledge in the areas of agent-based modeling, building engineering and architectural design. In general, the development of this proposal aims to:

1. Extend the notion of architectural complexity based on rigorous research and introduce a computational design methodology based upon behavioral modeling and the holistic consideration of building design.
2. Enable designers to explore larger solutions faster by establishing feedback loops between computational form generation and the analysis of design performance measures.
3. Introduce different types of agent ontologies in addition to swarm intelligence models. For instance, the belief–desire–intention agent model is used in this work to formulate behaviors that combine realistic constraints with design intentions into generative design routines.

1.7 How to read this book?

This book is organized into five parts that are self-contained. Part II is more theoretical and contains a literature review on the topics of complexity theory and complex adaptive systems (CAS). The underlying assumptions of complexity theory are reviewed, and a taxonomy of complexity definitions is provided. Concepts and tools for managing complexity derived from CAS are correlated with contemporary digital design practices and computational design models in AEC. Additionally, theory and applications of MAS

and the basic programming blocks of distributed artificial intelligence are briefly analyzed, along with some examples of biologically inspired evolutionary programming
approaches.

Part III is more technical and focuses on describing the components of the framework and the specifics of the design methodology. The internal structure of the proposed agent-based methodology is analyzed in combination with the types of agents
and the related design algorithms. Part IV is more design-oriented and features a set of
case studies, where the developed framework is used for design generation, analysis
and the evaluation of design alternatives. In the first design experiment, the MAS tool
implemented is applied to the development of a digital simulation by using data resulting from the physical implementation of a swarm robotic system. In the following design experiments, the agents are used to represent building components, and the tool is
used for (a) the generation and evaluation of facade designs and (b) the form-finding of
shell structures. The results of each of the design experiments are discussed separately
in each section. Lastly, in Part V, the results and findings of the experiments are summarized, along with the contributions of this thesis and suggestions for future work.

—

Part II: **Context**

A brief history of complexity theory

2 Background and related literature

This section reviews the literature relevant to this proposal, which includes established (a) methodological models of design, (b) digital design models in architecture, (c) complexity theory and (d) state-of-the-art multi-agent systems (MASs). The discussion begins by briefly describing methods of representing design processes computationally and specifically discusses existing digital design models in architecture. Next, there is a critical overview of complexity theory and its underlying principles, correlating them with contemporary building design. This section ends with a review of the existing MAS-related literature and a discussion on how the development of computational design frameworks, using a MAS approach, can help reduce the complexity of building design and enhance the designers' creativity in the future.

2.1 Design models for architectural design

Although design is an activity that can be found in all aspects of human endeavor, it is a relative recent arrival in academics; before the mid-twentieth century it was not understood as a process as widely as it is today [25]. Similar to complexity, there is no one unified theory of design, but there are multiple working models in every domain that sufficiently describe what design is.

The emphasis on theoretical design, as opposed to methodological description of design, has created a lack of clarity with respect to the nature and contributions of digital design methods. The increasing use of computers has created the need to represent design problems to computers and therefore forced both theoreticians and design researchers to formalize the design process and explicitly describe its theoretical aspects. A brief overview of theoretical design models in general is provided in Section 2.2.1 and in architecture in particular in Section 2.2.2.

2.1.1 Models of design

Design has been modeled as a language (semiotics), as a logic system [26], as a problem-solving process [27], by procedural methods [28], by reference to typologies [29] and by simulation and optimization techniques. Despite the many approaches that have been developed around design, it can be generally agreed upon that the design as an activity is (a) purposeful, (b) goal-oriented (c) and creative. Purposeful means there is always a purpose behind the design (i.e., we design a structure to provide shelter). Without a purpose, it is impossible to establish design goals. Beyond having a purpose, the design aims to achieve goals that involve searching for solutions that best meet the criteria described in the design goals.

Last but not least, design is a creative activity that seeks new solutions. If there is no creativity, we would have to fall back on existing solutions to tackle problems or

https://doi.org/10.1515/9783110797435-002

offer no solution at all. Creative design is sometimes defined as a novel combination of old ideas, but creativity also implies exploring uncharted territory and venturing into the realm of untested ideas. Literature shows that a large number of digital tools that focus on the first two characteristics of design (purposeful and goal-oriented) have been developed but fewer tools have focused on the creative aspect of design [30]. The following section offers a list of design models that are particularly relevant to developing computational design tools, namely (a) design as process, (b) design as simulation, (c) design as optimization and (d) design as space state exploration (Figures 2.1–2.4).

2.1.1.1 Design as process

Much effort has been applied to developing a method that describes design as a prescriptive process. By modeling design as a series of steps, the designer is provided with a method that not only gives structure to the activity but also helps sharpen the general understanding of what design is. Herbert Simon also considered the design activity as a problem-solving process and distinguished between well-structured and ill-structured problems. He based this distinction upon whether or not the problem at stake can be solved by general problem-solving procedures (i.e., analysis). Simon claimed that design problems are ill-structured because the problem space is not fully specified at the outset of the process [31]. In fact, some parameters of the problem may only "occur" to the designer after considerable work and research. Simon outlined a procedure for solving ill-structured problems by breaking them down into smaller, tractable problems and by following a cyclic approach, where the designer alternatively solves part of the problem while assimilating new information. Specifically, with regard to architectural design, Simon stated that:

> there is no definite criterion to test a proposed solution, much less a mechanize-able process to apply such criterion. Additionally, the solution space is not defined in any meaningful way, for a definition would have to encompass all kinds of structures the architect might at some point consider (i.e., a geodesic dome, a truss roof, arches . . .), all considerable materials (wood, metal, concrete, carbon fiber, . . .), all design processes and organizations of design processes (start with floor plans, start with list of functional needs, start with façade, . . .). Such a problem definition would make the problem impossible to compute. [31]

Similar to Simon, Asimow [28] has tried to formalize design as a process based on its programmatic description and proposed a three-step cycle, which includes (a) analysis, (b) synthesis and (c) evaluation. The design process in this case can be described as a kind of computer system, as a series of interlinking cyclic processes in which the introduction of new knowledge or new events can cause regression to a previous point from any point in the process (evaluation). These cycles, repeated in every phase from designing to building a structure, are: (a) the conceptual phase, (b) design development phase, (c) construction document phase and (d) construction [32] (Figure 2.1). As the project advances from concept design to construction, the introduction of new constraints and new information in every phase results in the decrease of design freedom and the increase of design knowledge in each of the cycles [33].

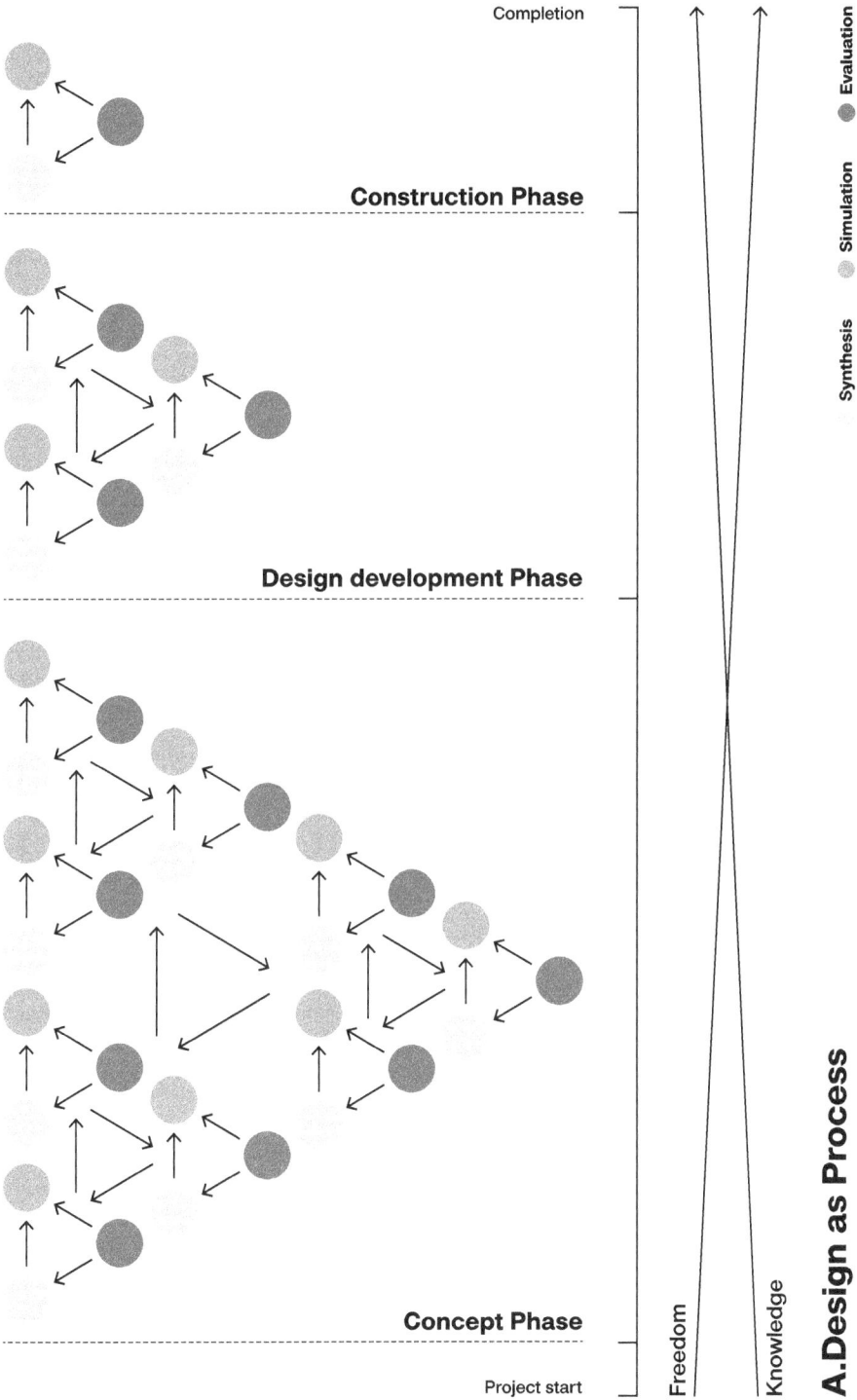

Figure 2.1: Methodological models central to computational design: design as a problem-solving cyclic process (a). Diagram based on Plesk's and Cryer's definition.

2.1.1.2 Design as simulation

When design is considered as a simulation, the aspects of the problem that needs to be taken into account are described with a certain degree of abstraction into a model. This model may be physical, mathematical or even an abstract thought model. Physical models have a long tradition of proving ideas and design concepts in architecture, unlike mathematical models that are a recent introduction, mainly due to the wide application of information technologies. In a simulation model, it is assumed that if enough of the important parameters are incorporated into the model, the simulation will serve as a testing ground for the system, in which a large amount of information could be observed during a short period of time. This allows designers to test design ideas faster, facilitating the decision-making process. Digital simulation models have been used in architecture extensively, either as a way to simulate form-finding and dynamic behaviors between particles [34] or as a way to simulate daylight and energy consumption [35], pedestrian flows [36] or build up construction processes (4D simulations) [37] (Figure 2.2).

2.1.1.3 Design as optimization

Designers strive to provide solutions that best match a set of design criteria; therefore, all designs can be seen as a form of optimization. This explains why optimization methods have received so much attention in different design-adjacent fields. Mathematical methods of optimization include many techniques, the best known of which is linear programming [30, 38]. Common to all such methods is the necessity to describe the design parameters as mathematical variables. As a result, design problems are usually converted to "well-structured" analysis problems for optimization by fixing the decision variables. The formulation of most optimization methods can be described in the following terms: decision variables set the context for the design, performance variables define particular solutions within a design space and objective functions define the particular solution of interest in the design space.

Thus, only objectives that can be written as functions are considered, and these are usually simplified by writing their objective functions [39]. Thus, the "optimum" found does not pertain to the real problem, but the limited and simplified version of it. Nonetheless, suboptimal solutions found this way can still give satisfactory results if a sufficient amount of the primary objectives are reasonably defined. Consider, for example, the following structural design problem: find the correct size of members to minimize the overall weight of a given structure. In this case, we can easily formulate a single objective optimization problem. The decision variables are the width and height (w, h) of the element, the performance variable is the overall weight (W) and the objective function is $f(w)$, subject to w,h. In architectural design, however, the aim is to satisfy multiple design objectives that might even be conflicting, so forming objective functions that lead to desirable design solutions becomes key. Recently, researchers have been focusing on multiobjective optimization routines for addressing the complexity of building design [38, 40] (Figure 2.3).

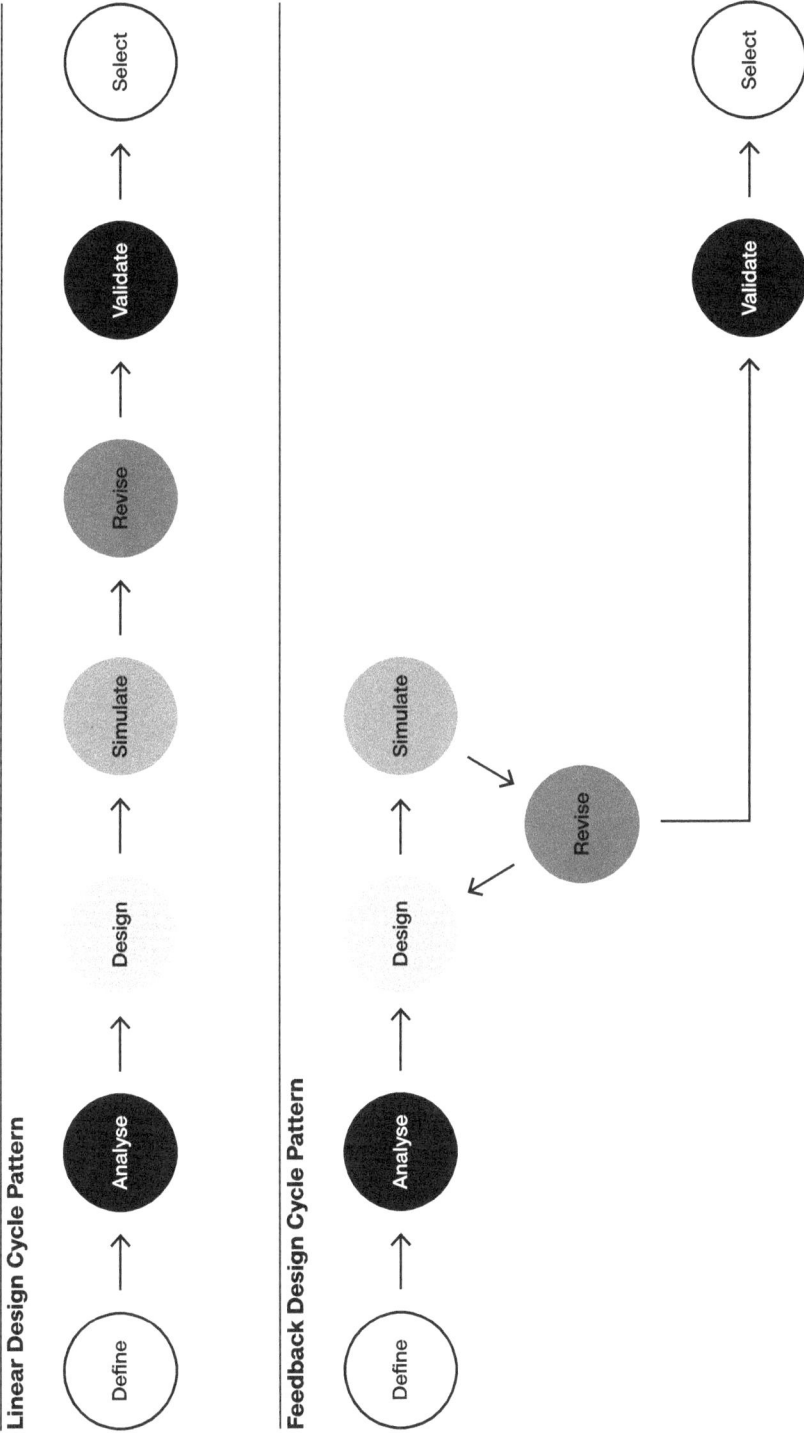

Figure 2.2: Methodological design models central to computational design: design as a simulation.
Diagram based on Koberg and Bagnal's design cycle patterns (b).

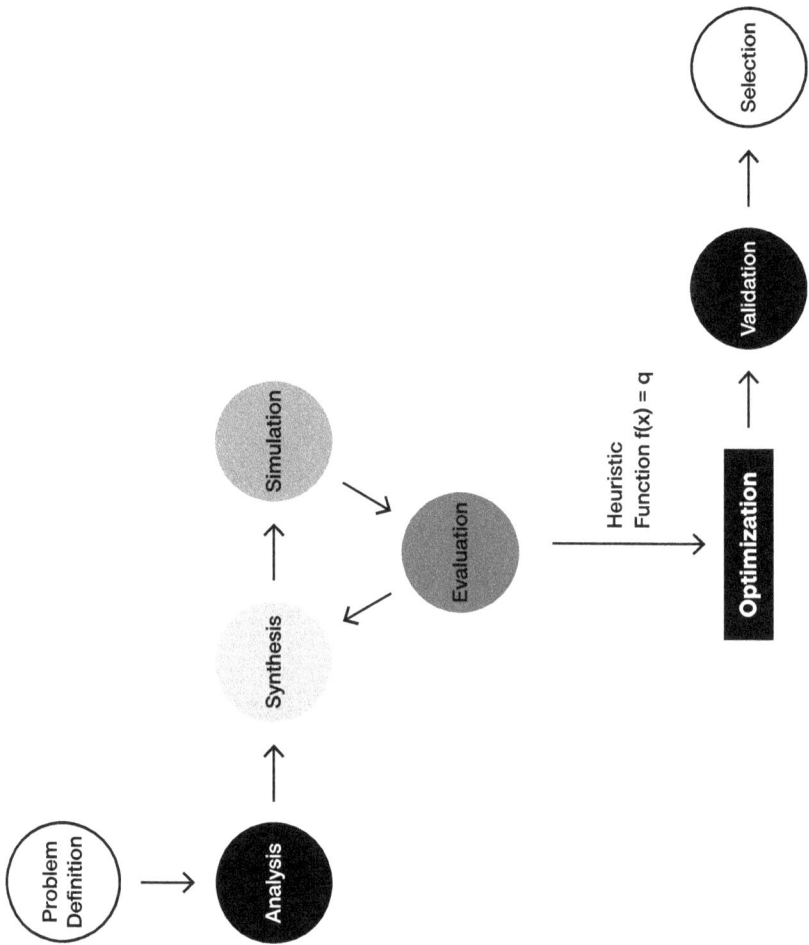

Figure 2.3: Methodological design models central to computational design: design as an optimization. Diagram based on definition developed by Asimow (c).

2.1.1.4 Design as space state exploration

Design has also been regarded as a decision-making activity, not solely as a problem-solving process [9]. This approach is related to decision theory and takes into account the fact that design goals are not always defined at the outset of the design process. In fact, in many cases, determining the goals is part of the design process itself.

In this approach, design is described as a "forward-looking" search activity which is (a) exploratory, (b) constrained (computable), (c) includes decision-making and (d) learning. Search is the common process used to inform decision-making, which implies choosing, and choices are framed by parameters that can be considered as variables. This set of variables relates to the problem decomposition and the design context. In ordering these variables, Gero [41] represents design as being comprised by three state spaces: function, behavior and structure. Function refers to the definition of an object's purpose/teleology or what it does. This can be seen as a result of behavior and it is distinct from purpose (the "why it does"). Behavior (performance space) refers to how it does something. In other words, it is the description of how the design behaves in a particular environment, and structure (decision space) or "what is." Thirdly, structure represents the physical object itself as described by material, topology, geometry and physical characteristics (Figures 2.4).

Despite the fact that heuristic/stochastic search methods can be used to find the values of variables in a specific state, what is also critical to design is the determination of the state space within which to search. This is called exploration of the state space and is akin to changing the problem space within which the decision-making occurs. Learning implies the restructuring of knowledge based on the iterative cycles of the analysis–synthesis–evaluation steps [41–43]. Considering the creative aspect of design as a problem with multiple space states allows us to formulate it as a Markov decision problem (MDP) [44].

An MDP is defined as a tuple/vector <**S, A, P, R**>, in which
S is a set of states (**Si**) that represents the state of the world (function, behavior, structure); **A** is a set of actions (**A¡**) available to the agent in states of the world (i.e., generate design and analyze); **P** is a transition function/table **P(S'|S,S)** probability of state **S'** given action "**A**" is taken in state "**S**"; and **R** represents reward/R (s) is the cost or reward of taking any action **A** in state **S**.

MDP provides a mathematical framework for modeling decision-making in situations where outcomes are partly random and partly under the control of the decision-maker. This makes MDPs a fitting framework for design problems where decision-making involves the designer, as well as other decision-making entities that might not be known beforehand (legislators, occupants, stakeholders, etc.). Rather than trying to find optimal solutions to a simplified problem, one might better seek satisfactory solutions to multiple subproblems [45]. This is the approach taken with stochastic/heuristic techniques used in artificial intelligence (AI) methods. The key characteristic of formulating design problems as MDPs is that it allows the decomposition of the problem into subproblems or design agencies. Each agency is comprised of a number of autonomous rule-based programmable entities (agents), which have heuristic search

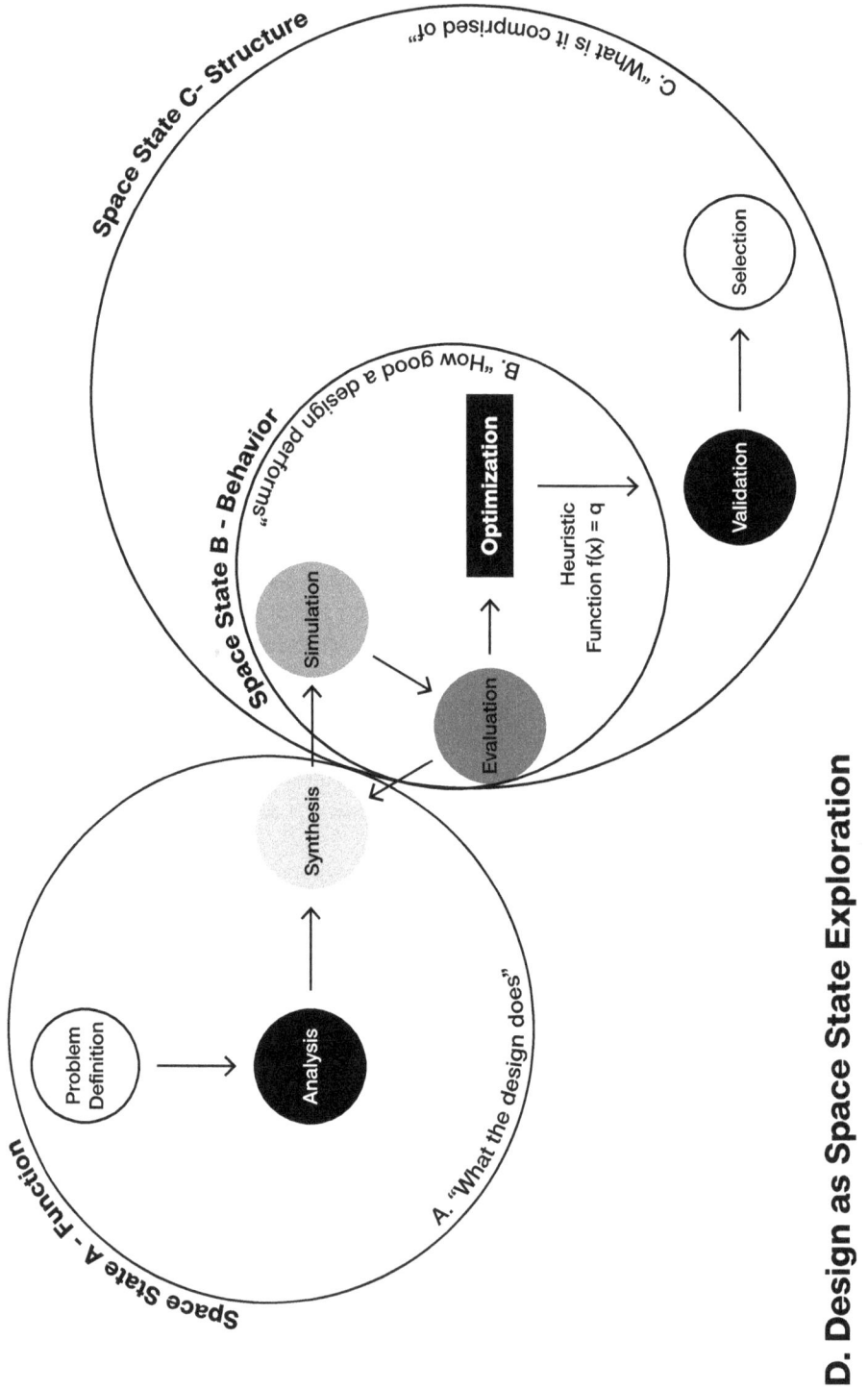

Figure 2.4: Methodological design models central to computational design: design as a state space exploration. Diagram based on definition developed by Gero and Sosa (d).

objectives encoded into their structures and must choose a sequence of actions in each state based on that. At each iteration, agents receive a utility due to the whole sequence of actions they have performed.

By interacting, the agents can decide the probability of whether an answer to a subproblem is true or false. By doing that iteratively for every subproblem and communicating their decision to other agents, they can inductively lead to a solution to the global problem. The MDP formalism assumes that in order to solve a problem, agents receive information through their sensors and decide the probability for a state S' based on the current state and not the history of states that the agent had. Based on the selected state, the agent performs one or a set of actions and evaluates whether it improved the world state or not.

2.1.2 Digital design models in architectural design

In the last 20 years, the wide application of digital design models has rapidly transformed the design and construction of our buildings from being the materialization of paper-based drawings to the materialization of digital information. The realization of buildings, such as the Guggenheim Museum in Bilbao, established a precedent for the bridging between design and construction, as an increasing amount of the projects that followed have been designed, documented, fabricated and assembled with the assistance of digital means [2, 46].

This was only made possible thanks to the increasing computational capacity and advancements in the field of computer graphics, paired with the availability of parametric design tools and advanced manufacturing processes [47]. In more recent years, a growing interest in performance-based models and data-driven design, along with the rapid integration of building information modeling (BIM), indicates a maturity in the use of digital tools beyond formal complexity and toward more performative design solutions [12, 48]. The integration of multiple design models through collaborative BIM platforms has proved to be beneficial for reducing building costs and for improving the design process, but has also called attention to the lack of existing methodologies to deal with complexity [5].

Even today, despite the great level of sophistication in the development of computer-aided design (CAD) tools, the dominant mode of design remains manual and rarely relies on computation [49]. Regardless of the sophisticated underlying infrastructure of current design tools, as long as the modeling process is done manually, we cannot regard current design tools as either computational or intelligent. Therefore, current design tools can be considered computational only in a narrow sense: that of being computer programs whose infrastructure relies on computers for the creation of digital geometric objects. Only during the last two decades, we have seen an increasing number of design tools based on computational design methods, and the majority of which deal with form-finding [34, 50–52]. To be clear, this work defines as noncomputational the design models that make use of computer programs to mostly support manual modeling processes, and as computational the design models that rely on computational procedures for the essential part of the modeling/designing process [53].

It is also important to clarify that one impact of the "computerization" of architectural design processes during the first digital turn was the automation of geometrical transformation and the mechanization of producing drawings [13]. On the other hand, the impact of using "computational" design processes, which we are currently experiencing, can lead to the reconsideration of architectural design as a whole, the complete formalization of the design process into code and the exploration of multiple solution spaces using stochastic techniques. This distinction becomes particularly important if we take into account that the complexity of the (manual) modeling process in a digital environment increases proportionally, if not exponentially, with the size and complexity of the geometric object in question. However, this is not true in an algorithmic modeling process.

In an attempt to categorize digital design models according to the various relationships between the designer, the conceptual content, the design processes applied and the design object itself, Oxman distinguishes between five paradigmatic classes of digital design models, namely CAD models, formation models, generative models, performance models and integrated compound models. In Oxman's approach, CAD models are descriptive and isomorphic to paper-based design methods, and therefore offer little in design thinking. Formation models, which can be developed using parametric modeling or animation tools (i.e., Autodesk Maya, 3d Studio Max, Rhinoceros 3d, Generative Components, etc.), are considered to be topological, as geometric modeling requires the designer to follow a formalized digital design process. In formation models, the geometric properties of objects can be manipulated digitally by the designer via a high level of interaction and parametric control (Figures 2.5). In generative models, the formation of geometry is associated with a computational mechanism (algorithm), and most of the time a fitness function is controlling a generative process. Genetic algorithms are a typical example of generative design models that have been extensively used as a design-driving mechanism with multiple applications [54–59].

Performance-based design models are a process of formation or generation driven by a desired performance. This is achieved by defining a set of measures and goals that must be satisfied [12]. Kotnik provides a more general classification and states that digital design models can be considered using the broad sense of computer, that is, a machine (hardware) that manipulates data according to a set of instructions (software). In such an approach, every piece of data included in a design model can be coded as a natural number N. This way, a piece of software (s) that operates on a computer can be considered as a function (f) that operates on a subset of inputs (in), which belong to the natural numbers (N) [47]. The function describes the relationship between the inputs and the set of all possible solutions/outputs (out) such that

$$f(\text{in}) = p(\text{out})\text{in}, \text{out} \in N.$$

Based on that assumption, Kotnik distinguishes between three general classes of design models, namely representational, parametric and algorithmic. His formalism encapsulates Oxman's classification and therefore CAD and formation models fall under the representational class, while generative models, performance-based and integrated compound models are all subclasses of the parametric class. The algorithmic class includes design models, where the concept of computational functions is implicitly coupled with defining relationships of architectural elements. Through this coupling, a computational generalization of geometry is achieved, which allows for new forms of creative expression but also requires a methodological formalization of design thinking.

This ties back to the general design models described in the previous section. In this thesis, we are interested in the algorithmic class. In order to better understand it, I describe below the basic algorithmic approaches and classify them into three different categories, namely feedback-based design models within the tradition of cybernetics and systems theory (adaptive, generative, form-finding); rule-based design models within the tradition of transformational grammars and axiomatic designs (rule-based and grammar-based); and intelligence-based design models within the tradition of AI (emergent/cellular and logic based).

The growing interest in algorithmic models during the latter years has resulted in the development of multiple paradigms and has also emphasized the lack of understanding of the underlying principles that have shaped digital design tools, such as information theory (IT), topology, cybernetics and complexity theory, to name a few. Although reviewing these fields goes beyond the scope of this work, a review of those topics and how they relate to design complexity is provided in Section 2.2. The next sections provide a short description of fundamental algorithmic models.

2.1.2.1 Control/feedback-based models

Control-based design models are computational models whose unified system of procedures governs the creation and manipulation of spatial forms and whose design method allows and controls this unified system as a whole. Control-based design models are exclusively a top-down modeling paradigm affirming the concepts of control automation, regulation and optimization, thus following the tradition of cybernetics.

Based on the implementation of the feedback mechanism, we can distinguish between adaptive, parametric and generative models. Control-based parametric models were implemented in commercial software packages, such as the CATIA software from Dassault, which has been the archetype for almost all current parametric design tools [5].

Form-finding is currently one of the most prominent research fields in feedback-based design models. It is a design approach rooted in the physical modeling of structures and the long tradition of funicular shells (i.e., domes). Already an extension of purely analytical processes to structural design, form-finding has been used to generate geometric models that couple form configuration with the forces in a closed feedback loop. In their simplest setup, form-finding methods, both computational and noncomputational, generate catenary curves and funicular shapes, as a direct outcome of the force distribution [16]. However, unlike civil structures where, for instance, the design of a bridge is conditioned solely by structural performance, in architecture most design problems require the consideration of multiple parameters, such as local resources, historical context, occupant behavior, environmental impact and constructability. Current form-finding methods focus on structural design goals

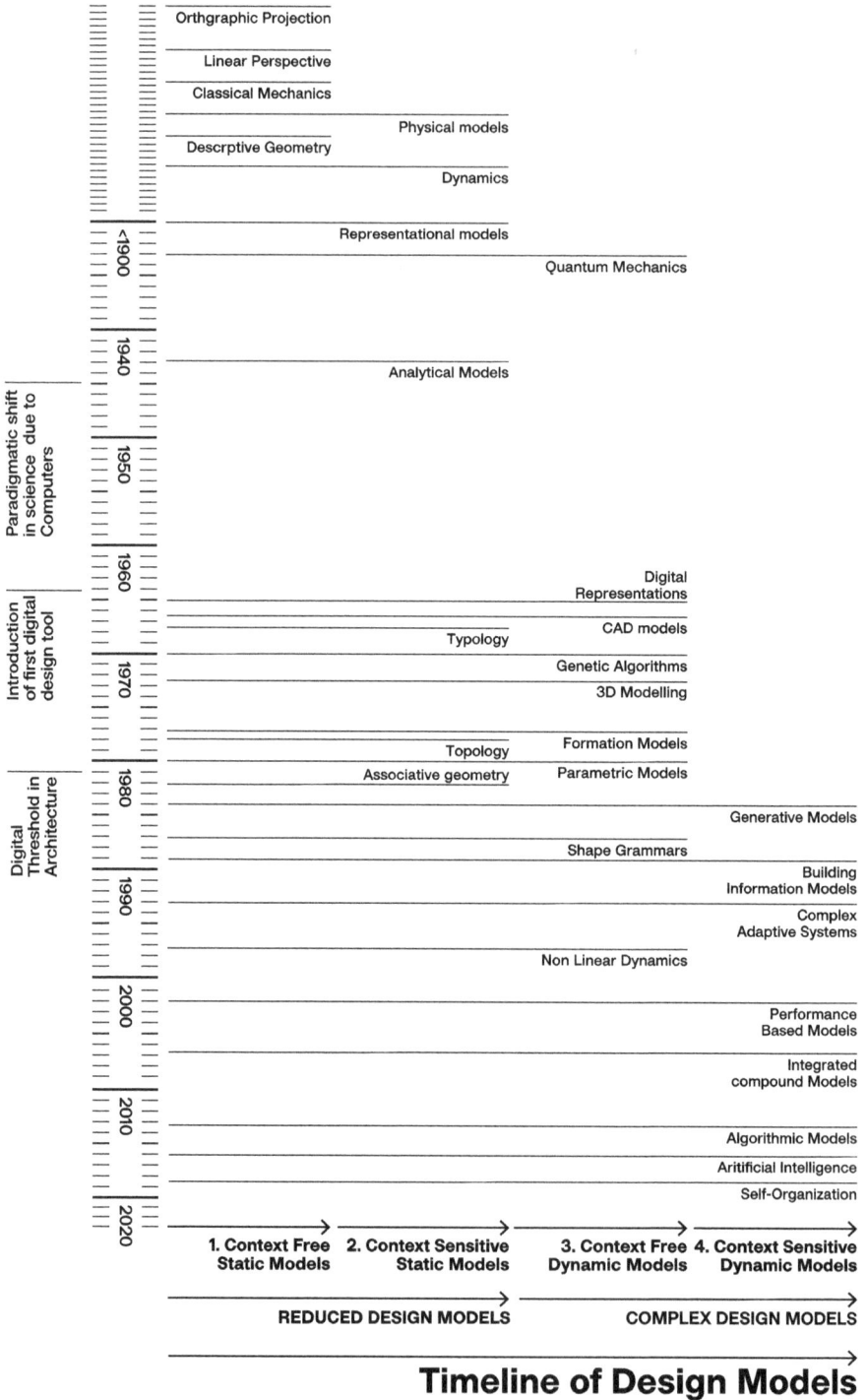

Timeline of Design Models

Figure 2.5: Timeline showing the evolution from reduced to complex design models.

but do little to consider the combination of different design goals, like environmental performance [60].

2.1.2.2 Rule-based models

Rule-based design models incorporate a set of rules whose application to a particular case helps to evolve the design of the object. The structure of such models involves two distinct levels and the dependence between them. The first level is the definition of the rules, and the second level is the form of application. Rule-based design methods include logic-based, grammatical/recursive and emergent/cellular models and were among the first methods to be researched in computational design.

Logic-based models utilize formal logical reasoning as the principal drive for design. Such models have used set theory to codify design decisions in propositional forms, which are then translated into geometry. The first prominent example of logic-based models can be encountered in Christopher Alexander's book, *Notes on the Synthesis of Form* [61]. Alexander developed the abstract idea of a "pattern" as a separate entity that encodes a set of relationships independent of all others. Following this approach, the design process can be divided into (a) defining the requirements of design, (b) identifying patterns that can address the requirements and (c) applying the patterns in ways that can lead to a design solution that satisfies the requirements.

Grammatical/recursive models are another type of model, whose design was extensively researched by G. Stiny through the perspective of "shape grammars." In his approach, the geometric rules of an architectural style can be described to a machine, which can regenerate them by the recursive manipulation and transformation of two-dimensional and three-dimensional (3D) shapes [62]. Stiny's shape grammars are based on Lindermayer's research of modeling plants and Chomsky's attempt to encode the complexity of language using simple rules [63]. Although shape grammars have been successfully used to generate design based on existing typologies, such as Frank Lloyd Wright-style prairie-like villas, this approach was limited to simple examples by the combinatorial complexity involved. A more recent implementation is the sequential shape grammar approach, which was developed at ETH in Switzerland. This approach has been implemented in a software called City Engine, and has been used for the procedural modeling of architecture and cities, mainly applied in gaming [64].

Emergent design models are based upon the local interaction of simple, rule-following programming entities (automata agents) and follow the tradition of Von Neumann's research into cellular automata (CA) and Turing's work on morphogenesis [53]. The design processes in such design models are divided into two levels. At the first level, a number of autonomous and modular code blocks called agents are programmed to interact and follow a set of local behaviors. At the second level, the accumulated behavior of the agents is considered within a specific environment and the consideration of all interactions globally leads to emergent phenomena. Such emergent phenomena are usually rendered into geometry as points, trajectories or 3D solid objects. A prominent example of emergent-based modeling is the implementation of the swarm behavior model by Reynolds [65]. Additionally, Frazer [66] used bottom-up concepts to define a computational model based on emergence and evolutionary programming, which he called "evolutionary architecture." Following Neumann's legacy on CA, Frazer's approach investigated morphogenesis in the natural world and used that as an abstraction for developing fundamental form-generating processes in architecture. However, Frazer's "evolutionary architecture" did not manage to efficiently describe how such morphogenetic natural processes can be transcribed and appropriated in order to develop emergent architectural design models.

2.1.2.3 Artificial intelligence (AI)-based models

Models inspired by AI were based on the idea of creating design systems that can develop some sort of "intelligence." The concept of developing computing machines that can be programmed to think on their own was established by Turing in his paper *Computing Machinery and Intelligence.* Minsky [67], who invented artificial networks, defined AI as "the science of making machines do things that would require intelligence if done by man." Greatly inspired by these advancements, Negroponte believed that the formalization of communication between humans and computers can transform the design process into one where intelligent machines (digital design tools) learn how to adapt to the designer and his style and, at the same time, can learn some objective truths via induction [68]. On a similar but more practical path, Sutherland's Sketchpad, the first digital design tool, was quite progressive as it was embedded with an elementary notion of intelligence, which had been totally missing in design tools up until then. Sketchpad allowed the designer to roughly draw a circle by adding points with a light pen, and the software would recognize the designer's intention and transform it to an exact circle.

Beyond the work of the researchers mentioned above, who set the foundations and envisioned how AI-based models could be applied in architectural design, there was a big gap of almost 40 years during which research efforts stagnated due to the fact that AI-based models had failed to deliver significant results. Only after the 2000s has a resurgence of interest in AI models been observed due to the wide availability of data and increased computational capacity [69]. Case-based reasoning (CBR), for instance, is an approach based on the assumption that in order to solve a certain problem one should refer to similar problems that have already been solved. This requires the existence of a big database of existing cases, which was not possible during the early days of computers. However, nowadays, thanks to the developments in the field of machine learning and the availability of BIM, this is no longer a problem as algorithms can be trained to perform feature searches. AI approaches similar to CBR, such as distributed constrained reasoning, have been recently applied in game theory and office building design [70].

2.1.3 Summary

In this proposal I am investigating rule-based models, particularly emergent models, and combining them with AI-inspired models and principles. The brief description of design models above shows that although logic and grammar-based models were the first types to be developed, they have not been successfully integrated into design tools. On the contrary, models based on control/feedback have received a lot of attention and a number of CAD tools have been developed based on these approaches. Emergent design models have only been studied recently in architecture, and therefore their use is still limited. However, emergent design approaches have a longer tradition in engineering where their modularity, robustness and distributed character have made them appropriate for solving complex problems, especially nowadays when there is an abundance of computing power. Moreover, emergent design models have been shown to be appropriate for developing systems that account for dynamic changes in their environment and can afford design complexity [71].

To understand how emergent design models can be used in the field of architectural design, we need to have a deeper understanding of complexity theory and complex adaptive systems (CASs), as well as their underlying assumptions. In the next section, I will attempt to define the term "design complexity" by bringing together definitions from domains that range from architecture, engineering and construction (AEC) to IT and general systems theory (GST). Distinguishing between different types of complexity is considered to be essential, due to the fact that complexity in architectural design has been mainly related to geometric complexity, and has not been considered from a complexity theory perspective.

2.2 A brief historical review of complexity and its relationship to architectural design

One of the main characteristics of the digital age is its diversity and complexity, and it is essential to recognize that there is no one-size-fits-all solution to how we process analyze and manage the swaths of data that we dealing with daily. Complexity is intrinsic to almost every significant building from the antiquity until our times whether it relates to its geometry, its materials or its program. From an architectural point of view, it is remarkable to observe the evolution of building construction in recent years. Next to iconic stone cathedrals and buildings of centuries past, now stand prismatic and freeform steel structures equipped with building systems capable of responding to their environment in a dynamic fashion (Figures 2.6). The introduction of responsive building components and sensors that track occupant activity within buildings and the creation of their digital twins are rapidly transforming our built environment from complex physical systems to cyber-physical ones [3, 4]. Thus, our environment has become increasingly complex and therefore complexity theory and computation have been radically influencing research in nearly all disciplines, in both the sciences and the humanities [72]. A good example is the paradigmatic shift in sciences, such as physics and biology, which is the result of studying complexity by adopting computers as primary tools for simulating and modeling natural processes. Reductionist models have been successively modified or replaced as the predominant paradigm of research in recent years.

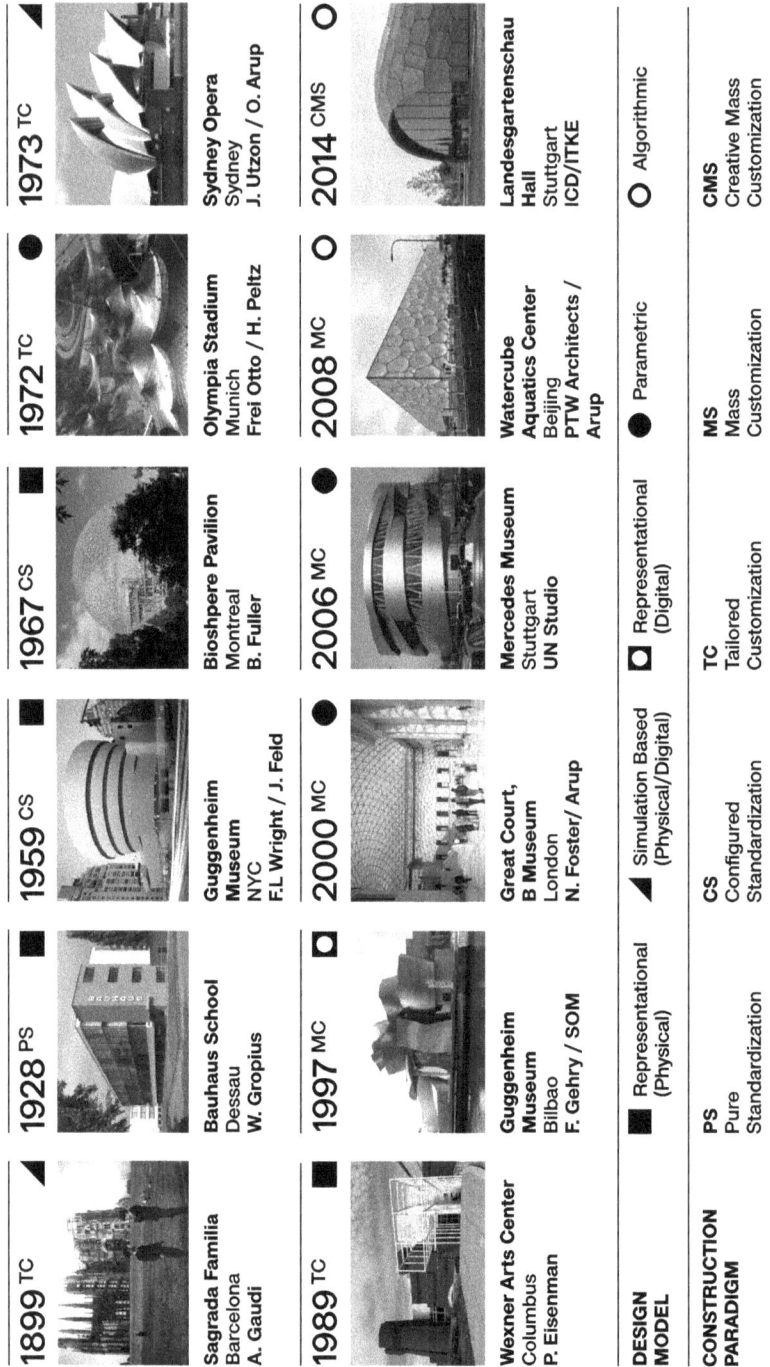

Timeline of Iconic Buildings' Design and Construction Approach

1899 TC	1928 PS	1959 CS	1967 CS	1972 TC	1973 TC
Sagrada Familia Barcelona A. Gaudi	Bauhaus School Dessau W. Gropius	Guggenheim Museum NYC F.L Wright / J. Feld	Bioshpere Pavilion Montreal B. Fuller	Olympia Stadium Munich Frei Otto / H. Peltz	Sydney Opera Sydney J. Utzon / O. Arup

1989 TC	1997 MC	2000 MC	2006 MC	2008 MC	2014 CMS
Wexner Arts Center Columbus P. Eisenman	Guggenheim Museum Bilbao F. Gehry / SOM	Great Court, B Museum London N. Foster/ Arup	Mercedes Museum Stuttgart UN Studio	Watercube Aquatics Center Beijing PTW Architects / Arup	Landesgartenschau Hall Stuttgart ICD/ITKE

DESIGN MODEL	■ Representational (Physical)	▲ Simulation Based (Physical/Digital)	◻ Representational (Digital)	● Parametric	○ Algorithmic
CONSTRUCTION PARADIGM	PS Pure Standardization	CS Configured Standardization	TC Tailored Customization	MS Mass Customization	CMS Creative Mass Customization

Figure 2.6: Iconic building structures correlated with the design model and building paradigm used for their realization. Images derived from photos by Baldomer G. I Roig, S. Drakopoulos, Right Cow Left Coast, J. C. Benoist, R. Roletschek, M. Eiskalt, B. Spagg, T. Hisgett, A. Dunn, J. Herzog, E. Pantazis, J. Nebelsick via Wikipedia, arch2o.com and author's own archive (Creative Commons License).

To further clarify this, the mechanistic worldview of nature, which relies on the continuous top-down reduction of a whole into its parts, is being replaced by the correlation of local interactions and the identification of patterns that can bring the parts into an equilibrium as an emergent property of the overall system [73]. For example, scientists and biologists have closely investigated CASs, such as termite colonies, and by tracking how termites forage and collectively build their habitats scientists have developed mathematical models in order to understand how the complex geometry of the habitat (termite mounds) is related to the environmental conditions, the termites' method of locomotion and the locally available materials [74]. Unlike the fields of biology and physics, architectural design thinking has not been greatly affected by computational thinking, even though information technologies have radically changed the way we design and construct buildings. With a few exceptions, in its early manifestations, digitally mediated architecture focused on geometry and approached complexity in a diagrammatic instead of a scientific manner. Digital design approaches remained rooted in representation-based design paradigms, instead of developing a deeper understanding of complexity and reconsidering the design process in the light of computation [6].

To approach a topic as broad as complexity theory and draw conclusions that can prove useful to the design computing community, this work implements the methodology illustrated in Figures 2.7. The literature review includes research papers from scientific fields that go beyond the fields of AEC and range from biology and physics to complexity theory.

The bibliographic research (ca. 250 publications) was organized in two levels: on a "local" level, the literature within the architectural CAD communities (Cumincad, CAADria, eCAADe and ACADia) was queried based on keywords relating to complexity (i.e., complexity theory, design complexity, architectural complexity, etc.), and main references and key terms were extracted. On a "global" level, the largest available corpus of digitalized books (ca. 10,000 publications dating from 1910 to 2010) was queried using N-Gram viewer for different types of complexity (i.e., architectural complexity) and related key terms (i.e., feedback, emergence and self-organization), which were extracted from the local bibliographic search. N-Gram Viewer is an online graphing tool that implements a probabilistic Markov model for predicting the combination of characters on a database (collection of books), given an input sequence of characters and charts, annual word counts and possible combinations (i.e., database = GoogleBooks, input sequence = design complexity, n-grams: design, complexity, design complexity, etc.) [75].

This "global" search was used as a form of validation to highlight how the types of complexity and related terms extracted from publications in architectural communities appear in the global literature. The results of my local analysis of the literature are illustrated in Figures 2.8, which plots the appearance of the word complexity in publications from different disciplines in the past 100 years. The reader can clearly observe a constant trend of publications relating to complexity between 1935 and 2010 in a variety of scientific fields, including mathematics, physics and biology, as well as cybernetics, IT and computer science.

In architectural literature, apart from Venturi's famous book *Complexity and Contradiction* that addresses complexity as a reaction to the uniformity and reductionism

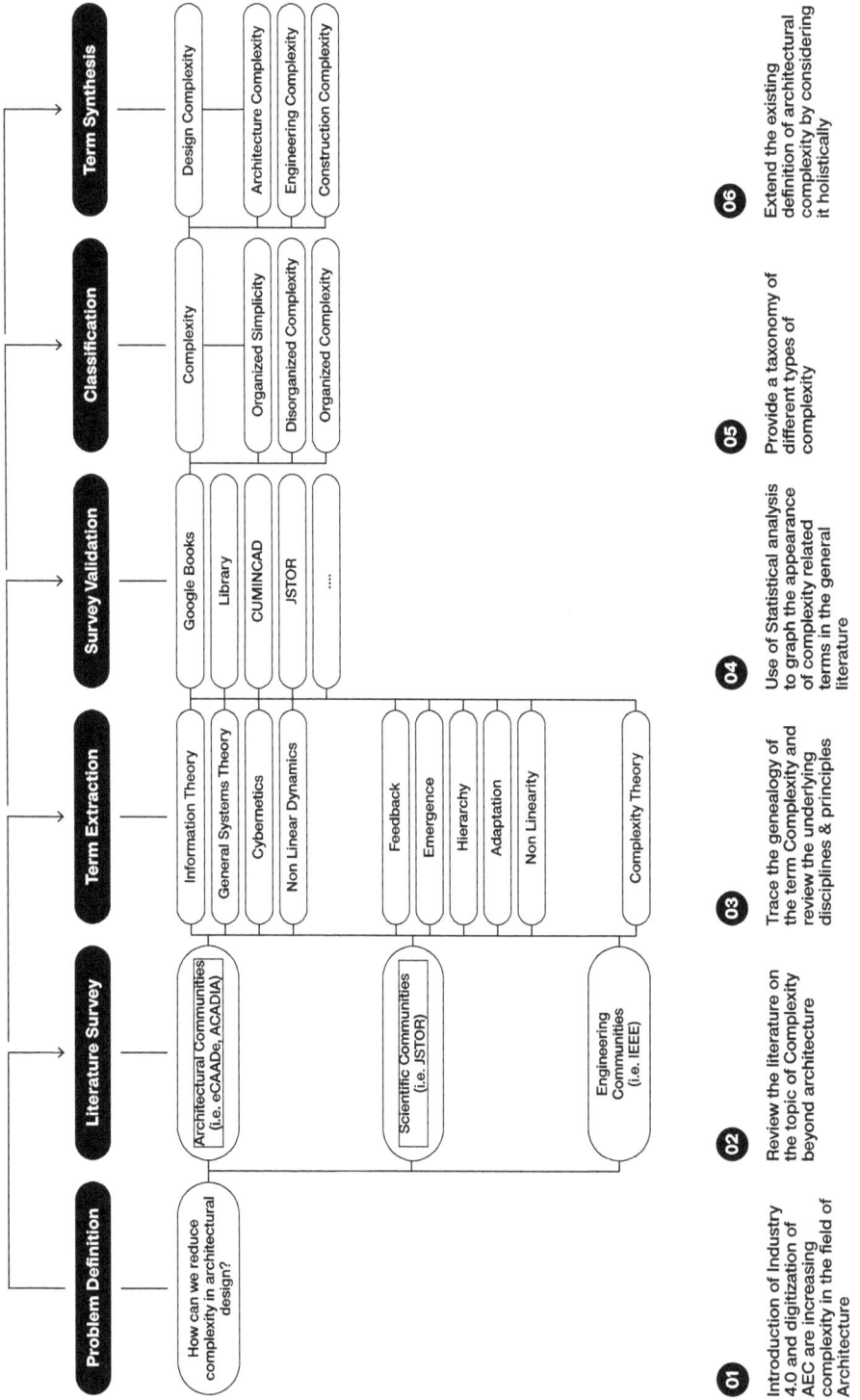

Figure 2.7: Illustration of the research methodology implemented for this literature survey.

imposed by modernism [76], there have been very few publications dealing with the topic in a rigorous and scientific manner. However, in the early 1990s, there was a significant increase in the number of publications dealing with complexity in architecture and engineering, as shown in Figures 2.8.

The review of the literature suggests that in architecture this was a consequence of the wide application of digital design tools, while in engineering and science it was the consequence of the first validated results coming out of the Santa Fe Institute (SFI). The SFI was founded in 1984 and was the first institution dedicated to providing common ground for researching complexity theory. It demonstrated how computational methods and complexity theory can be successfully applied across disciplines to solve real-world problems [77]. In Figures 2.9, the key terms relating to complexity are extracted and plotted as they appear in the global literature. The graph shows there has been a steady increase in the use of these terms since the 1940s. In Figures 2.10, the appearance of different types of complexity in the literature is plotted. It is striking that in the first half of the twentieth century, most of the terms – except for structural complexity and adaptation – were almost absent from the literature.

In the following sections, a brief historical overview of the evolution of the term is provided, and the underlying principles of complexity are described in order to better understand the term. Based on this analysis, a taxonomy of different types of complexity is devised, and measures developed to manage it within the contemporary architectural context.

2.2.1 Theoretical framework for approaching design complexity

Everyday language has included terms for complexity since antiquity; however, the idea of treating it as a coherent scientific concept is quite new [1]. Nonetheless, in the late nineteenth century, scientific progress supported by the technological advancements brought by the industrial revolution questioned the linearity and reductionism of the Newtonian paradigm, which existed in traditional sciences, such as mathematics, biology and physics [1, 78]. The establishment of new theories in the twentieth century provided researchers with new tools for studying how living organisms evolve (i.e., molecular biology) and how CASs behave (i.e., a beehive) and put complexity in the scientific landscape [103]. In the 1930s, Alan Turing was the first to associate complexity with the amount of information needed to describe a process, offering a different perspective [79]. This led Shannon to the formulation of IT in the 1940s by associating the amount of information exchange between the feedback mechanisms of different systems for the accomplishment of a given task [80]. In 1950, Bertalanffy [81] introduced GST, which dealt with systems holistically and considered their complexity in relation to the number of their parts and their relationships.

Jon Von Neumann [82] mathematically described the logic and structure of automata and considered communication systems as stochastic processes for solving complex problems. In the 1960s, Wiener introduced cybernetics, and focused on analyzing the complex behaviors between systems that operate across multiple domains

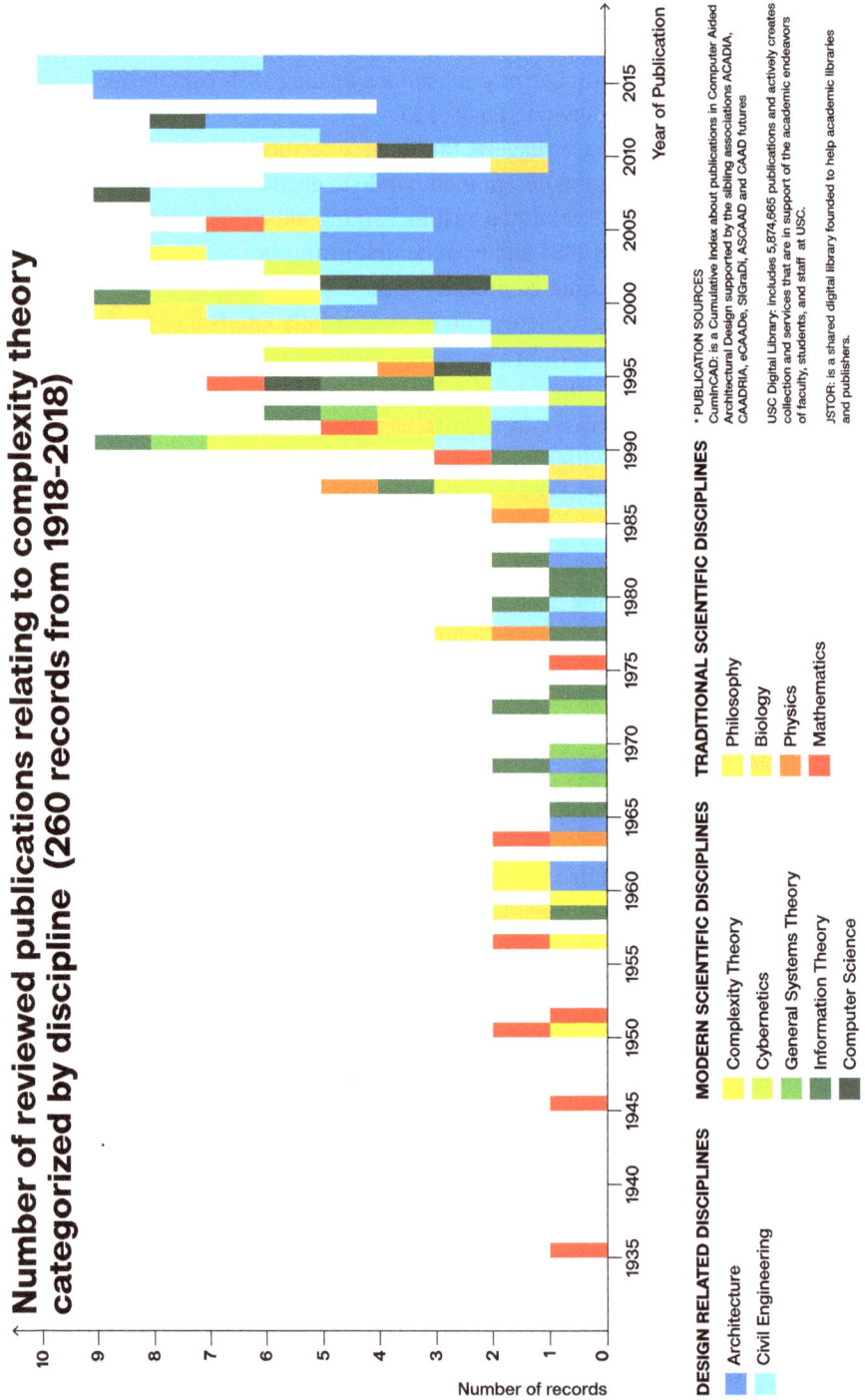

Figure 2.8: Number of publications including the term "complexity" in architecture and other scientific disciplines. The database is comprised of 260 publications which are found in academic libraries by querying the term "complexity".

such as biology, physics and architecture [96]. From the 1970s onward, complexity theory started to formalize as a separate discipline due to the incapacity of existing models to explain how biological organisms and CASs function [77]. In more recent years, the emerging field of software engineering and systems management brought about an interest in defining different types and measures of complexity [83, 84] relating to the amount of computing resources and steps needed to perform computational tasks.

The multiple definitions of complexity are an impediment to developing a clear understanding of complexity and indicate the lack of a unifying framework before the founding of the SFI. In fact, many complexity definitions reviewed in this paper represent variations of a few underlying schemes [84, 85]. A historical analog to the problem of defining and measuring complexity is the problem of describing electromagnetism before Maxwell's equations. In the context of electromagnetism, factors such as electric and magnetic forces that arose in different experimental contexts were originally considered as fundamentally different [86].

It is now understood that electricity and magnetism are in fact closely related aspects of the same fundamental quantity, the electromagnetic field. Similarly, researchers in architecture, biology, computer science and engineering have been faced with issues of complexity but have naturally considered them within the context of their own discipline. To date, the most comprehensive body of work relating to complexity relates to the research at the SFI (Crutchfield 1994). Since it was founded, the SFI has laid the foundations for most topics relating to the study of complexity theory, such as evolutionary computation and agent-based modeling, to name just a few [78, 87, 88]. The literature survey indicates that the existence of a common body of work on complexity, such as the SFI, has fostered research in different fields so it would only make sense to focus additional efforts on the development of a more unified framework for understanding complexity.

2.2.2 Sources of complexity

The main source of complexity is undoubtedly nature, which produces complex structures even in simple situations and can obey simple laws even in complex systems [89]. Complexity arises whenever one or more of the following five attributes are found in a system: (1) the existence of many parts, relationships and/or degrees of freedom; (2) multiple states/communication; (3) broken symmetry (differential growth); (4) emergent properties; (5) nonlinearity; and (6) a lack of robustness [90]. Consequently, the number of components in a building system, the tight coupling of all connected elements on multiple levels (social, structural, functional and geometrical) and the establishment of specific hierarchies among the elements significantly increase the complexity of such a system.

Living systems such as organisms, communities and coevolving ecosystems are the paramount examples of organized complexity [91]. For example, the genomic systems of a higher metazoan cell encodes on the order of 10,000 to 100,000 structural and regulatory genes, whose joint orchestrated activity constitutes the developmental program underlying the ontogeny of a fertilized egg [73]. However, apart from exam-

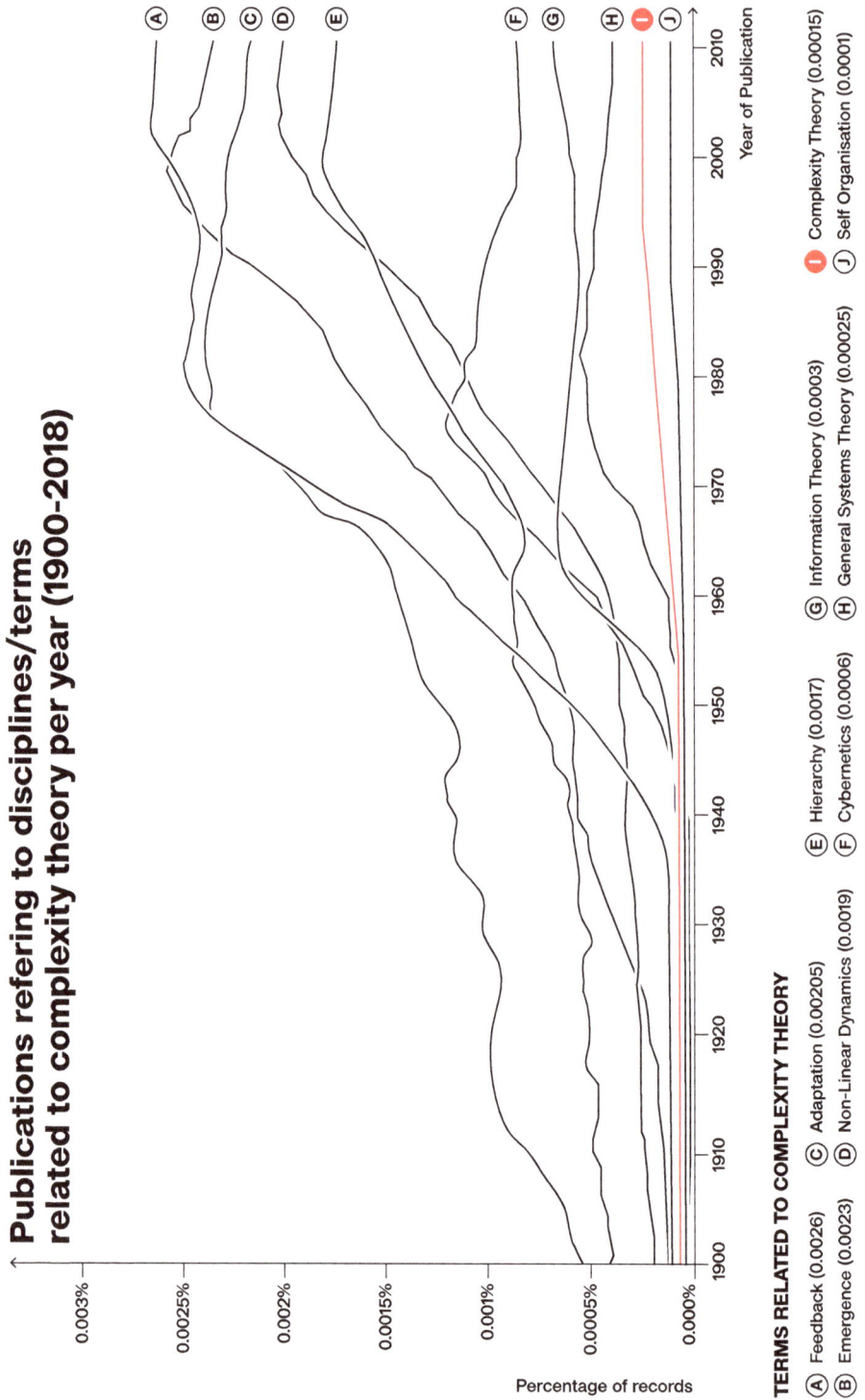

Figure 2.9: Graph plotting the appearance of key terms relating to complexity in the Google Books library.

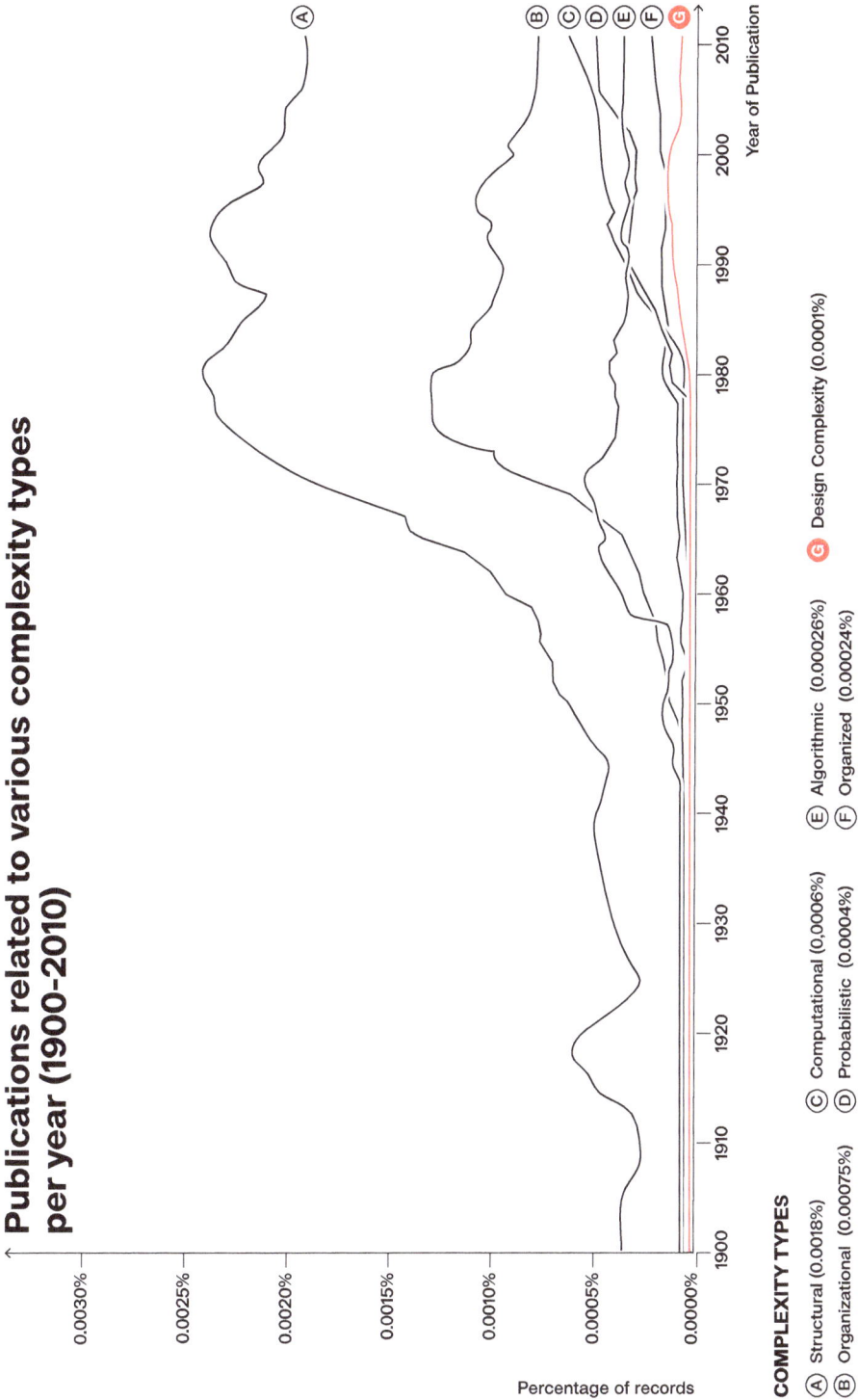

Publications related to various complexity types per year (1900-2010)

COMPLEXITY TYPES

Ⓐ Structural (0.0018%)
Ⓑ Organizational (0.00075%)
Ⓒ Computational (0,0006%)
Ⓓ Probabilistic (0.0004%)
Ⓔ Algorithmic (0.00026%)
Ⓕ Organized (0.00024%)
Ⓖ Design Complexity (0.0001%)

Figure 2.10: Graph plotting the appearance of different types of complexity as n-gram among 250,000 publications.

ples in nature and human life (e.g., behavioral, social and environmental sciences), instances of systems with characteristics of organized complexity are also abundant in applied fields, such as architecture and engineering [92]. Jane Jacobs [93] states that an essential quality shared by all living cities is the high degree of organized complexity, while Gordon Pask considers buildings not as "machines for living" but as complex environments within which the inhabitants cooperate and perform their mental processes. Thus, he considers architects as "system designers" and was one of the first researchers to identify a demand for a systems-oriented thinking in order to respond to the complex nature of architectural design. It is important to point out that there is another level of complexity within systems, such as buildings and cities, which goes beyond analyzing and understanding how they function, and which relates to the complexity of creating something that does not exist.

A number of researchers have recently adopted a more systemic approach and suggest that living cities and inhabited buildings should be considered as complex holistic systems [94]. In Figures 2.11, a building is represented as a complex system in which tightly interacting subunits are composed and assembled on many different levels of scale with hierarchies that transcend socioeconomic and cultural relationships down to geometrical forms and the basic structure of materials. In Figures 2.12, examples of physical systems (i.e., mount hill) and designed abstract systems (i.e., buildings) are graphed based on their design complexity and scale.

2.2.3 Underlying assumptions of complexity

To be able to study design complexity beyond the context of architectural design, the association of complexity theory with a number of underlying assumptions outside the classic scientific paradigm have to be taken into account [85, 95]. Classic science is based upon the assumptions that (a) an entity can be divided into component parts and that cumulative explanation of the parts and their relations can fully explain the entity (reductionism); (b) phenomena can be studied objectively (objectivity), which means that if different observers look at the same phenomena in the same way they will create similar descriptions; and finally (c) there is linear causality between phenomena, which means a cause leads to one or multiple effects in a linear fashion from the initiation to the finalization of a process [95].

The seminal work of Alan Turing and J. V. Neumann laid the foundation of complexity theory by relating complexity to the bulk of information exchange, which was defined as the length of the shortest algorithmic description for executing a given task [79, 82]. Shannon formulated a general theory of communication, as it relates to the amount of information exchanged between the feedback mechanisms of different systems for the accomplishment of a given task [80], and is considered to be the father of IT.

Along this path, both Wiener [96] and Fischer [97] viewed communication systems as stochastic or random processes and helped define IT mathematically by introducing the concepts of disorder and entropy from thermodynamics. Stochastic processes have since been central in modeling and solving complex problems with unknown structures

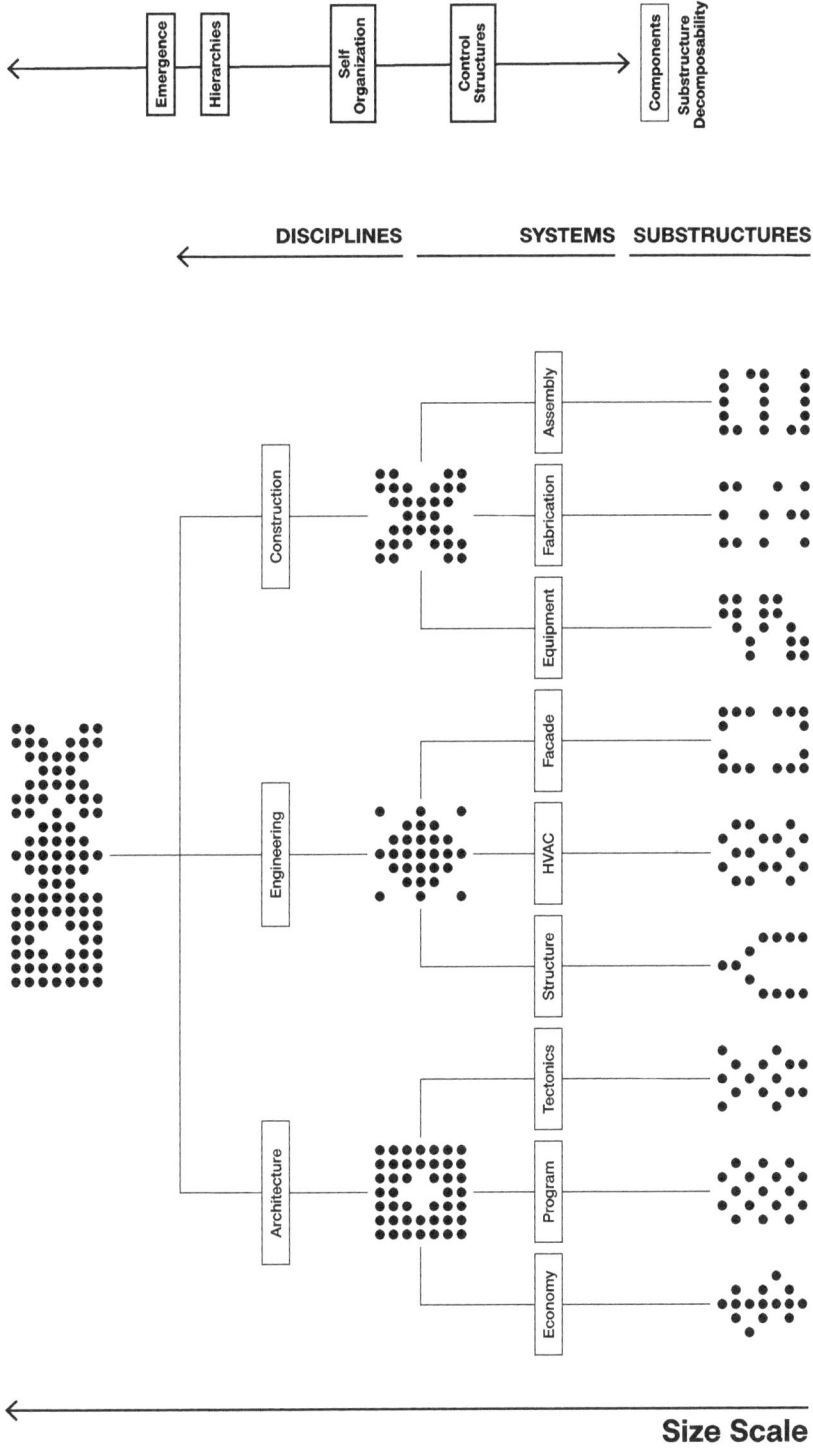

Figure 2.11: Illustration of the main characteristics of complex systems with regard to building design.

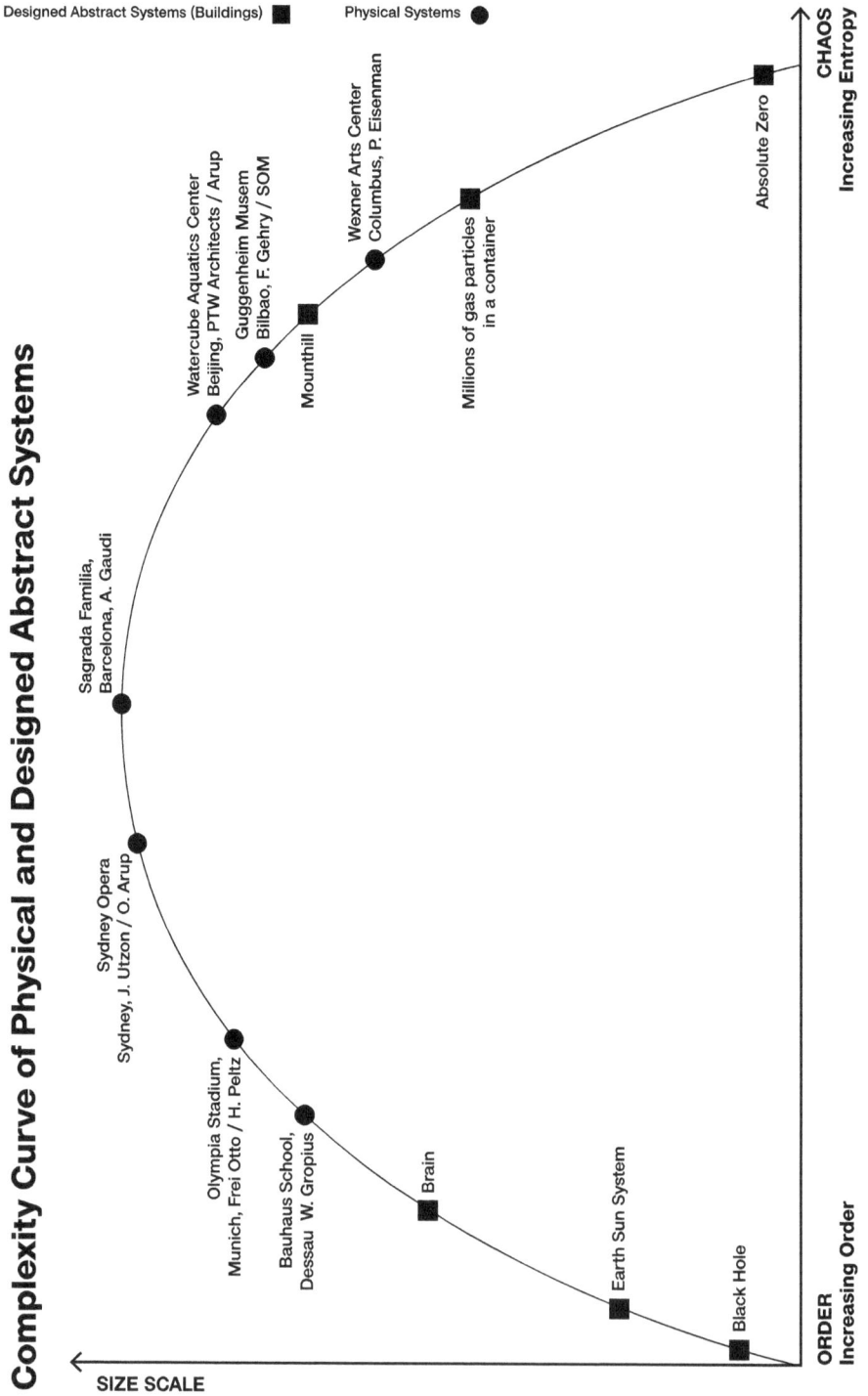

Figure 2.12: Graph mapping examples of physical and artificial systems based on their level of complexity. (ranging from total order to chaos).

and boundaries, they are of great interest for design exploration and evolutionary computation [12].

GST acknowledges the similarity of principles that apply to systems, regardless of the nature of their parts or the relations and "forces" between them [98]. Following this principle, "gas particles in a container" is a clear example of a physical system, while self-organized assemblies of organisms, such as a beehive, an anthill or a human community, can be considered typical examples of a CAS. GST defines a system as some circumscribed portion of the world that can be recognized as "itself," in spite of the fact that its constituent parts are subject to perpetual change [99]. Different systems can be characterized based on four fundamental components: structure, behavior, communication and hierarchy (levels of organization) [100].

Wiener [96] focused on the relationships among system components and the manipulation of hierarchies that exist within them, rather than analyzing each one of the components in isolation. Notions, such as feedback and control, were more central to the discipline than any law of traditional physics or mathematics. By embracing nonlinearity via circular causality (feedback) and by introducing concepts such as "forward-looking search" in system design, cybernetics contributed to the holistic understanding of complex natural phenomena [78]. It offered a new framework on which all individual systems may be ordered, related and understood based on concepts such as behavior, feedback and hierarchy and consequently its contribution to the field of complexity has been tremendous [101].

These "younger" scientific disciplines, which mainly appeared in the second half of the twentieth century and became the cornerstones of complexity theory, introduced an alternative paradigm to that of classic science, one that is nondeterministic and nonlinear [25]. In Figures 2.13, the term "complexity" is decomposed into multiple levels based on this worldview. Unlike classical scientific approaches, this new worldview is based on three basic assumptions. First, an entity is best understood when considered in its entirety (holism) and has characteristics that belong to the system (as a whole) and not to any of its parts "individually." Second, the observer is not independent from the phenomena and the observer's experiences add to the perceived reality (subjectivity). Lastly, there exists circular causality (feedback), which means that cause and effect in different phenomena is not always linear and that there is a dynamic (nonlinear) exchange between action and experience [102].

2.2.4 Defining complexity

Based on the set of assumptions described in the previous section, Weaver [103] identifies three types of complexity: organized simplicity, disorganized complexity and organized complexity. Figures 2.14 provides a taxonomy of the different classes and subclasses of complexity that has been surveyed in the literature. Organized simplicity applies mostly to "designed physical systems," such as the ones engineers were modeling in the nineteenth century (i.e., the mechanical loom). Disorganized complexity applies to both physical and artificial systems, whose behavior is almost impossible to predict (i.e., the

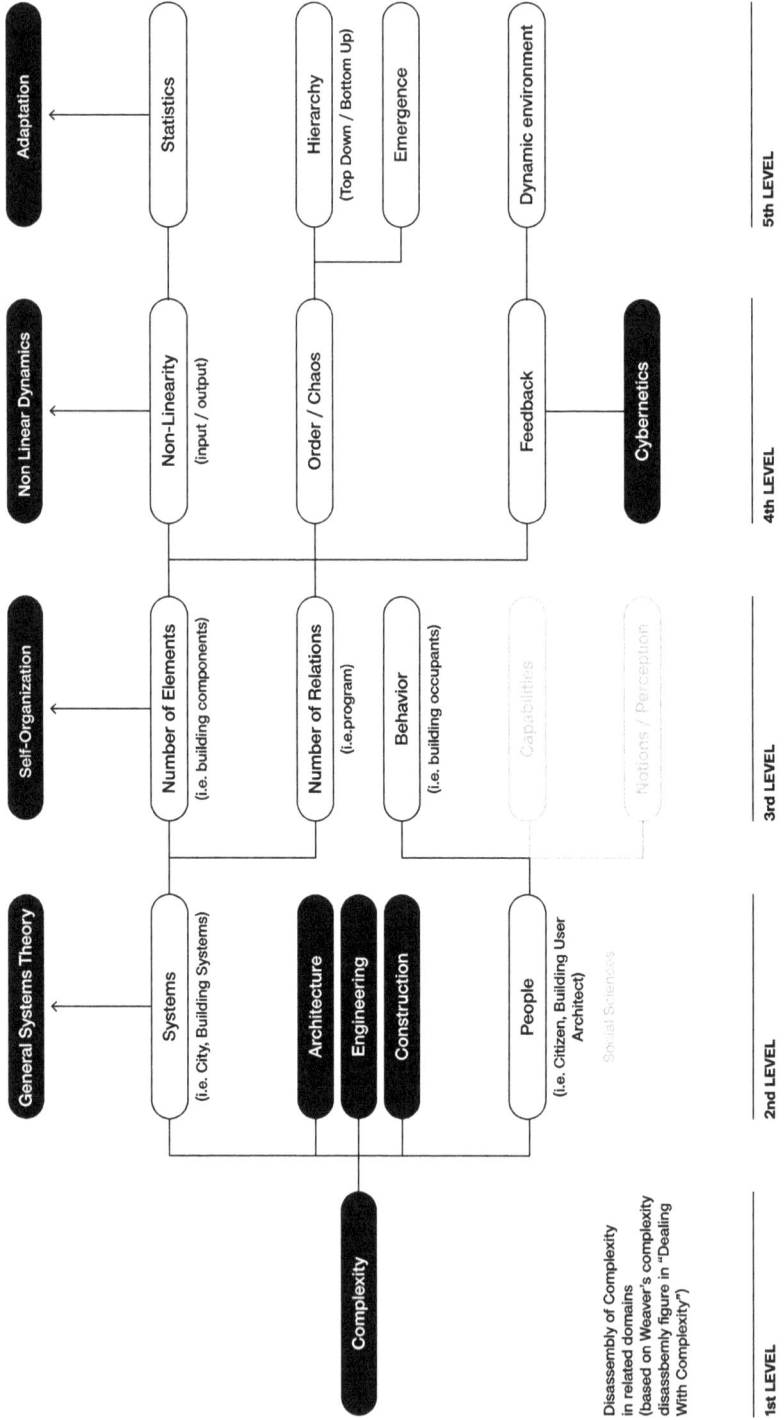

Figure 2.13: The term "complexity" is disassembled into different levels based on where it applies (second level), where it arises (third and fourth levels) and key properties in addressing and managing it (fifth level).

motions of a million particles in a gas container) [92]. Organized complexity is encountered in CASs (i.e., a beehive) and in designed abstract systems (i.e., a building) and is therefore interesting to designers, architects and engineers. By surveying the scientific and architecture literature, it is possible to distinguish four main types of organized complexity which are typical in such systems, namely structural (also organizational), probabilistic (also deterministic), algorithmic and computational [104–106].

2.2.4.1 Complexity as an absolute quantity

In biology, a living organism can be classified as structurally complex, because it has many different working parts, each formed by variations in the implementation of the same genetic coding [89]. If we consider an organism as a system, probabilistic complexity is the sum of the interrelationships, inter-actions and interconnectivity of parts within the system and between the system and its environment [107].

Based on the definition of a Turing machine, Kolmogorov defined algorithmic complexity as the length of the description provided to a computer system in order to perform and complete a task [108]. This highly compressed description of the regularities in the observed system, also called a "schema," can be used to define the complexity of a system or AI computing machine [67]. Algorithmic complexity is also called descriptive or Kolmogorov complexity in the literature, depending on the scientific community, and is defined as finding the universally shortest description of an object or process [109]. If we consider a computer with particular hardware and software specifications, then algorithmic complexity is defined as the length of the shortest program that describes all the necessary steps for performing a process, that is, printing a string.

Algorithmic complexity, in many cases, fails to meet our intuitive sense of what is complex and what is not. For instance, if we compare Aristotle's works to an equally long passage written by the proverbial monkeys, the latter is likely to be more random and therefore have much greater complexity. Bennett introduced logical depth as a way to extend algorithmic complexity and averaged the number of steps over the relevant programs using a natural weighting procedure that heavily favors shorter programs [83, 110]. Suppose you want to do a task of trivial algorithmic complexity, such as print a message with only 0 s, then the depth is very small. But if the example above with the random passage from the monkeys is considered, the algorithmic complexity is very high, but the depth is low, since the shortest program is Print followed by the message string. In the field of mathematical problem solving, computational complexity is defined as the difficulty of executing a task in terms of computational resources [111].

In computer science, computational complexity is the amount of computational effort that goes into solving a decision problem starting from a given problem formulation [112]. Within this classification, non-deterministic polynomial time (NP) complexity is one of the most fundamental complexity classes and is defined as the set of all decision problems for which the instances where the answer is yes have efficiently verifiable proofs that the answer is indeed "yes" [113, 114]. In other words, computational complexity describes how the time required to solve a problem using a currently known algorithm

Taxonomy of Complexity Types

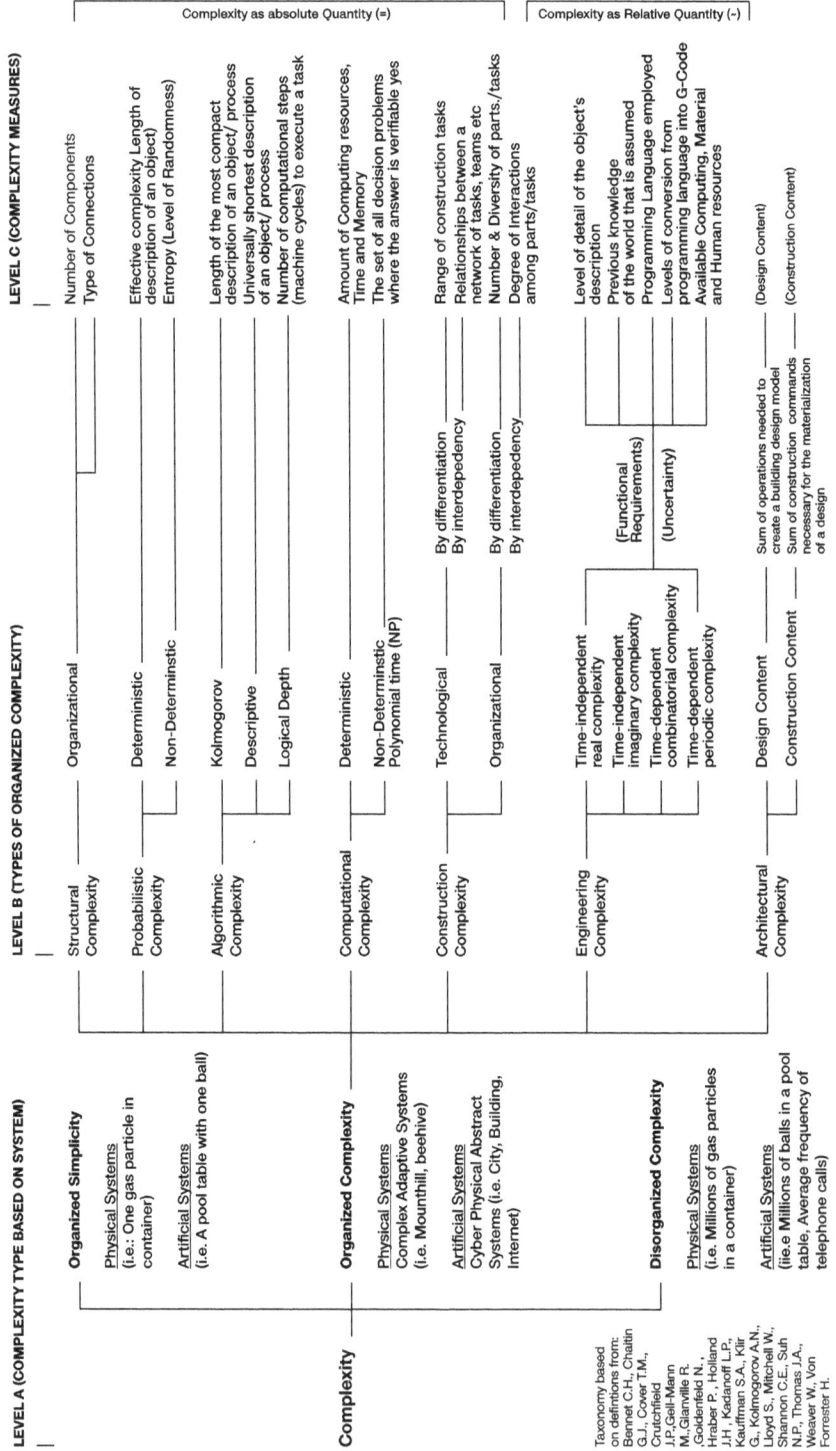

Figure 2.14: A taxonomy of different complexity type.

increases proportionally to the size of the problem. Depending on that relationship, problems are classified as polynomial time (P), NP, NP-complete or NP-hard, which describes whether a problem can be solved and how quickly. For NP-complete problems, for instance, although a solution can be verified as correct, there is no known way to solve the problem efficiently [115].

2.2.4.2 Complexity as a relative quantity

Mitchell defined architectural complexity in a digital context as the ratio of added design content to the added construction content [2]. In Mitchell's definition, design content is defined as the joint product of the information already encoded in a CAD system and the information added in response to conditions and the requirements of the context at hand by the designer. The construction content of a building is defined by Mitchell as the length of the sequence that starts with the fabrication description of a component and ends with the assembly of the whole building [116]. Per the above definitions, Mitchell's definitions overlap with that of algorithmic complexity. Design content refers to the length of the description necessary for describing to a computer system a set of instructions to create a 3D geometry. Construction content refers to the length of description necessary to generate toolpaths for the fabrication and on-site assembly of the design content. In Mitchell's definition, the designer – by operating a CAD system – handles the complexity of defining the architectural shape, and therefore his definition does not appropriately capture the computational complexity of creating "a design" (i.e., decision-making during the design process).

In engineering, Suh introduced axiomatic design and divided complexity into two domains, namely functional and physical. The functional domain includes a set of constraints, attributes and desires coming from the user as well as a set of functional requirements (FRs) that a design object needs to fulfill. The physical domain includes a set of design parameters and a set of fabrication and construction processes [105, 106, 117]. In the physical domain, the complexity of an object is related to the coupling of design parameters and the available construction processes and therefore can be described as an absolute quantity. Within the functional domain, complexity is regarded as a measure of uncertainty in achieving a set of goals defined by a set of FRs, which are coupled with a set of design parameters. According to Suh, a design is considered complex when its probability of success is low: that is, when the information content required to satisfy a number of FRs by a number of design parameters is high. With this definition of engineering complexity, Suh provided a tool to view the complexity of designed and engineered systems from a scientific, rather than a purely empirical approach, and aimed to create a higher level of abstraction in order to enable designers to synthesize and operate complex systems without making them overly complex, per se.

Lastly, in the field of construction, complexity is defined as a function of the size and uncertainty of the project on the one hand, and the combination of organizational and technological complexity on the other [118]. These two types of complexity are classified in terms of differentiation and interdependency, and thus organizational complexity by differentiation refers to the number and diversity of parts involved in a construction process, while organizational complexity by interdependency refers to the degree of interactions between a given project's elements [119]. Technological complexity by differentiation refers to the range of construction tasks, while complexity by interdependency refers to the relationships between a network of tasks, teams, technologies and construction activities.

2.2.5 Measuring complexity

As can be observed in Section 2.2.4, contemporary researchers in IT, biology, and engineering and computer science have developed different definitions and measures of complexity, but there seems to also be an overlap as they were asking the same questions of complexity but within their own disciplines. For an extensive analysis of complexity measures, see [120, 121]. However, by reviewing the literature we can conclude that the most frequent questions that appear in the literature across disciplines for quantifying complexity of an object, an organism, a problem, a process or even an investment are:
1. How hard is it to describe?
2. How hard is it to create?
3. What is its degree of organization?

The difficulty of description (i.e., logical depth) can be measured in bits (i.e., effective complexity), while the difficulty of creation (i.e., design content) can be measured in time and energy (i.e., entropy). Lastly, the difficulty of organization can be subdivided into two groups: one which measures the difficulty of describing an organizational structure and another that relates to the amount of information shared between the parts of a system as a result of its structure.

2.2.6 Decomposing complexity in the AEC

Admittedly, the design, construction and management of a building is indeed a challenging problem involving multiple disciplines and therefore it is hard to create an absolute definition, as well as a measure of complexity. Following Mitchell's and Suh's definitions, design complexity in the AEC will be examined under the scope of two domains. The first is the virtual/functional domain, which is directly related to two levels of complexity: the complexity of design problems and the complexity of design processes. The second is the real/physical domain, which is directly related to the construction (fabrication and assembly) and building systems integration.

We will consider the design-to-construction process holistically and discuss how complexity arises in these subtopics, while examining the possible ways it could be addressed. In doing so, the aim is to clarify architectural complexity, and translate achievements from other fields for design purposes.

2.2.6.1 Complexity of design problem representation

In design, as well as in science, when given a specific problem one has to deal with many interconnected variables, often deriving them from FRs [122]. Contrary to science, in the design world problems are wicked, which means there is no clear formulation that contains all the information the problem-solving mechanism needs for understanding and solving the problem [122]. Unfortunately, the lack of clarity in many of these parts increases the complexity of this kind of problem. Through the act of designing, architects and designers face two different aspects of complexity [123, 124]. One aspect relates to the lack of complete information about the design problem, which makes the formulation of a universal design solution difficult [106]. The other aspect relates to the fact that the target is to create something new, which means that the solution is not specified. The paradox is that if a new and innovative design procedure can be specified, how can the resulting outcomes be innovative?

Architectural design has a long history of addressing complex programmatic requirements through a series of steps without a specific design target [125]. Unlike other fields, such as engineering, where the target is to solve a particular problem in the best possible way, architectural design problems, because of this novelty aspect, are open-ended, in a state of flux, uncertain and therefore ill-structured [122]. For instance, the task of designing a house leans toward the side of ill-structured problems; the amount of uncertainty involved makes the specification of the problem hard, so the solution becomes complex [126].

Simon supported the idea that the degree of complexity of any given problem critically depends on the description of the problem. Holland described optimization problems in domains as broad and diverse as ecology, evolution, psychology, economic planning and AI and, by abstracting from the specific field, he examined commonalities relating only to the complexity and uncertainty of such problems [66, 91]. Although designing a house is not an optimization problem, the design methodology required to approach it using digital means can share common features of adaptation and self-organization with an optimization problem in biology, such as the construction of ant hills [127].

2.2.6.2 Complexity in the design process

Although there is no well-formulated consensus model of the design process in architecture, a typical model that features the following has emerged: (a) the assumption that most design problems are ill-defined (wicked) by definition; (b) the recognition of the importance of prestructures, presuppositions or proto-models as the origins of solution concepts; (c) the emphasis on a conjecture–analysis cycle in which the designer and the other participants refine their understanding of both the solution and the problem in parallel; and finally (d) the display of the essential spiral and nonlinear characteristics [128]. Even though the use of computational tools offers an opportunity to formalize the design process, there are no formal architectural design methods that follow the above model to work in a systematic way.

It is largely accepted that during the design process the architect navigates through an ill-defined problem domain and employs various strategies to elaborate the problem description, iteratively generates and explores design alternatives and, after a number of iterations (e.g., when given a time constraint), the architect proposes a solution [43]. In the digital realm, this design process can be described as a structured and exploratory activity. Decision-making in this case is more complicated as the architect has to describe to the computer explicitly the decomposed problem, the set of variables that relate to this problem decomposition as well as the context (i.e., requirements from the city or building code). In order to reach a decision, exploration of alternatives via variable instantiations is the common process. Design exploration, in this case, is akin to changing the problem space within which the decision-making occurs. Learning implies the restructuring of knowledge based on the presupposition-conjecture-analysis-evaluation cycle [42].

The ill-structured nature of the design problems, the changing context and the engagement of the human factor do not allow the clear definition of the solution space to be explored and therefore increase the complexity of the design process. Nonlinearity and the amount of interconnected design parameters between the conjecture–analysis cycles also increase the complexity. In an attempt to improve the latter, both research and professional practices have focused on automating traditional manual methods of production using CAD and algorithmic design tools [43, 129]. Current parametric design systems have facilitated the design and management of nonstandard geometries and, at first sight, seem to reduce the complexities of the design process – at least in terms of algorithmic complexity. This is easily measured if we consider that the printout of a code for a parametric model, together with a table of all the parameter sets, is much shorter than all the workshop drawings [10].

However, there are complexities relating to the description of the problem and the definition of efficient design strategies, which remain largely unresolved. For instance, in biological systems, the blueprint of an organism, that is its genetic code, is considered to be a set of instructions that are dependent on a particular environmental context for its interpretation and manifestation and is subject to evolution and adaptation. In architecture, digital design tools were developed to streamline the production of the blueprints of buildings, and thus focused less on formalizing the encoding process where the designer interacts with the computer in order to manifest his/her ideas. Although the architectural design process has been computer-based for more than 20 years, only recently has there been rigorous research leaning toward the adaptation of computational design techniques for design exploration [30]. In order to leverage the power of computation, more emphasis should be placed on researching how design abstractions can be formally described to computers. That way, similar to biology, evolutionary and learning mechanisms can be used to extend the cognitive capacity of designers and help them explore new design schemes or evolve existing ones based on previous knowledge and/or experience.

2.2.6.3 Complexity in construction processes

Construction projects are invariably complex and are becoming increasingly more so because of the fragmented nature of the industry and its capability to both generate and collect a large amount of data [130, 131]. Building construction is typically characterized by the engagement of multiple, separate and diverse groups such as architects, engineers, consultants and contractors for a finite period of time. On a higher level, organizational complexity in a project arises when the number and level of differentiation and interdependencies of all the contributing organizations increase [132]. The differentiation can be either vertical, referring to the level of detail the activities of a project might entail, or horizontal. Horizontal differentiation refers to the number of formal units involved, such as departments and groups, or to the way the tasks are structured in terms of labor subdivision and the level of specialization required for each task [133]. For instance the organizational complexity of a project can increase if the number of different occupational specialization utilized to accomplish a project is high. Another example is when during the different phases of a project, architects, engineers and different specialists are working from geographically dispersed offices at different time zones and within different social structures. The dynamic and distributed character of the construction environment increases complexity as a result of the required amount and types of information exchanged between all contributing parts (i.e., designers, engineers and contractors) [134]. The multitude of different disciplines, the lack of integrated frameworks and the reliance on classification methods conducted by human protocols hinder this communication exchange and have caused inefficiencies, project cost overruns and time overruns. Furthermore, the quality and maintainability are reduced, the design intent is diminished and the efficient access to objects and information in a timely manner is inhibited [21, 135].

On a lower level, complexity in construction occurs when dealing with the fabrication of nonstandard geometries and nonrepetitive assembly methods. The fabrication process relates to the manipulation of raw material for the production of discrete elements, while the assembly process refers to the combination of discrete elements into systems [2]. Complexity in fabrication can be described as the length of the translation of a specific geometry or shape description into a sequence of instructions for a computer numerically controlled machine or a robotic arm that will fabricate the geometry in question. Additionally, complexity in the assembly process can be described as the number and diversity of steps required to combine discrete fabricated elements into a single structure.

Consequently, if expressed in terms of algorithmic complexity, the number and descriptive intricacy of elements and/or the steps needed for their fabrication and assembly increase construction complexity. However, the dynamic environment of the construction site and errors in the construction process result in nonlinearity and high levels of uncertainty, thus making it harder to describe construction processes in terms of algorithmic complexity; they can be better described, relatively speaking, in terms of entropy.

2.2.6.4 Complexity in building (control) systems

The twentieth century has included the introduction of many building systems technologies, such as electrification, air-conditioning, fire protection systems, active structural damping, automatic control systems, computer networks and high-performance glazing, to name a few [136]. These technologies have transformed buildings from simple structures providing shelter into complex material systems that react to environmental parameters through automated facade systems and, to a certain extent, respond to their user's needs [137]. Although it is remarkable how new technologies have been sequentially incorporated into the building construction process, there are still issues that need to be addressed, such as the interactions between different building systems, processes and occupants [138]. The general vision is an intelligent building, which is responsive to the requirements of the occupants, environment and society by being functional and productive for occupants in terms of energy consumption and CO_2 emissions [139].

Due to the complexity and diversity of behavioral patterns and preferences, the influence of the occupant's behavior is considered only in simulation [21]. Moreover, the level of sophistication of the system's components, combined with the fragmented nature of the construction industry, has not allowed the integration of different systems in the early design phase or later on in a building's life cycle [136]. Consequently, a great challenge remains: to what extent do building systems perform to the level they are intended, both in terms of occupant satisfaction and in terms of energy consumption?

2.2.6.5 Extending the definition of architectural design complexity

Based on the types of complexity described above, it is clear that in order to holistically encapsulate the complexity of architectural design in a digital context it is important to consider the definition of architectural complexity beyond the realm of architecture. This is done by introducing concepts from the field of engineering, and specifically by adopting Suh's concept of engineering complexity. Suh divides complexity into two domains, namely a functional and a physical one. This classification is integrated with Mitchell's definition of complexity in order to further analyze the concepts of design and construction. Following Suh's definition of engineering complexity, the design content lies in both the virtual and functional domains and can be further subdivided into architectural design and engineering design content. Architectural design content includes constructing a design model that combines the constraints, environmental conditions and design intentions (design approach) set forward by the occupants/stakeholders/decision-makers with a set of FRs.

This design model maps a set of constraints and attributes to the set of FRs and couples them with a set of design parameters by considering process variables. The engineering content includes finding the shortest description for coupling the functional parameters with design parameters by considering process variables. Construction content includes the mapping of design parameters to process variables, such as available resources (material), building technology, construction activities, time and cost. Therefore, the complexity lies in the physical domain.

The complexity of construction content is extended to include the number of operations necessary to realize a design and also to engulf (a) the level of differentiation of tasks, (b) the interdependency among the function of the tasks, (c) the degree of labor skills each task requires and (d) the level of uncertainty in completing a task. Risk, and the uncertainty of how design parameters are coupled with a set of process variables (building paradigm), can be considered a measure of complexity in the physical domain.

The complexity of the design content in the virtual domain depends (a) on the amount of design decisions required to formulate a design model and (b) on the uncertainty of satisfying the FRs given a set of design parameters. Moreover, entropy can be perceived as a measure of randomness of operations in the design process and can be used for measuring the potential of generative design systems to generate novel solutions based on a set of design decisions [9] (Figures 2.15 and 2.16).

2.2.7 Tools for managing complexity

The survey of the literature across different disciplines indicates that the main research tools for managing complexity include (a) abstraction, (b) modularity and (c) the idea of scalability. Abstraction can be used as a tool to compare data by treating them as generic entities that can be compared, encapsulated and drawn generalizations from. Modularity can be used as a concept that enables the development of functionally specific components that specialize in solving particular aspects of problems. Lastly, multiscalability is a concept that allows one to formally express features and principles (rules) that may be present across different levels but may have completely different effects according to the specificities of the scale. For instance, a fundamental rule in one scale may, on a much larger scale, reveal itself to be a frozen accident. In order to emphasize the importance of abstraction, modularity and scalability, let us consider an example from physics by P. Ferreira [140]. Suppose you want to describe the number of particles of a given entity with the mathematical function:

$$\{- \sum_i (\vartheta^2 \nabla^2_i)/(2m_e) - \sum_i (\vartheta^2 \nabla^2_j)/(2m_n) + e^2/(4\pi\varepsilon_0) \sum_{i1,i2} 1/|r_{i1} - r_{i2}| +$$
$$+ z^2 e^2/(4\pi\varepsilon_0) \sum_{j1,j2} 1/|R_{j1} - R_{j2}| - ze^2/(4\pi\varepsilon_0) \sum_{i,j} 1/|r_i - R_j|\}\psi = E\psi.$$

This equation, which describes matter at atomic levels, i and j, ranging between 1 and 1,020, represents the number of particles in a human body. It is therefore complicated to solve this equation for ψ, the wave function for the particles. Actually, it is impossible to solve this equation analytically for the atom of helium ($i = 2$ and $j = 1$). So how can we proceed when faced with such problems?

Although we are unable to solve this equation analytically, physicists use abstraction to explain the dynamics of larger particles. Physicists consider the problem at different levels, for example, at the molecular level, and develop models from there,

Phases of design to construction process in relation to architectural complexity

DIGITAL DESIGN APPROACH

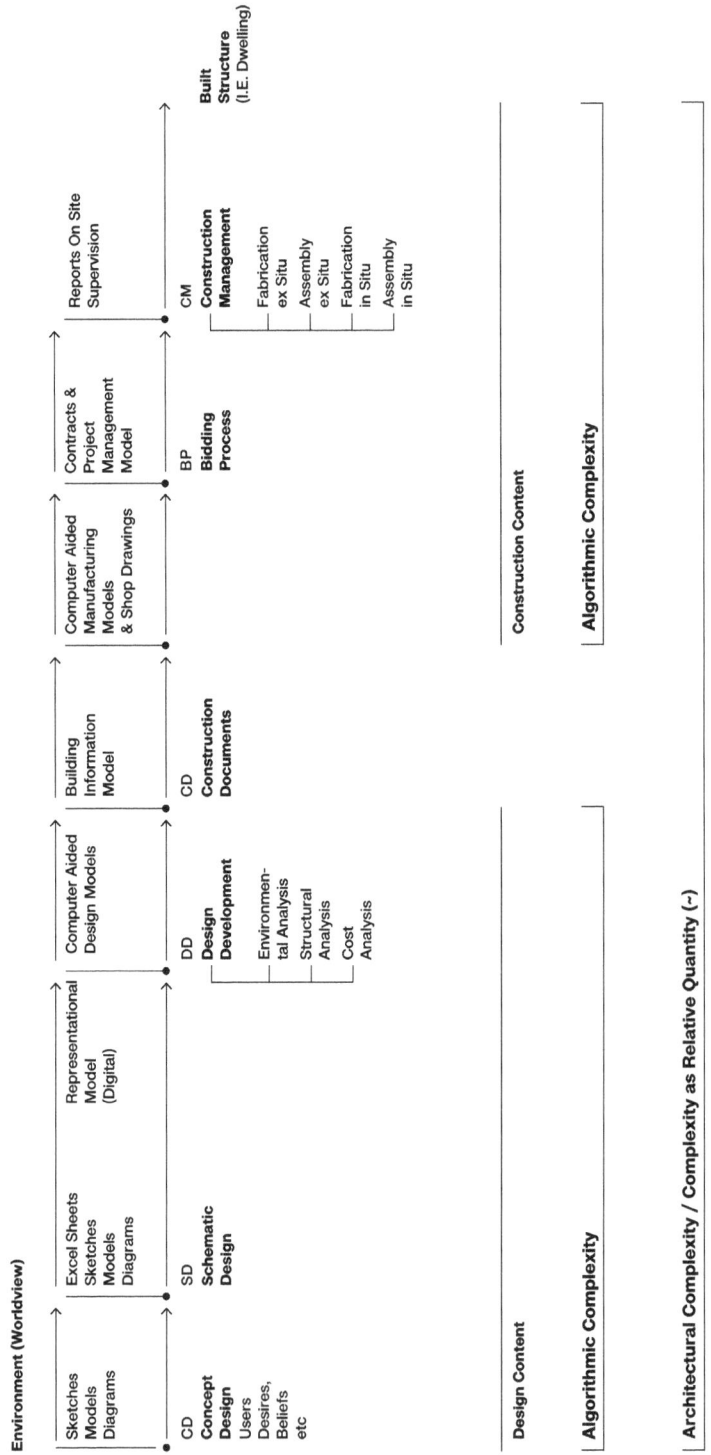

Environment (Worldview)

Sketches Models Diagrams	Excel Sheets Sketches Models Diagrams	Representational Model (Digital)	Computer Aided Design Models	Building Information Model	Computer Aided Manufacturing Models & Shop Drawings	Contracts & Project Management Model	Reports On Site Supervision	**Built Structure** (I.E. Dwelling)

| CD **Concept Design** Users Desires, Beliefs etc | SD **Schematic Design** | | DD **Design Development** Environmental Analysis Structural Analysis Cost Analysis | CD **Construction Documents** | | BP **Bidding Process** | CM **Construction Management** Fabrication ex Situ Assembly ex Situ Fabrication in Situ Assembly in Situ | |

Design Content

Algorithmic Complexity

Construction Content

Algorithmic Complexity

Architectural Complexity / Complexity as Relative Quantity (~)

Figure 2.15: Design-to-construction process in based on W. Mitchell's definition of arch. complexity.

Computational Design Approach

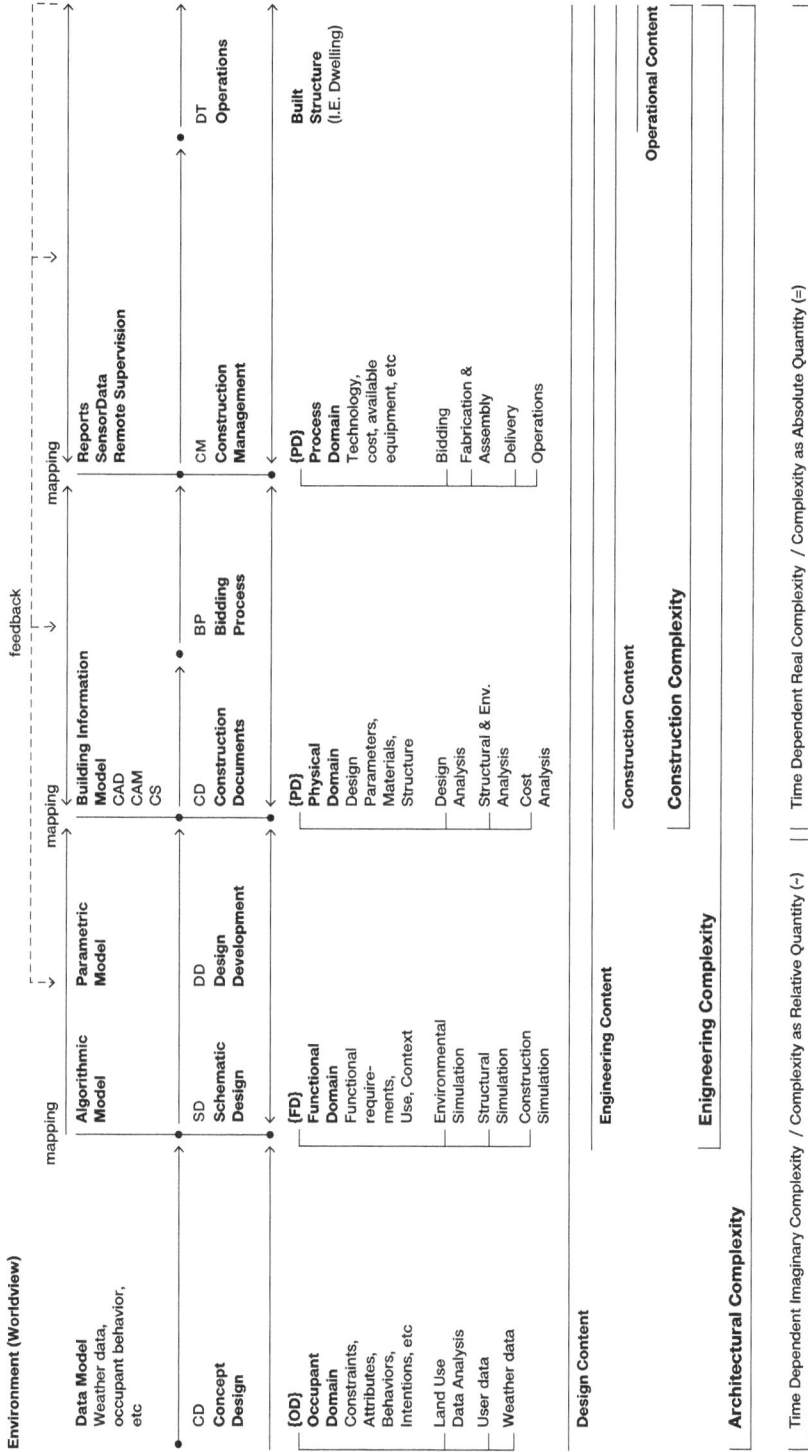

Environment (Worldview)

mapping →   feedback   mapping →

mapping →   mapping →

	Algorithmic Model	**Parametric Model**	**Building Information Model** CAD CAM CS	**Reports** **SensorData** **Remote Supervision**

Data Model
Weather data,
occupant behavior,
etc

CD
Concept
Design

SD
Schematic
Design

DD
Design
Development

CD
Construction
Documents

BP
Bidding
Process

CM
Construction
Management

DT
Operations

[OD]
Occupant
Domain
Constraints,
Attributes,
Behaviors,
Intentions, etc

Land Use
Data Analysis
User data
Weather data

[FD]
Functional
Domain
Functional
require-
ments,
Use, Context

Environmental
Simulation
Structural
Simulation
Construction
Simulation

[PD]
Physical
Domain
Design
Parameters,
Materials,
Structure

Design
Analysis
Structural & Env.
Analysis
Cost
Analysis

[PD]
Process
Domain
Technology,
cost, available
equipment, etc

Bidding
Fabrication &
Assembly
Delivery
Operations

Built
Structure
(I.E. Dwelling)

Design Content

Engineering Content

Construction Content

Operational Content

Architectural Complexity

Enigneering Complexity

Construction Complexity

Time Dependent Imaginary Complexity / Complexity as Relative Quantity (-)	Time Dependent Real Complexity / Complexity as Absolute Quantity (=)

Figure 2.16: Design-to-construction process based on the author's definition of arch. complexity.

considering the characteristics of molecules they can observe, such as mass, charge and polarity. Equations and simulation models in physics are often solved by approximation, using large amounts of computer resources and power.

With the introduction of computers, physicists were able to computationally simulate and predict molecular behaviors that could not be observed otherwise. Thus, in many scientific disciplines, computation has been considered a neutral rather than an exploratory tool, and has advanced the discipline's mathematization [47]. In architecture, despite the widespread availability of digital design tools and the increasing computational capacity, which allows the development of complex building models across scales, the design approach is in most cases traditional and rather geometry centered. Designers tend to create descriptive design models by making assumptions for the occupants. These models are then rationalized in a top-down manner for environmental and structural parameters or construction constraints. However, information technologies supported by data-driven design approaches enable the development of multiscalar predictive models that integrate multiple design parameters with multiple simulations (structural behavior, occupant behavior, etc.) and can generate geometric configurations in a bottom-up manner. These generated designs can then be evaluated and refined based on performance parameters and real-time occupant feedback.

At this point, it is important to emphasize the difference between computer-based tools and computational tools. Up until the early 2000s, the majority of digital design tools were computer-based, automating and mechanizing data handling within the design process. Recently, we have seen the introduction of computational tools that promote design exploration and an attempt to extend the designer's intellect by correlating data in novel ways [30]. The generalization of geometry via computational tools and methods requires a higher level of formalization of design thinking but also provides new forms of creative expression. Suh, who introduced the axiomatic design approach, argues that via design formalization and by systematically incorporating scientific principles in design, there is the potential to inform design with empiricism and intuition and evaluate the complexity of a design problem holistically [117]. Along the same lines, Kotnik suggests that the consideration of digital design as computable functions offers an opportunity to systematize design knowledge and compare existing methods [47]. This is due to the fact that mathematical functions make the governing of cause and effect explicit, and therefore connecting methods of digital design with the concept of computational functions offers a platform for directly transferring formal mathematical concepts into architecture.

However, there are two approaches regarding this transfer of mathematical concepts onto contemporary digital practices. One approach is related to performance-based design techniques, which are directly related to optimization problems. This approach, although quite prevalent nowadays, can easily direct design thinking toward the parametric manipulation of optimization routines. The other approach is based on the very idea of computation and the algorithmic description; it offers the possibility of precisely controlling the relationships of FRs and design parameters between architectural elements in unique ways, rather than providing optimal solutions. The formalization of design thinking cannot replace the process, but it can act as a framework for exchanging knowledge between fields of science and design and a more systemic examination of contemporary design practices.

The tools for managing complexity outlined in the section above can offer a high-level framework for managing the multiple levels of complexity included in building design using digital design methods. Following the axiomatic design approach, design can be broadly defined as the creation of a synthesized solution that satisfies a set of perceived needs through the mapping of processes between FRs in the functional domain and design parameters in the physical domain. Through this perspective, concepts such as self-organization, autonomy, topology, holism and entropy from the field of complexity theory can be used to correlate the multiple levels of complexity across the different fields of AEC.

2.2.8 Complexity and complex adaptive systems

One of the most important characteristics of complex nonlinear systems is that "the sum is more than its parts." In other words, complex systems cannot be successfully examined by breaking them down into parts and then partially studying the characteristics of every part that comprises the system in isolation. Instead, it is necessary to look at the whole system, even if that means taking a crude look, and then allow possible simplifications emerge from the work.

This makes it clear that the study of CASs has a lot in common with the design process. CASs are particularly interesting for managing design complexity because it has been proven that a model or schema in relatively few dimensions is exploring a gigantic strategy space far from any optimum or equilibrium [23]. Think, for example, of a computer learning to play chess. In the not-so-distant past, chess was an unsolved problem, as the game "Go" still is today, and the adaptive computer learning necessary for chess was not yet available. Nowadays, reinforcement learning models have made great progress in developing models of computer Go games. Similar approaches could prove helpful for design problems as well. In order to better understand how CASs can be utilized in the design process, we need to ask a number of questions and relate CAS to other complex phenomena that do not share the same properties.

How does a complex system operate? How does it engage in passive learning about its environment, in the prediction of future impacts of the environment and in the prediction of how the environment will react to the computer's behavior? Another question that arises is how it differs from a system like a turbulent flow? Turbulent flow in fluid dynamics is a complex phenomenon where an eddy is created because of disruption in the flow (i.e., obstacle). The initial eddy gives rise to smaller eddies, and so on. Certain eddies have properties that enable them to survive in the flow and have offspring (new eddies), while others do not and die out. Why is a turbulent flow not regarded as an evolutionary and adaptive system?

The answer lies in the way information about the environment is recorded. In CASs, information is not merely listed in what computer scientists call a lookup table. Instead, the regularities of the experience are encapsulated in highly compressed form as a model or theory or schema. Such a schema is usually approximate, and sometimes wrong, but it may be adaptive if it can make useful predictions, including interpolations, extrapolations and sometimes generalizations to situations that are very different from those previously encountered. It is important to note that the adaptive process need not always be extremely effective in achieving apparent success at the genotypic level, in biological evolutionary terms. It is more important that the adaptive process is effective at the phenotypic level, which is a combination of the expression of an organism's genetic code (its genotype) and the influence of environmental factors that can affect the characteristics and behavior of an organism. A number of examples stemming from research on CASs demonstrate that simple organisms have the capacity to improve their predictions and behaviors over time through the appropriate encapsulation of information related to their environment [141].

Today, most researchers in computational design research and academic circles have yet to take a crude look at the whole. Instead, a lot of computational design research is focused on specialization (i.e., performance-based design and robotics) and it is taken for granted that serious work can be done only by looking at one or few aspects of a building (i.e., geometry). Yet every architect needs to make decisions while pretending that all aspects of a design scenario have been considered, including all the possible interactions among them (design engineering and construction). However, if there are only disconnected specialists to consult, the collation of their opinions and information does not always reflect a fair picture of the whole, and the final result will not always be well orchestrated. If the complexity that arises from the interaction of these specializations itself is not managed, the potential of digital media and computation in architectural design is not fully leveraged.

2.2.9 A holistic design approach for managing complexity

The analysis in the previous sections shows that the complexity of creating a building design description (added design content) lies in the functional domain and is conditioned by the (a) time, (b) information exchange (bits), (c) energy (i.e., entropy) required to come up with a design proposal and (d) the level of uncertainty (i.e., percentage of probability) of fulfilling a set of design goals defined by FRs and selected design parameters.

On the other hand, the complexity of constructing a given design description (construction content) lies in the physical domain and can be measured as an absolute quantity with field-specific dimensions. Therefore, the construction content defined by Mitchell is extended to include the (a) description of a sequence of fabrication and construction activities, as well as (b) the range of construction activities, (c) the level of interdependency between activities and existing resources, (d) the technological sophistication required and (e) the level of risk in delivering it. By embracing design complexity on multiple levels, particularly in the physical and functional domains, we can determine the complexity of a design-to-construction process by considering not only geometric complexity and the available technologies (i.e., construction methods and digital fabrication) but also environmental parameters (location, orientation and azimuth), building performance data (heating and cooling loads), dynamic user behavior (space occupancy) and specific construction tasks.

Practical application of complexity theory may be found in many disparate disciplines and has included modeling approaches with both systemic scientific and utilitarian objectives. It is not operationally meaningful to view complexity as an intrinsic property of an object (i.e., a building); instead, it is better to consider complexity holistically and assume that the level of complexity arises from, or exists in, abstractions of the world. In the field of software engineering, for instance, by breaking down complexity into structured (computable) situations, engineers have managed to deal with both complex problems and new technology.

2.2.10 Summary

To summarize, this section has provided a critical review of the key assumptions and definitions of complexity in architecture and other disciplines. The review of the literature shows that the lack of a unifying theory for complexity has hindered research efforts and has resulted in researchers from different disciplines creating similar definitions of complexity, because they were considering complexity only within the bounds of their domain. The review also shows that in the field of digital architecture, the lack of scientific approaches has resulted in approaching complexity diagrammatically and solely through the perspective of geometry.

The main sources of complexity and their underlying assumptions have been clearly laid out. Additionally, a taxonomy of complexity definitions was presented along with a classification of complexity measures that have been developed to help design researchers situate the complexity of buildings on the general map of complexity theory. The term "complexity" in design has been decomposed in the domains of the AEC, and the existing definition of architectural complexity in digital design [2] has been extended with notions of complexity from the fields of engineering and construction management.

From the discussions in this section, we can conclude that understanding complexity is important in order to develop computational methodologies that reach beyond geometry and to help designers develop new kinds of design abstractions based on a deeper understanding of natural form-making processes (e.g., nest building), rather than the figurative replication of a form (e.g., the free-form shape of a nest). There is a need to develop design tools that can account for multiple aspects of design (structural, material, environmental) early in the design stage and handle complexity without the tools being overly complex themselves. To achieve this, the decomposition of design problems into subproblems is vital to help develop MASs that can help deal with the problems. Last but not least, the investigation of mechanisms that govern CASs is considered crucial to the development of custom MAS for design.

2.3 Multi-agent systems (MASs)

A MAS is defined as a computerized system composed of multiple interacting agents within an environment. Agents can act together to achieve more complex goals than the ones an agent can achieve on its own. Minsky termed as "agency" a group of agents acting together, which later evolved into the field of MAS [5, 67]. The main advantages of MAS approaches are that due to their modularity and distributed nature, they are able to solve complex problems. MASs are considered an example of distributed AI (DAI), and such systems have shown capacity to develop intelligence via heuristic search methods and reinforcement learning.

Even though there is a considerable overlap between agent-based modeling and simulation (ABMS) and MASs, it is important to point out that a MAS is not always the same as an ABMS. The main difference between a MAS and an ABMS is that the latter is used to search for insight into the collective behavior of physical agents (e.g., bees and termites) which, by simply following local rules and without overall (top-down) coordination, leads to complex global phenomena. Classic examples of ABMS can be encountered in CASs and include, among others, flocking behaviors and ant colony optimization models. Most of these models fall under a more general research area known as swarm intelligence [23]. On the contrary, research on MASs is targeted toward solving specific engineering problems, and the agents' behaviors and structure can be modeled according to the problem and not according to a natural system [142]. The ABMS terminology tends to be used in scientific fields such as biology, while MAS is more common in engineering and technology. Typical applications of MAS research range from online trading and disaster responses to manufacturing or modeling pedestrian flows. The next section discusses how the study of CASs can help designers develop MAS. Section 2.3.2 contains a list of different models and ontologies of agents, the basic programming unit of MAS.

2.3.1 Complex adaptive systems and MAS for design

A popular analogy used to describe the potential use of MAS approaches to solve complex problems is that of social insects. Termites, for instance, have been colonizing a large portion of the world for several million years and collectively building sustainable structures that utilize available resources. The success of social insects and other creatures (i.e., beavers) in building their own habitats [143] can serve as a starting point for developing new abstractions in architecture and engineering using a MAS approach.

Following the founding of SFI, there has been a growing interest in developing models to simulate CASs that can be used to reproduce some features of the natural system, or to apply the model to another system to predict its behavior [23]. Necessary steps toward achieving this goal are (a) understanding the underlying assumptions, (b) understanding the mechanisms that generate collective behaviors in nature and (c) developing metaphors that show how similar mechanisms can be applied to architectural design. Regarding the underlying assumptions, the previous section includes an analysis of different assumptions and how they can be used to form abstractions. Regarding the mechanisms, this is where modeling plays an important role.

Modeling an artificial system, such as a building or a city, is very different to modeling an ant colony. In the former, the model is used to represent an idea and manifest its constructability, while in the latter case the model is used to uncover what happens in a CAS and use that knowledge to make testable predictions. Digital models in architecture have been successfully employed for formal explorations, but only in a few cases have they been used to make predictions for other types of structures or for studying the construction process and performance of the structure [144]. However, as building complexity increases, developing predictive models based on concepts, such as DAI and emergence, is considered helpful for avoiding inefficiencies resulting from centralized top-down planning. Such concepts form the basis of MAS and can serve as a platform to develop new kinds of abstractions for design.

Emergence is a set of dynamic mechanisms whereby structures appear at the global level of a system from interactions among the system's constituent units. Actions/ tasks are executed on the basis of purely local information without reference to the global pattern, which is an emergent property of the system rather than a property imposed upon the system by an external ordering influence [23, 145]. To better understand how emergence occurs in MAS, there is a list of the basic MAS characteristics below.

2.3.1.1 MAS characteristics

A community of agents, whether digital or physical, is generally referred to as a MAS. There are multiple definitions of MASs and Weiss provides an extensive review of agent ontologies [146]. According to Kasabov and Kozma [147], in order for a MAS to be considered intelligent it should have the capacity to:

1. accommodate general problems' rules incrementally;
2. adapt in real time to its environment;
3. analyze itself in terms of behavior, error and success (state awareness);
4. learn and improve through its interaction with the environment (embodiment);
5. learn by processing large amounts of data;
6. use memory for storage and/or retrieval; and
7. represent short-term memory, long-term memory, age, etc.

2.3.2 Models of agents

Agents are often described as an abstract functional system such as a computer program that can carry out tasks on behalf of its users or physical systems. This could be an autonomous vehicle, like a drone or a robot, that acts based on signals it receives from its sensors. Due to this, software agents are often referred to as abstract intelligent agents to distinguish them from the physical implementations, for example, computer systems, robotic systems and biological systems. In the field of DAI, an agent is an autonomous entity that can "receive" perceptions from its environment through sensors and act upon the environment using actuators.

This work will focus on intelligent agents as a way to lay the foundation toward developing DAI-driven design tools based on MASs. Agents can vary from very simple to very complex. An agent interacts with other agents in its environment and can adjust its actions to achieve a goal (Figures 2.17). In the case of an "intelligent" agent, the agent can learn or use knowledge (i.e., existing databases) to achieve goals. An example of a basic, yet intelligent agent is a reflex machine such as the thermostat in a building.

2.3.2.1 Mathematical structure of agents

A very simple agent program can be defined mathematically as an agent function that maps every possible percept sequence to a possible action the agent can perform or to a coefficient, to a feedback element, to a function or even a constant that eventually affects the action:

$$f{:}P * \rightarrow A.$$

An agent function can incorporate various principles of decision-making, such as the calculation of utility of individual options, deduction over logic rules, probabilities [148] or fuzzy logic [149]. An agent is therefore an abstract concept, which makes it particularly interesting for design purposes as it can be used to express multiple things, including requirements and constraints. The program of an agent maps every possible percept the agent receives into an action. The word "percept" is used to refer to the agent's perceptional inputs from its environment at any given moment, usually coming from a set of sensors (i.e., light sensor, pressure sensor, etc.). The following sections briefly discuss basic agent architectures. Simply put, an agent can be anything that perceives its environment through sensors and acts upon that environment via actuators.

2.3.3 Agent ontologies

Weiss [146], who has studied DAI and especially MASs extensively, classifies agents into four main categories:
1. **Logic-based agents:** The decision about what action to be performed is made via logical deduction.
2. **Reactive agents:** Decision-making is implemented in some form of direct mapping from situation to action.
3. **Belief–desire–intention (BDI) agents:** Decision-making depends upon the manipulation of data structures representing the beliefs, desires and intentions of the agent.

4. **Layered architecture (LA) agents:** Decision-making is realized via various software layers, each one more or less explicitly reasoning about the environment at different levels of abstraction.

2.3.4 Agent classes

From the aforementioned categories, depending on their complexity and degree of perceived intelligence and capability, the BDI and LA agents can be further divided into five main classes, according to Russell [150]. A brief overview of these five main classes is described below to provide a better understanding.

2.3.4.1 Simple reflex agents

Simple reflex agents act only on the basis of the current percept, ignoring the rest of the percept history. The agent function formulated above is based on the condition–action rule: if a certain condition exists, then take a certain action. This agent function only succeeds when the environment is fully observable. Some reflex agents can also contain information on their current state, which allows them to disregard conditions whose actuators are already triggered.

If the environment is only partially observable, infinite loops are often unavoidable for simple reflex agents. Such issues can be avoided by randomizing the actions of the agents at certain intervals, to enable agents to escape infinite loops. Weinstein et al. [151] have successfully used simple reflex agents for sorting text databases and for making them appropriate for data mining.

2.3.4.2 Model-based reflex agents

Unlike simple reflex agents, model-based agents can handle partially observable environments. Their current state is stored inside the agent, maintaining some kind of structure that describes the part of the world that cannot be seen. This knowledge of "how the world works" is called a model of the world, hence, the name model-based agent. A model-based reflex agent has a type of internal model that is related to the sequence of percepts the agent receives and therefore reflects some of the unobserved aspects of the current state. Input (percept) history and impact of the agent's actions on the environment can be determined by using its internal model. The agent uses its internal model to choose an action based on the impact of its action or the history of percepts it has received.

2.3.4.3 Goal-based agents

Another important class is that of goal-based agents. This further expands the capabilities of the model-based agents by using goals or targets. Goals can be described as desirable situations, which allow the agent to choose between different behaviors by selecting one that reaches a goal state. Search and planning are the subfields of AI

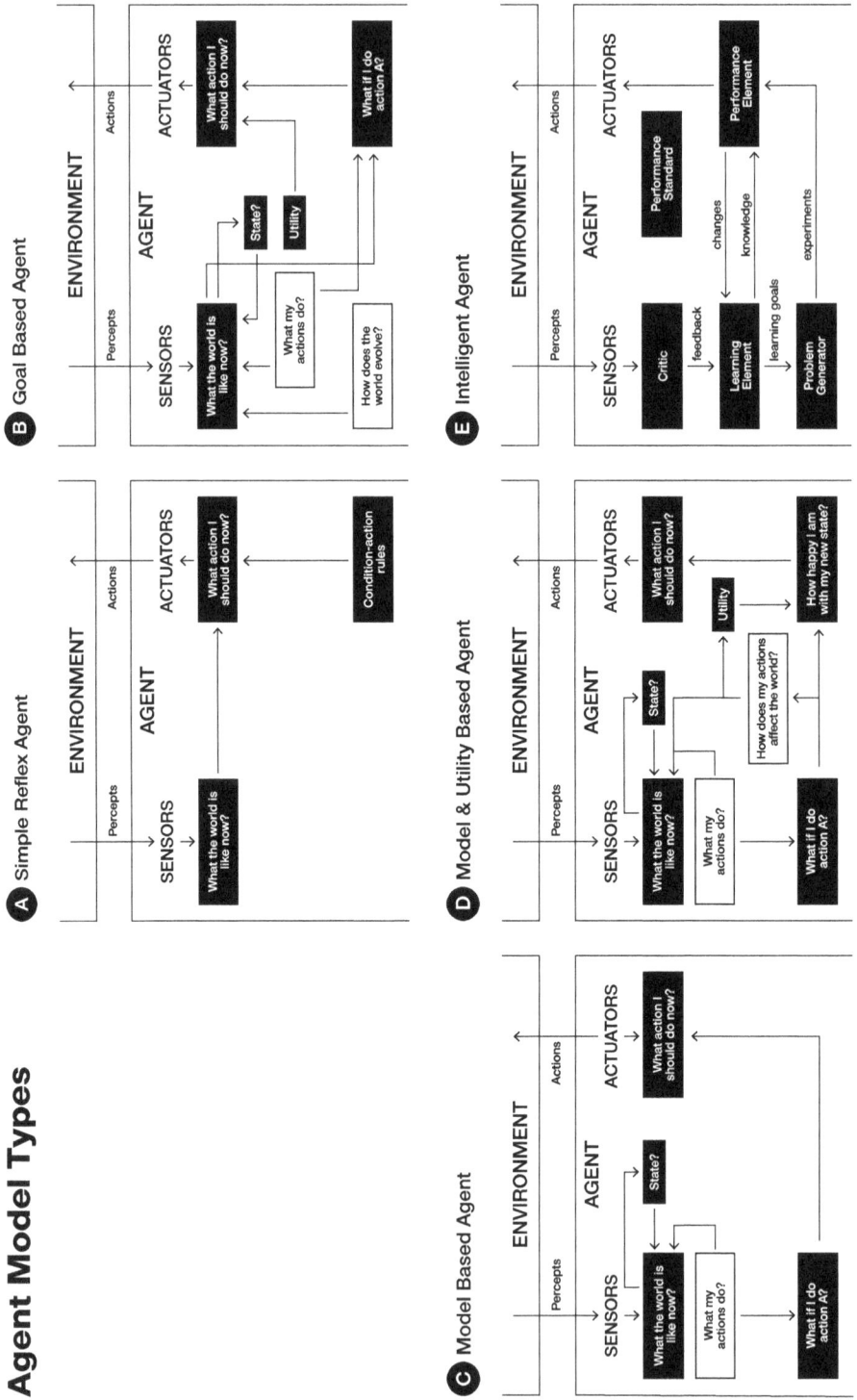

Figure 2.17: Types of agent models (ontologies) with an increasing level of complexity (a–d).

that study the decision-making process of selecting actions that lead the agent to achieve its goals. The advantage of this class of agents is its flexibility; the knowledge that drives the decision-making is represented explicitly and can be easily modified. A typical and widely used example of a goal-based agent is the distributed flocking model developed by Reynolds [152].

2.3.4.4 Utility-based agents

Goal-based agents only distinguish between goal states and nongoal states. It is possible to define a measure of how desirable a particular state is. This measure can be obtained through the use of a utility function that maps a state to a measure of the utility of the state. A more general performance measure should allow a comparison of different world states according to exactly how well the agent would attain a goal state.

A rational, utility-based agent will choose the action that maximizes the expected utility of the action outcomes, that is, what the agent expects to derive, on average, given the probabilities and utilities of each outcome. A utility-based agent has to model and keep track of its environment, tasks that have involved a great deal of research on perception, representation, reasoning and learning.

2.3.4.5 Intelligent/learning agents

Learning has the advantage that it allows agents to initially operate in unknown environments and to become more competent than their initial knowledge alone might allow. Learning agents have a more complex structure, and apart from percepts, actuators and states they also have (a) a learning, (b) a critical, (c) a performance and (d) a problem generator layer. The most important distinction is between the learning layer, which is responsible for making improvements, and the performance layer, which is responsible for selecting external actions. By using feedback mechanisms about the current state of the agent, the critical layer determines how the performance component should be modified to do better in the future. The performance component is what we have previously considered to be the entire agent; it takes in percepts and decides on actions. The last component of the learning agent is the problem generator. The problem generator is responsible for suggesting actions that will lead to new and informative experiences.

2.3.4.6 Hierarchies of agents

In order to actively perform their function and achieve their goals, intelligent agents today are normally organized in a hierarchical structure containing many subagents. Intelligent subagents process and perform lower level functions. Intelligent agents and subagents are able to form a MAS that can accomplish difficult tasks or goals with behaviors and responses that display a form of intelligence.

2.3.5 Applications of MAS in the AEC

Research has shown that the complexity and uncertainty often encountered in design problems can be effectively addressed with distributed computation and AI [78, 142, 153]. The nature of design problems is ill-structured and therefore, computationally, designers must engage in defining abstractions in order to design, explore and optimize [31, 122]. The capacity of distributed systems, in this case MASs, to abstractly model requirements as agent goals and to adapt to local conditions has rendered them appropriate for solving a large class of real-world problems in a number of domains, including software engineering, financial markets, security and game theory [146, 154, 155]. MASs are also closely related to CASs, which are characterized by their ability to self-organize and dynamically reorganize their components in different ways and across multiple scales [156]. This process allows the agents to negotiate, survive and adapt within their environments [142], but it also requires a basic understanding of evolutionary and generative mechanisms in nature and the development of mathematical models that simulate physical processes [23]. There exist a number of properties, such as aggregation, nonlinearity, flows and diversity, and mechanisms, such as planning, tagging, internal models and building blocks, which are common to MASs and serve as a reference for designing and developing agent-based models that can be synthesized to form a MAS [142, 146, 148, 157]. Due to their modularity, MASs are adequate for producing portable, extensible and transferable algorithms, with better integrated development environments and more applications [142, 158].

The application of MASs in the AEC industry has been less pervasive. Beetz et al. [159] classified MASs in the AEC under three domains of design generation, namely knowledge capturing and pattern recognition, the simulation and performance of building designs and collaborative environments. In the fields of engineering and construction, researchers have been exploring the applicability of MASs from different perspectives, such as collaborative design, construction scheduling and structural optimization, to name a few [159–162]. ABMS has been used in digital fabrication and building construction for its ability to abstract, adapt and simplify real-time complexities into simple basic rules [129]. Additionally, there has been significant research in developing MASs for autonomous collective construction at the level of algorithms and also at the level of hardware [24, 163].

In the field of architecture and computational design, the focus of the research on the use of computational methods to date has been mostly on design generation (form and aesthetics) and simulations [149, 164]. Approaches to design generation can be classified as linear or nonlinear based on algorithms that operate either in top-down or bottom-up manner [42, 165, 166]. Many researchers have argued that top-down approaches offer control instead of enough design flexibility, as they operate on fixed design topologies that are sequentially decomposed [167].

On the other hand, bottom-up algorithms can be challenging to apply for design purposes and often exhibit a lack of control in the design outcome [168]. In the architectural literature, a majority of the research using the bottom-up design approach has focused on the generative aspect of agent-based simulations and performance models and has mainly implemented swarm or boid algorithms [161, 169–174]. Snooks argues that "swarm intelligence" can enable the encoding of design requirements either into agent behaviors of different populations that belong to interrelated subsystems or within a population with adjustable or differentiated behaviors of one system [161, 175, 176]. The distributed nature of agent-based models enables the mutual negotiation of relationships between different design parameters, such as program and form or structure and ornament [175]. Focusing more on pattern recognition and the representational aspect of design problems, Achten has proposed a MAS framework for graphic unit recognition in technical drawings. This approach suggests that singular agents may specialize in graphic unit recognition, and MASs can address problems of ambiguity through negotiation mechanisms [177]. Menges uses swarm-based agent models in order to establish communication across different design environments (architectural design and structural design) and/or different hierarchical levels (global geometry and material structure), allowing for the uninterrupted flow of information from input parameters to multiple design constraints [17, 178]. Contrary to other researchers, Menges and the Institute for Computational Design in Stuttgart have used agent-based models to realize a number of prototypical structures and empirical data to develop an interactive, agent-based framework for integrative planning in architectural design [179].

In the field of structural engineering, Soibelman et al. implemented an agent-based reasoning model to enable designers to more rapidly explore conceptual structural designs for tall buildings. In their approach, a MAS system (M-RAM) provides the designer with previously adapted solutions for evaluation. The solutions are generated by a distributed multiconstraint reasoning mechanism [180]. Dijkstra and Timmermans [181] have created a custom platform, AMANDA, to simulate pedestrian flows in buildings and urban environments.

Meissner [182] has used agent-based simulations for the support and integration of fire protection engineering in the planning process, while Klein et al. [183] have used MASs in combination with MDPs in order to develop alternative building management and control systems in relation to occupant habits and preferences. However, according to Anumba et al. [160, 184], the encoding of the design requirements (i.e., building design requirements) into agent behaviors and the definition of an agent upon decomposition of a given design problem are most often highly complex and consequently hard to achieve.

Trying to address that issue, a novel approach is presented by Marcolino et al., based on the development of collaborative MAS environments; it combines alternative agent models (social choices) with number theory. Instead of using an agent-based modeling approach to generate geometry, in this case, the MAS is applied to optimize efficient building designs generated in a parametric environment (Revit) using a genetic algorithm [185]. This approach presents teams of uniform and diverse agent populations with different design and performance goals. The system developed aggregates the agents" opinions, which relate to a predefined range of design requirements, in order to provide designers with a larger number of Pareto optimal design solutions.

2.3.5.1 Gap analysis using MAS in AEC

The review of the literature shows that there are generally two critical impediments to the furtherance of MASs, particularly in architecture: first, there is a lack of methodology to enable (software) designers to clearly specify and model their applications as MASs, and second, there is a lack of widely available MAS toolkits that support designers in effectively exploring larger solution spaces. Thus, more research efforts are required to develop MAS toolkits that combine generative processes with analytical processes and user-related data (i.e., light) in a distributed fashion, as well as tools that enable designers to use and adjust more sophisticated agent-based algorithms, such as MDPs and team formation models.

The lack of toolkits has led to duplicate efforts, as a lot of work in the field of architecture uses different implementations of ABMS methods, specifically boid and swarm intelligence algorithms. In addition, apart from a few exceptions, most of the agent-based modeling approaches have not coupled the agent models with actual building components [179] or databases of existing buildings [180].

Even though there has been a number of examples of agent-based models being used to simulate user behavior, such as circulation patterns, the AEC industry has done little to adapt techniques to accurately incorporate the end-users' behavioral and performance information during the design phase of buildings [48, 166]. Studies have shown that if buildings are designed according to their users' needs, behavior and preferences, there is potential to reduce the total energy consumption of the building during its operational phase [186]. Additionally, research within other domains and industries, such as security [187, 188], economics [189] and game theory [190], has shown that user-centered designs could significantly increase the efficiency of building systems.

A common approach used to incorporate user-related information during the design phase is through the use of behavioral models to simulate users' movements, interactions and responses within the designed environment. Such simulations are also designed to estimate the building's energy consumption more accurately based on its occupants' possible interactions, comfort levels and preferences [191, 192]. Although such simulations have been promising and provide a more user-centered analysis of a building's operations, in many cases, due to the complexities of human behavior (e.g., preferences and personalities), they do not provide accurate and realistic representation of the actual occupants' behavior during the operational phase; therefore, in some cases, the building could be less energy efficient and not accommodate the occupants' needs [193].

Therefore, the holistic approach is often lacking. Using a holistic approach would offer designers the capacity to integrate generative design rules with (a) user-related information (i.e., preferences); (b) multiple analyses and performance criteria, such as environmental and lighting analysis; (c) building and material constraints; and (d) performance optimization functions.

Part III: **Methodology**

Synthesis, analysis and evaluation

3 Research methodology

Until recently, most existing 3D modeling design tools could only do what the designer instructed them to, and many optimization tools needed a lot of data in order to operate properly. However, architects are often called upon to make decisions and evaluate design models without having all the parameters fixed. Performance-based parametric tools can be used to help search through large design solution spaces, but this can prove largely inefficient and time-consuming. Beyond enabling geometric and information modeling, design tools should also serve to extend the capability of architects to solve difficult problems, but in order to do that they require capabilities covering the following five areas: search, pattern recognition, learning, planning and induction.

The suggested methodology and corresponding framework have three purposes: first, to enable architects to build design models that can accommodate changes as the project progresses, instead of needing to rebuild it at every phase (as is often the case with parametric design models); second, to provide a structured environment in which to develop agents and behaviors that can be applied to architectural problems, and to develop a computational apparatus that facilitates agent-based design research on the one hand, and comparison of results with existing design approaches on the other; and finally, to allow the reduction of solution spaces using heuristic methods and improving the efficiency of traversing through multiple solution spaces by combining the designers' input with pattern-recognition techniques. By using planning methods, we may obtain a fundamental improvement by replacing the solution space with a much more appropriate solution landscape.

This proposal investigates and develops an agent-based framework for the computational exploration of design alternatives in the early design stage. Despite being field-specific, it combines concepts from areas that range from architectural design, structural and environmental engineering to complexity theory and multi-agent systems (MASs) and is adaptable enough that it can be used by a variety of disciplines that involve geometric modeling in design or engineering.

The recent advancements in the architecture, engineering and construction (AEC) with the increasingly fast application of building information modeling [37] and performance-based design approaches [12], as well as the introduction of integrative planning methods [194] and robotic construction platforms, such as a digital construction platform [195], indicate that research into methods and tools that holistically deal with the design-to-construction process is necessary.

3.1 Generic design problem solvers using MASs

Developing algorithms and computational tools that solve specific problems is a major area in both engineering and computer science, and there are a number of algorithms able to find optimal solutions to problems. Although such approaches can be very efficient in specific engineering problems (i.e., structural optimization), they are

https://doi.org/10.1515/9783110797435-003

applied after a design has been formulated. Therefore, up-to-date computational design tools have not dealt with the creative aspect of design in the early stages, but rather focus on the later stages and rely on a given design space after the problem is well defined. Although these tools have proven invaluable for finding optimal solutions for specific subproblems, they provide little design intuition and feedback for addressing bigger problems.

The growing interest in complex adaptive systems is due to the fact that by observing mechanisms that exist within communities of social insects, like termites, researchers have developed algorithms that can be applied to a variety of problems and domains. These algorithms can be considered generic problem solvers and are of particular interest to architectural design, where almost every problem is unique. Creating agent-based models and developing algorithmic solutions for addressing design problems are relatively new research topics in architecture, mainly due to their computational complexity. This proposal specifically focuses on the application of new agent models for combining different fields in the AEC. The following section describes the main elements of our methodology.

3.1.1 Design problem decomposition

As mentioned in Section 2.2.6, design problems are often described as ill-defined, and in terms of their computational complexity one could classify them as NP-hard. NP-hard is a type of problem in which:
–   it is not known how to generate a correct solution,
–   it is not known how to test the correctness of a proposed solution and
–   it is possible to compare two proposed solutions and select the more correct one.

Based on the problem classification, a number of computational methods have been developed for solving them (solvers). Following the categorization of problems, algorithms are also categorized into (a) greedy, (b) deterministic, (c) stochastic, (d) exact, (e) approximate, (f) progressive, (g) adaptive, (h) specific, (i) generic and (j) open [39].

This work implements fundamental algorithms that fall under the broad categories of generic, stochastic, adaptive and open algorithms, and applies them in the field of design using agent-based modeling. Since there are multiple classifications and descriptions of algorithms, brief descriptions are given below for each of those categories to provide clarity.

Stochastic algorithms include a random component, and the result can be predicted probabilistically. We should note that on digital computers, all processes are inherently deterministic, but pseudo-randomness is sufficient to classify an algorithm as stochastic. Hill climbing (HC), simulated annealing (SA) and swarm algorithms are fundamental stochastic algorithms.

Adaptive algorithms can operate on a changing set of constraints and inputs. These algorithms run continuously within a dynamic environment. An example of such algorithm is self-organizing maps.

Generic algorithms are designed to tackle a wide variety of problems. This flexibility is accompanied by a significant drop in performance. Divide and conquer is an example of a basic generic algorithm.

Open algorithms allow external entities (be they human beings or other algorithms) to participate in the solution process. Seemingly unpromising lines of inquiry can be investigated upon request by an external agent. Research suggests that generic solvers are more appropriate for design, as they are capable of solving almost any problem. But what enables a generic solver to deal with many problems? Doesn't that require a large amount of intelligence or data?

It certainly does, but the important aspect is that the intelligence (knowing) does not need to exist within the solver itself but rather in the way the problem is decomposed, and the solver is configured. If the solver is open, it can be extended by another algorithm or interactively by the user, who can fill in knowledge gaps. For example, a generic solver does not need to "know" anything about geometry in order to find if an opening is properly placed; it only needs a collaborating algorithm that can take care of geometric representations (i.e., the Non-Uniform Rational B-Splines (NURBS) environment). By implementing open design systems and dividing the "knowing" (i.e., design generation, design knowledge and intuition) and the solving aspect (analysis and evaluation) into disjointed algorithms, it becomes easier to develop generic problem-solving approaches. Due to this decoupling, such algorithms are also easier to repurpose. Communication between the algorithms becomes key and the interaction among them defines the problem-solving process. The process can be briefly described as follows.

The generic algorithm generates alternative design solutions based on a problem specification. The companion algorithm processes the solution and assigns a quality rating (e.g., cold, warm and hot). The generic algorithm is responsible for interpreting the messages, and the collaborator algorithm is responsible for computing the "fitness" of each solution. The algorithms communicate via exchanging numbers and characters (messages), and their communication can be described mathematically:

$$f(x) = q, \tag{3.1}$$

$$f : t \rightarrow R. \tag{3.2}$$

Equation (3.1) is called the fitness or heuristic function and defines the language between the two algorithms mathematically. The output of f, which is labeled as q, describes "fitness" as a numeric value.

Equation (3.2) shows the mapping of this function, which specifies the type of data that goes in and out of such a function. The mapping notation (3.2) states that f must consume data in the form of t and return data as a real number (R).

The combination of heuristic function and mapping creates the solution space (landscape) for a given problem, which can be explored with the aforementioned type of algorithms. The process includes iterative runs during which the generic solver(s) search(es) and find(s) their way around these landscapes and converge on high ground as quickly as possible [39]. The designer's responsibility becomes developing meaningful problem spaces and setting up relationships between design elements and design targets (performance) using one or more fitness functions.

3.1.2 Hypothesis

The agent-based framework revolves around the hypothesis that designers will need to develop new types of abstractions (i.e., mathematical models) that will be represented as agents used to describe design problems in generic problem-solving algorithms in order to be able to deal with the increasing complexity of building design. Figure 3.1 presents an agent-based design framework for computational morphogenesis, suggesting the modeling of design requirements from different design domains into agent behaviors. The framework focuses on the early design stage, and the objective is to enable designers to couple geometry with different types of numerical analyses in an agent-based fashion and automatically generate and evaluate design alternatives using principles of evolutionary programming. The implementation of custom agents allows designers to traverse the solution space and extend the existing form-finding methods, such as particle spring systems, and couple them with analytic data via heuristic functions. Strategies for exploring the solution space can be achieved via appropriate problem decomposition and by introducing task-specific agents, their behaviors, hierarchies and the heuristic functions between them. The fundamental novelty of this methodology is that different types of agents are implemented for each aspect of the design cycle, namely synthesis, analysis and evaluation. To allow for extensibility, the framework is composed of a set of class libraries organized around a core agent library, which is described in the following section.

3.2 Framework development for integrating multiple design phases

For an agent-based computational design tool to be able to generate design solutions, the design problem needs to be described in a number of agents and represented in a tractable way. This is a particularly difficult task due to the fact that problem requirements are not formalized until the later stages of the building design process. In architecture, the majority of agent-based design approaches have focused on adapting the basic behaviors of swarm intelligence models developed by C. Reynolds in order to fit the context of specific design problems (i.e., simulation) [152].

This work focuses on situations with multiple types of agents in which the variables and constraints are distributed among the agents so that no single agent controls all the variables. In such situations, the design problem is defined as a distributed constraint satisfaction problem, and each agent may interact with only a few agents in the system. Local interactions become a feature of the agent; therefore, the agents within such networks can be part of a team and must cooperate with each other to achieve a design goal or they can each have individual targets and goals.

Multi Agent Systems' (MAS) Design Approach

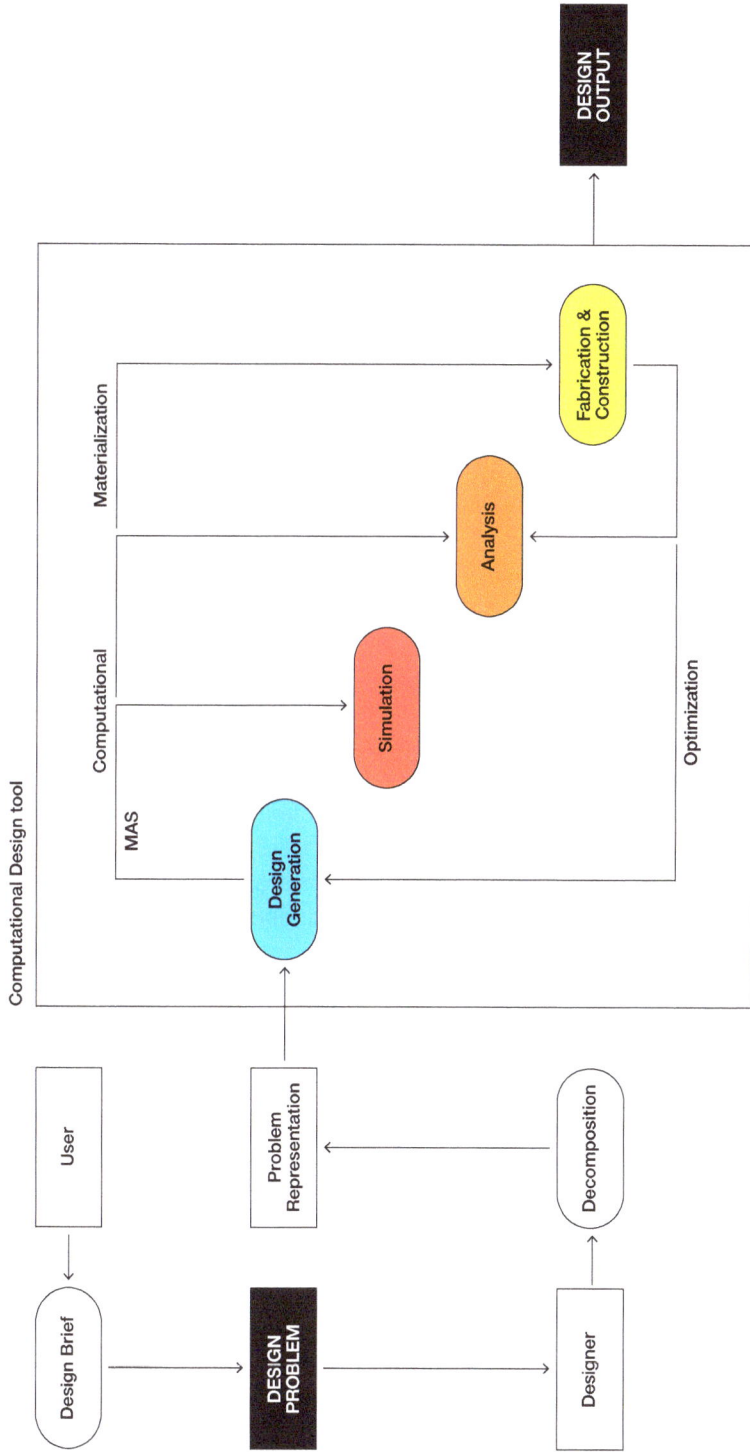

Figure 3.1: Diagram illustrating the multi agent system's design approach.

To provide clarity, an agent is denoted as a software-based programming block and/or computer system that shares the following properties: (a) an agent exists within an environment and responds to it, while interacting with other agents. Therefore, the agent is situated, in that its behavior is based on the current state of its interactions with both the population of the agents and the environment; (b) an agent may have explicit objectives that condition its behavior and are directly related to specific performance criteria, in which the goals are not solely targeted to maximize effectiveness but are used to assess and improve the decision-making process; (c) an agent can adapt and change its behavior based on a utility function that uses analytical data or the agents' own evolution and interaction history. In this case, individual adaptation requires agents to have sufficient memory to keep track of their actions, usually in the form of a dynamic agent parameter (utility), and therefore (d) an agent has resource parameters that indicate its current stock of one or more resources (energy, material and information) [142].

Each of the aforementioned properties is expressed as layers within the internal structure of an agent. The established typical agent structure in this work includes (a) an interface layer through which the agent communicates with their environment, (b) a definition layer that describes the set of states and goals of each agent, (c) an organization layer that decides the type of actions to be taken by the agent at a given time based on analytical data, (d) a coordination layer that keeps track of past and current decisions and finally (e) a communication layer that establishes that the agents are able to communicate among themselves. The key assumption is that any given architectural design problem can be distributed among agents that cater to different aspects of the problem in question. In particular, this model suggests the creation of separate agent classes with goals related to different steps of the design process, namely (a) synthesis: the generative agent class, (b) analysis: the specialist agent class, (c) evaluation: the evaluator agent class and (d) coordination: the coordinator agent class (core agents).

The interface layer defines a set of control variables and constraints that relate to the design problem (i.e., design a facade). These variables and constraints are connected to a number of actions the agent can perform (i.e., add opening and place a facade panel) in the definition layer. The plan and schedule of the agents' actions are defined in the organization layer, which also outlines rewards and penalties in order to be able to evaluate the outcome of each action. The communication layer defines the type and amount of data communicated among different agents, while the success or failure of each action is measured using rewards and costs. Using distributed constraint reasoning, a set of agents can start forming teams (i.e., generative with specialist) and cooperate by coordinating their actions, plans and schedules [188].

In the proposed MAS framework, the four generic agent types – each focusing on a specific goal – are applied to different design cases. Each agent type appends to different design domains, while goals are defined by the designer, based on available data and specific design intentions. In this work, case studies are presented, where agents are simple programming modules with the aforementioned structure and are linked to design software in order to perform different actions (i.e., generate geometry) based on their type and state. The agents implement core geometric functions that exist within the design software by accessing specific application programming interfaces (APIs), such as RhinoCommon [197], the Energy Plus API [198] and Karamba 3D API [199]. However, there are experiments where entire computer systems have been considered agents and the agents' actions are related to the implementation of specific commands within the system [70]. The parameters in the research include the decomposition of simple design problems into different design agencies that form agent networks; the generative capacity of the MAS framework to create unique and complex outcomes that improve with iteration and integrate geometry formation with environmental performance criteria as they relate to established standards and user preferences; and integrating geometry rationalization based on the coupling of structural performance with environmental performance criteria.

This thesis is a preliminary step toward developing an integrated MAS toolkit that utilizes computational methods to find solutions through discovery, rather than precise analysis. Figure 3.2 shows the proposed framework, emphasizing the main steps and processes involved in this methodology. In the following sections, we describe in detail the tasks involved for the development of such toolkit.

3.2.1 Task 1: development of agent classes

This class development is implemented in two stages: first, the generative aspect of design, in which agents act autonomously; and second, the optimization of generated outcomes, in which agents act collaboratively and negotiate to find optimal solutions. The system explores these complexly coupled relationships between the generative and analytical design processes. First, a set of generative agents and behaviors are modeled, based on a given design site's location and orientation, a building facade bounding context and designer-defined parameters that append to building components (length, width, thickness and type). Initially, the systems' agents act autonomously and develop design alternatives, which satisfy local rules and constraints from the geometric domain, avoiding specific areas that are reserved for window openings and views, and collision checking for constructability.

During a second loop, the designs are analyzed by a set of specialist agents, which communicate their data back to the generative agents in order to adjust parameters to regenerate design alternatives based on specific user preferences and performance goals. Five different classes of agents are modeled with actions, properties, states and goals (Figure 3.3). The agent classes are as follows:

1. **Generative agents** relate to the design intention and geometric properties of the building component and are responsible for generating facade panels that regulate the amount of light that enters the office space.
2. **Specialist agents** with a number of different subclasses (based on the types of analysis) for analyzing and evaluating the generated designs' performance.
3. **Simulation agents** are responsible for simulating analytical results and/or user preferences and presenting them to the designer.
4. **Evaluation agents** are responsible for collecting the available analytical data and are based on heuristic functions that evaluate and rank design alternatives.

MAS Generative Design Tool Framework

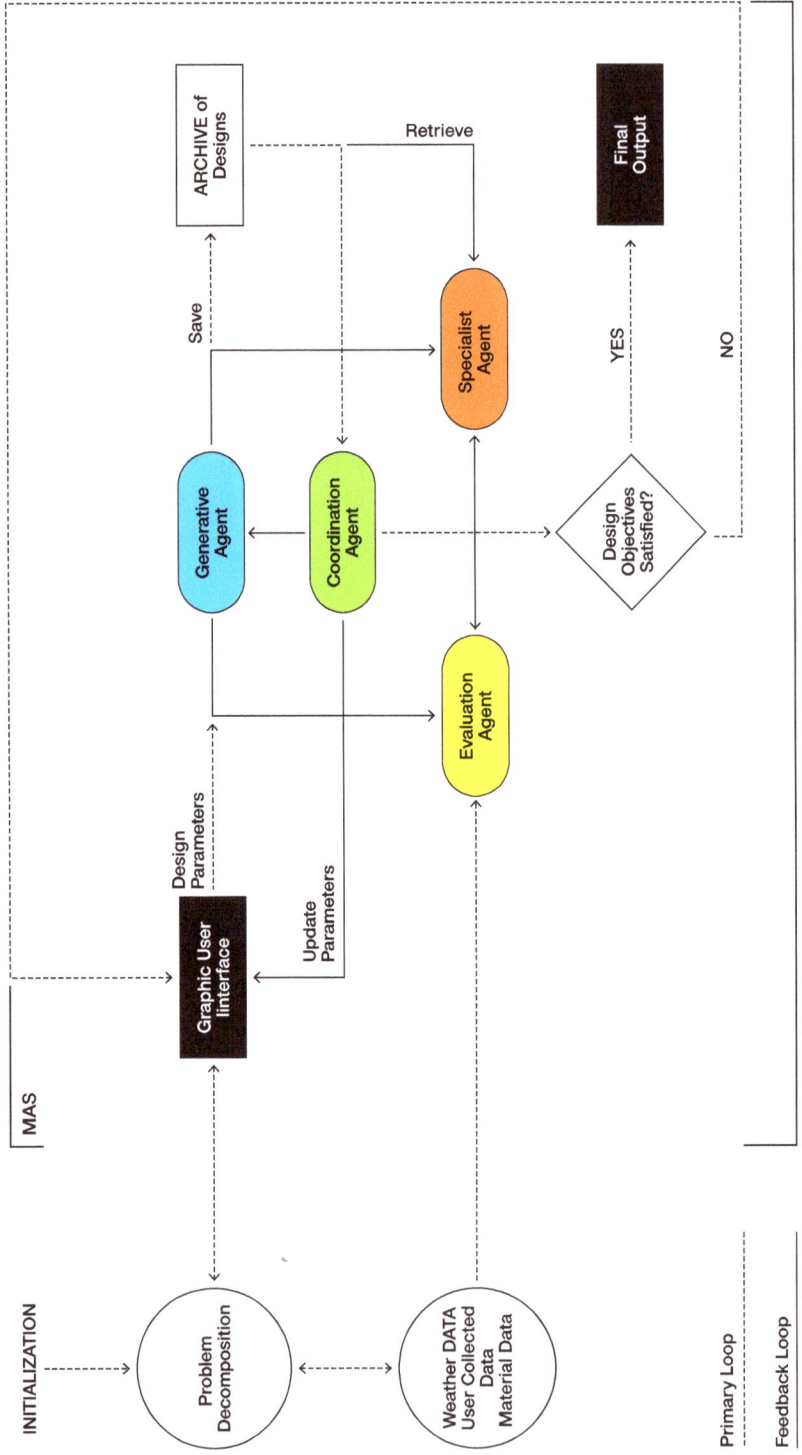

Figure 3.2: MAS framework diagram showing agent classes and interdependencies between agents.

Typical agent's structure

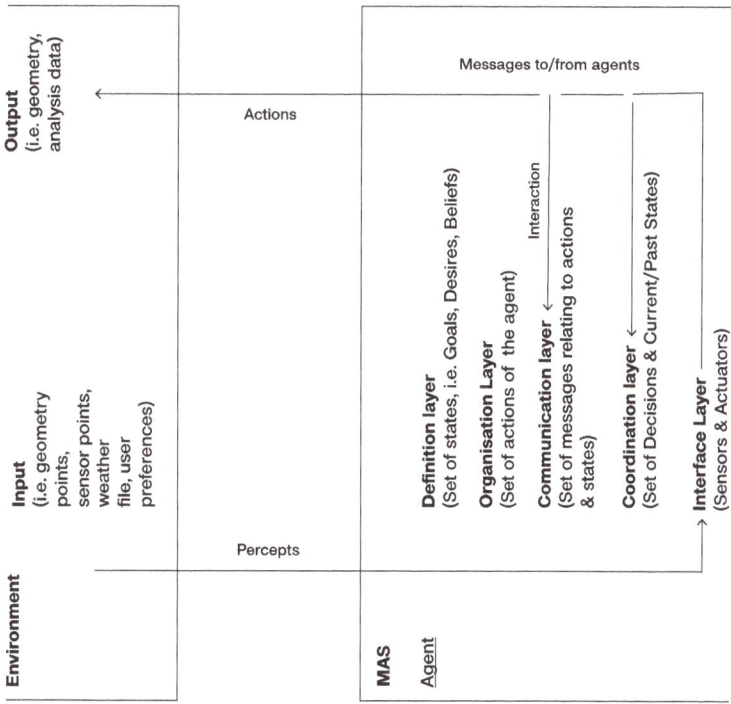

Agent hierarchy within MAS

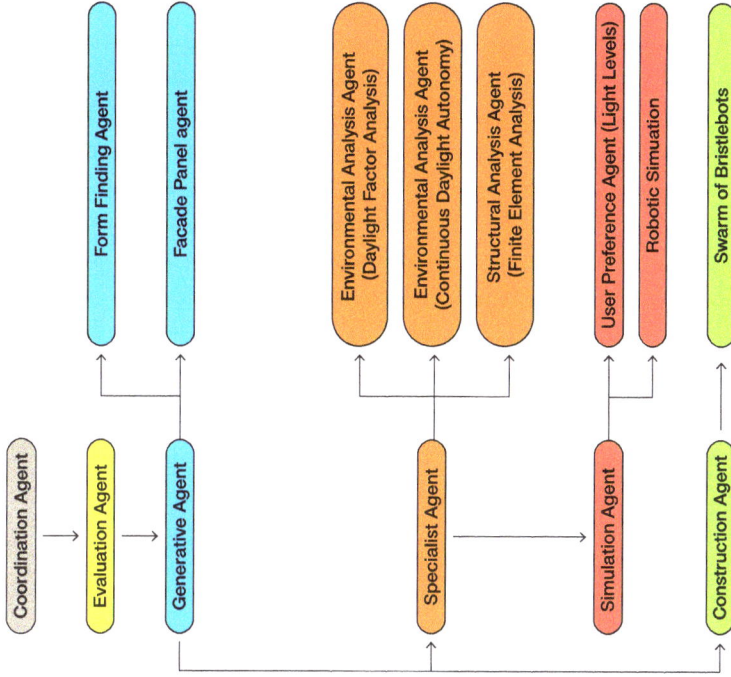

Figure 3.3: Diagram of typical internal agent structure, agent types and agent hierarchy within the MAS.

5. **A coordination agent** ensures that each agent is aware of the other agent states and is responsible for the communication and coordination of the different classes.

All agent types have the same number of layers, but their definition is based upon a specific domain (i.e., generative design, environmental or structural engineering) and a basic set of principles that are related to the specific domain and affect the agent's behavior. Behavioral rules may vary in complexity and the levels of information considered during the decision-making process. The level of information needed for each agent can be either based on established analytical methods for environmental and structural design or on the designer's input. For instance, the behavior of a generative agent that represents a facade panel is dependent on a set of input parameters that are coupled with basic principles found across environmental design methods such as (a) facade orientation, (b) boundary conditions, (c) Sun's positions, (d) view availability or (e) the level of daylight required with regard to the better use of that space.

3.2.2 Task 2: search space exploration using heuristic algorithms

As a first step toward steering the behavior of the generative agent for producing design alternatives, its input parameters were coupled with analytical values obtained by a specialist agent's stochastic search algorithms. By developing simple heuristic functions, the designer specified the relationship between design parameters with analytical values. This relationship was either based on the designer's experience or on a sensitivity analysis that allowed the designer to check which design parameters have bigger impacts on the specific analysis. For example, if the designer performs a daylight analysis, the specialist agent will collect a numerical value that reflects lux values [196], and if the designer performs a structural analysis, the specialist agent will collect a numerical value that reflects displacement [200]. The specialist agent passed the input parameters to a commercial analytical solver and collected the analytical values it communicated to the generative agent. At each iteration, the specialist agent was responsible for performing the analysis and communicating the analytical values to the generative agent (updating the values of said generative agent). In cases where there were multiple analyses, a weighted value was attached to each design parameter that indicated the impact of each design parameter to the analysis. For each of the parameters that were updated, a credit (e.g., +1) was attributed to each agent if the value obtained at the sensor point with the highest impact was closer to the target value (Figure 3.4). If the obtained value was further than the target value, the attribution was a penalty (e.g., −10). Based on this credit or debit, the generative agent decided in which direction to update a design parameter.

The experiment began by implementing four search optimization procedures and selecting the one that performed better in terms of time, design diversity and applicability on other problems. Initially, two basic local search algorithms were implemented: HC and SA, since they are fundamental heuristic algorithms that prove to be applicable and perform quite well for a wide range of problems. Two more complex

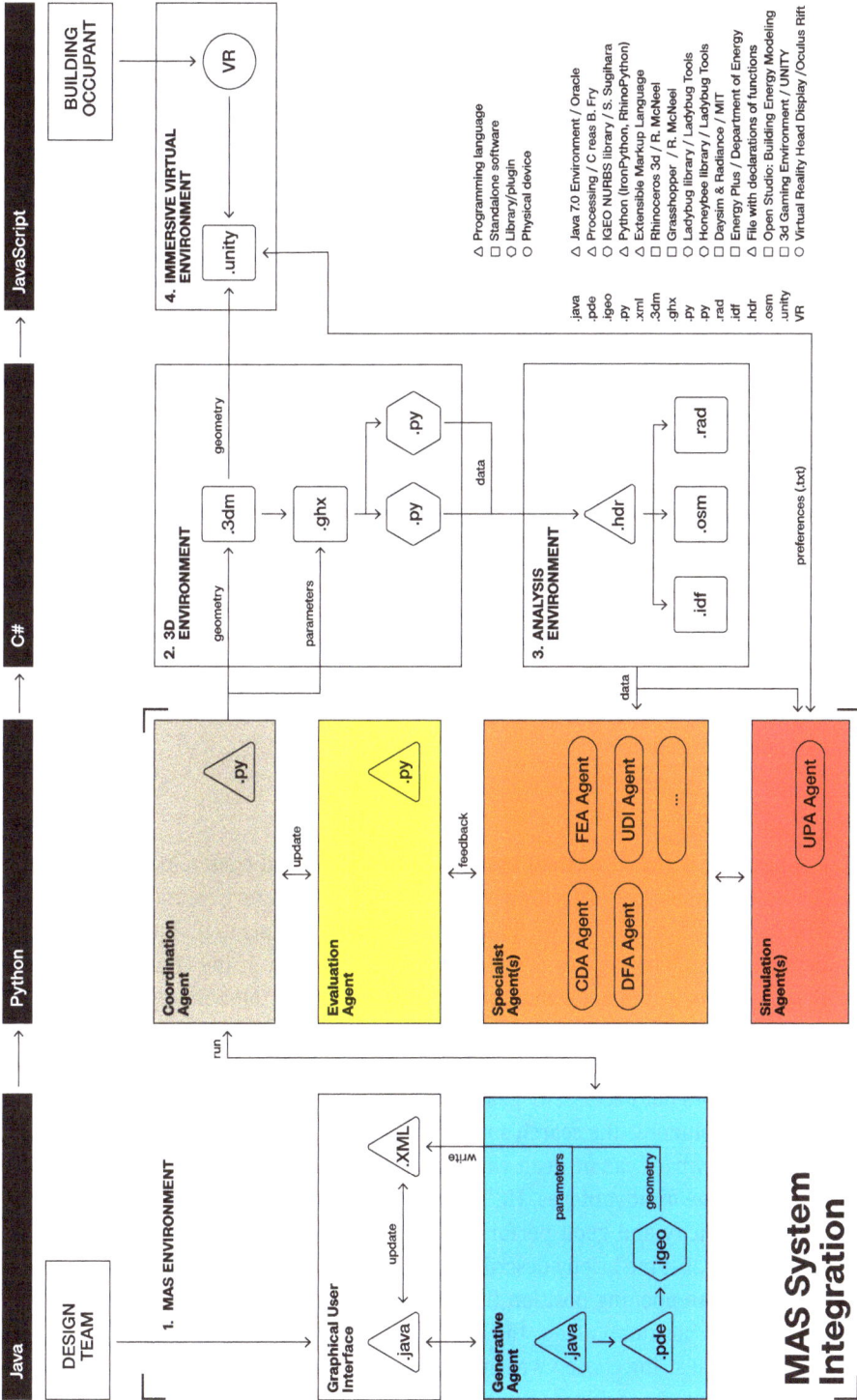

Figure 3.4: MAS System Architecture: agents types, softwares and data exchange.

evolutionary algorithms were implemented, namely a particle swarm optimization (PSO) algorithm, which is a distributed local optimization algorithm, and a stochastic search diffusion (SDS) algorithm. Unlike HC and SA, PSO does not use selection; rather, improved solutions appear via the interaction of the agent population.

Although the first two algorithms are fundamental in the artificial intelligence research, as noted by Minsky in his seminal paper *Steps Towards Artificial Intelligence*, and they have been widely used in the fields of engineering, little research has been done on their application to design problems [201]. Therefore, it was essential, as a first step toward building an intelligent MAS tool, to apply them to design problems and compare them to population-based approaches such as PSO and SDS.

For each of the approaches, a custom heuristic function was implemented that associates a design feature (i.e., window position on a facade) with one or more analytical results (i.e., the amount of daylight entering a space). An example of a simple heuristic with two analytical values was presented in the previous section. In order to be able to draw conclusions, we see how the different algorithms perform when (a) generating unique solutions each time they run or (b) operating on a small versus a large sample size. In order to be able to validate the results, the problem was reduced to finding the optimal position for an opening on a generic facade surface and searching linearly to map (brute force) the whole solution space.

The relationship of the parametrization of the design features was observed in relation to the analytical results, and the generated information was used to guide the search more efficiently, based on this relationship and the performance of each of the approaches discussed below.

3.2.2.1 Hill climbing

HC is a basic local search method best suited to convex and constrained optimization problems. It is an iterative algorithm that starts with an arbitrary solution to a problem (i.e., the position of an opening on a facade), and then attempts to find a better solution by incrementally changing a single element of the solution. If the change produces a better solution, an incremental change is applied to the new solution, repeating until no further improvements can be found. HC is a good method for finding a local optimum/optima (a solution that cannot be improved upon by considering a neighboring configuration), but it does not necessarily guarantee that this is the best (optimal solution) out of all possible solutions (the search solution space), as it can be easily trapped in local optima. This drawback can be dealt with by using repeated local searches (shotgun HC). Despite the above disadvantages, HC is an algorithm with a wide field of application due to its simplicity and good performance; therefore, I consider it an optimization search approach. Let me briefly describe the following algorithm:

Step 1: Choose an opening position (i.e., the x, y position of the window) at random. Call this position (string) the best evaluated (in the experimental case study the initial string chosen was evenly spaced openings).

Step 2: Choose another window position a step size away from the previous position (i.e., for two windows, there are a maximum of eight possible steps to search in). If the change of the window position parameter leads to an equal or higher value of the heuristic, then set best-evaluated to the resulting string.

Step 3: Go to step 2.

3.2.2.2 Simulated annealing

SA is a probabilistic technique for approximating the global optimum of a given heuristic function. It is a method suitable for problems with a large solution space, where searching for an approximate global optimum is more important than finding a precise local optimum in a fixed amount of time. It can be used for solving both unconstrained and bound-constrained stochastic optimization problems [202, 203]. To clarify, consider the example of placing openings onto a facade: the position of the window openings is coupled with the total radiation behind the facade. At each step, the algorithm changes the position and size of the window and measures the amount of light entering the space behind the facade, based on the heuristic mentioned in Section 1.3.1. The algorithm selects a new value from the parameter range, based on whether the obtained analysis was closer or further from the desired value (i.e., maximizing the natural light availability). The algorithm is briefly described below:

Step 1: Choose a new parameter value at random to generate a geometry. The distance of the new point from the current point or the extent of the search (budget) is based on a probability distribution with a scale proportional to the light distribution inside the room (heuristic function).

Step 2: Receive a numeric value from the analysis (i.e., environmental analysis).

Step 3: Accept all new points that increase the heuristic, but also decrease the probability of choosing such points over iterations, favoring points at a further distance from the objective (sub-optimal). By accepting points that decrease the objective, the algorithm avoids being trapped in local minima in early iterations and can explore globally for better solutions.

Step 4: Terminate when the budget is exhausted, or the maximum number of iterations has been reached.

3.2.2.3 Particle swarm optimization

PSO is a population-based stochastic algorithm for optimization. It consists of three basic steps: (1) generating the particles' positions and velocities, (2) updating the velocities and, finally, (3) updating the particles' positions. Here, a particle refers to a point in the design space that changes its position from one move (design iteration) to another based on velocity updates. Unlike evolutionary algorithms (i.e., a genetic algorithm) the particle swarm does not use selection, but rather evaluates the interactions between the design parameters and a defined heuristic and iteratively updates them aiming to improve the quality of problem solutions over time. In this respect, PSO is similar to the genetic algorithm but its main advantages are: (a) it can converge faster toward optimal solutions [204] and (b) it performs better computationally. The algorithm operates on a collection of design iterations called "particles" that move in steps throughout a surface domain. The steps are described briefly below:

Step 1: Begin by generating n initial particles with different opening positions and assigning different velocities (length of panel) and probabilities to each generation type.

Step 2: For each generated design, perform an environmental analysis and retrieve results.

Step 3: Evaluate the results based on the given heuristic function, determine the best (user defined-desired) opening positions, the new velocity, and the best generation probability for the agents.

Step 4: Update (iteratively) the opening positions (the new location is the old one plus the new probability, modified to keep the facade surface filled), probabilities of types and neighbor agents.

Step 5: Iterate through steps 2 and 3 until the algorithm reaches a stopping criterion, which could be a maximum number of iterations or a point where the designer is satisfied.

3.2.2.4 Stochastic diffusion search

Stochastic diffusion search (SDS) is a "parallel probabilistic pattern-matching algorithm" and is capable of rapidly identifying the best instantiation of a target pattern in a noisy search space [205]. Unlike stigmergic or swarm behaviors that rely on the modification of physical properties of a simulated environment, SDS uses a form of direct (one-to-one) communication between agents, similar to the tandem calling mechanism employed by one species of ants, *Leptothorax acervorum* [206]. In SDS, based on the problem decomposition, agents formulate a simple hypothesis that serves as a candidate solution to the search problem. The agents iteratively perform cheap, partial evaluations of their hypothesis and share information about them (diffusion of information) through direct one-to-one communication with other agents. As a result of the diffusion mechanism, high-quality solutions can be identified from clusters of agents with the same hypothesis. A brief description of the operation of SDS is given as follows:

Step 1: Populate a surface (i.e., facade) with agents who avoid specific areas (i.e., openings).

Step 2: Generate an initial population of solutions with different opening positions on the facade.

Step 3: Create an agent for each design solution. The agent maintains a hypothesis for the placement of the openings in the form of a message (i.e., window position will increase daylight).

Step 4: Perform analysis (i.e., daylight factor analysis) for each generated design and retrieve the results.

Step 5: Evaluate the results based on the given heuristic function.

Step 6: If the fitness criteria are satisfied, select another position randomly. Repeat steps 3 and 4.

Step 7: If the condition is not satisfied, report the position and ask another agent if it was able to satisfy the condition. If the answer is yes, the agent saves this position.

Step 8: Repeat until the majority of the generated solutions satisfy the fitness criteria.

3.2.3 Task 3: agent coordination, evaluation and negotiation

Communication and negotiation mechanisms among the agents are established to update the behavior of the generative agent(s) and improve their geometric results. Text file messages update values and/or actions to negotiate across different agents. At each iteration, the coordinator agent class checks the state of other agents, reports their state (i.e., analysis has finished), calculates their utility and predicts future actions based on the utility.

The effect of their actions and behaviors on other agents is dependent on the hierarchy established among the different agents (i.e., a generative agent is higher in the hierarchy than the specialist agent) or on the degree of importance of each agent's related behavior, which is expressed as a utility. The target of the negotiation for each agent is to satisfy its own goal while minimizing the negative side effects on the other agents. The satisfaction of each goal is measured by the increase or decrease of each agent's utility.

Hierarchy among the agents is established and applied by a coordinating agent that communicates and controls the rest of the agents. Figure 3.3 illustrates the basic structure of an agent and the established hierarchies among the system's agent classes. The designer is responsible for designing the interaction mechanisms between agents and the different input design parameters. The designer also couples different types of analyses with design parameters and establishes trade-off processes among the agents.

3.2.4 Task 4: prototype development (MAS design tool)

The system prototype is built on top of several open-source software platforms and tools as well as commercial software that offer open APIs. The core agent types/classes of the system are developed in Java on the Eclipse Platform, which serves as the common interface for diverse integrated development environment (IDE)-based products to facilitate our integrations [207]. The custom-programmed MAS utilizes libraries and classes from Processing, a Java-based programming language with its own IDE [208].

For the geometric adaptation and transformation of the building components, the IGEO library is implemented, which has been developed to offer automatic data management of NURBS-based geometry as agents, as well as method chaining for coding efficiency [168]. A custom java applet and a graphical user interface (GUI) are created to generate geometric configurations, which are then imported into the Rhino 3D NURBS design environment for further analysis [209].

Designs are analyzed and evaluated using Grasshopper, a visual scripting editor within Rhinoceros and, specifically, two Python-based environmental simulation plugins, Honeybee and Ladybug [35]. Automation functionalities are added via custom programming in Python, to obtain and simulate environmental analysis from specialized stand-alone software (Open Studio, EnergyPlus, Radiance and DAYSIM). Karamba 3D [210] is used to perform finite element analysis on generated designs, while Kuka|prc [211] is implemented to generate robotic construction simulations and evaluate designs based on their constructability. The data obtained is saved as .xml or .txt files, which are used to model and form the parameter bounds of an environmental set of agents.

The Python programming language is used to handle the calls of the different platforms, while a "federated" system architecture is used to relate multiple software environments. Extensible Markup Language (XML) is deployed to control and manage the agents' properties and states, because it provides a flexible and adaptable information identification method that allows designing a customized markup language for almost any type of document [212]. Figure 3.5 illustrates all of the system's architecture, a description of the platforms used and the corresponding relationships within this system.

3.2.5 Task 5: develop a GUI to allow interactivity between designer and MAS

The ability to interact with the optimization routines visually permits considerable control by the designer over the forms of solutions generated by a stochastic optimization method, and also helps to avoid the algorithm getting stuck in local optima [213]. Radford notes that designers can improve the performance of stochastic algorithms through the interactive manipulation of parameters. In order to be able to engage the designer and allow him or her to interact with the system, I have developed Termite, an applet that operates within Processing, but can also be called from the visual programming editor, Grasshopper, and allows for the easier building of agent-based models within the 3D environment of Rhinoceros 3D. The Termite toolkit contains some packaged agent classes that are programmed in Java and Python and have a wrapper for Grasshopper, which allows the designer to easily build agent-based simulations by combining different components and visualize them from the viewport of Rhinoceros (Figure 3.6). To allow for more integration beyond the bounds of a specific 3D environment, I have also started the development of a prototypical Java-based GUI, which is stand-alone and can call different types of software based on the design objective. The GUI allows the designer to (a) define input geometries, (b) select available analytic solvers and (c) define heuristic functions that correlate geometry generation with the analysis using XML files. The designer can then initialize the search process and evaluate the results by visually assessing both the geometry and the corresponding performance. By integrating Parallel Line and Pareto Front plots in the definition of the simulation agent, the designer can select which analyses she/he wishes to correlate and choose which way to visualize it. By showing correlations between the geometry of the generated designs and their impact on the analytical results graphically, designers can gain insights into how the adjustment of design parameters affects the performance of the design. After gaining insights into the parameter range, the designer can call one of the heuristic algorithms to run iteratively and present design alternatives that meet the predefined design performance targets.

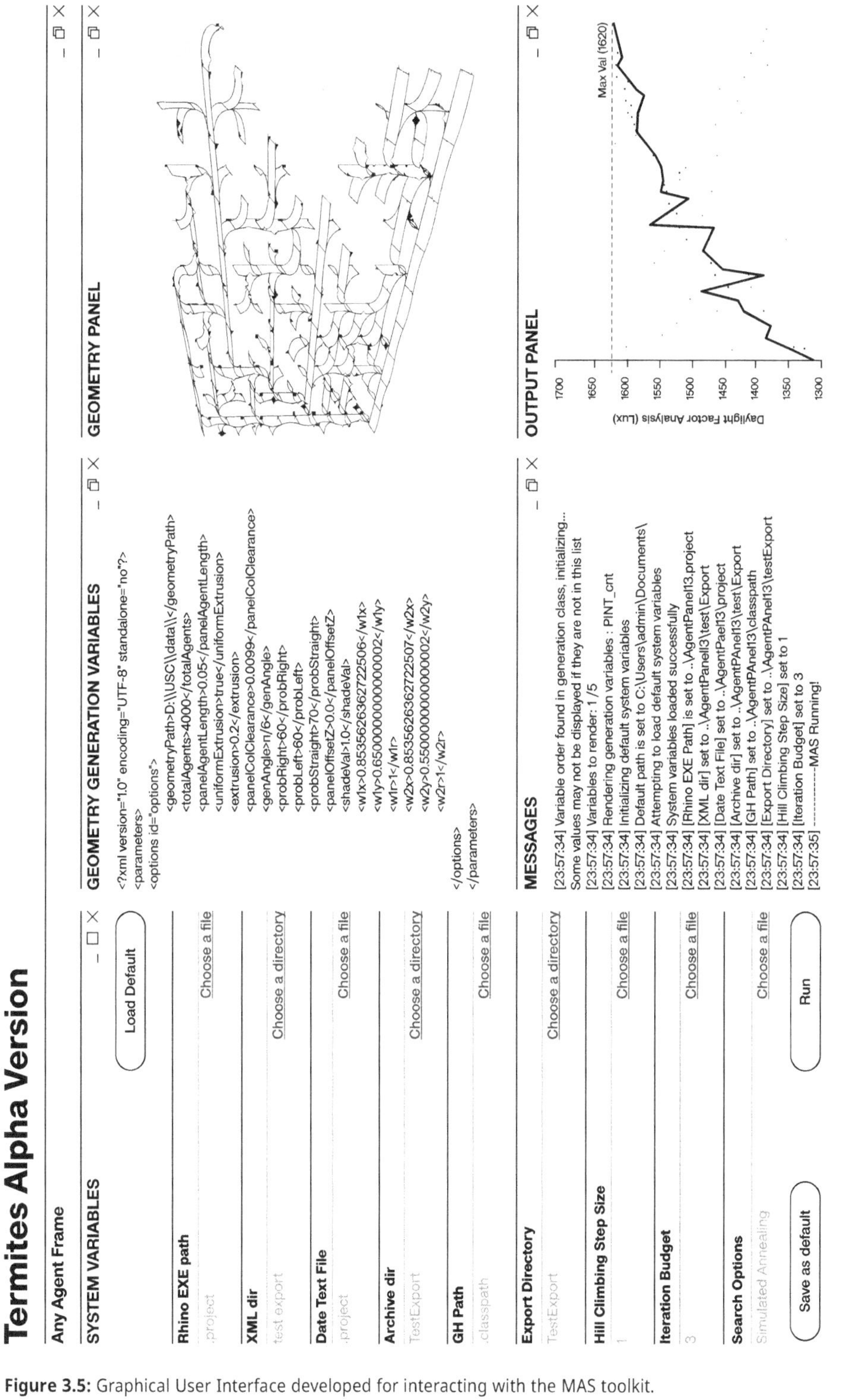

Figure 3.5: Graphical User Interface developed for interacting with the MAS toolkit.

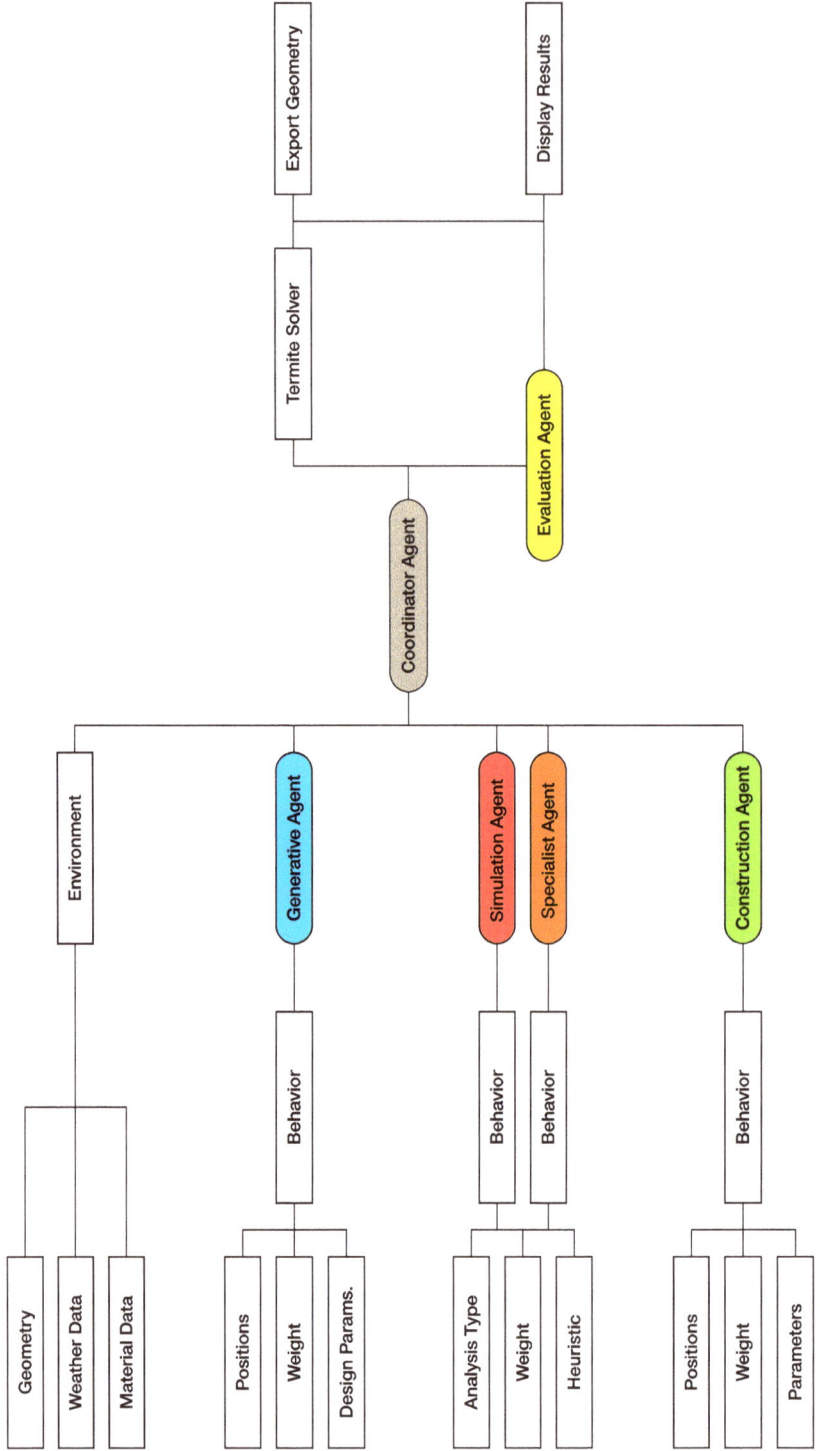

Figure 3.6: Description of an agent based design approach as a VPL graph.

Part IV: **Case studies**

Designing with behaviors

4 Design experiments

The best way to demonstrate, test, measure and iterate upon the multi-agent system (MAS) framework is to pursue a series of design experiments. The design experiments in question are built around the literature review and primarily address two crucial gaps. First, although swarm intelligence methods have been used in architectural design to generate bottom-up design simulations, there are very few examples of how such techniques can be employed to develop and coordinate swarms of robots in the physical world in ways that lead to emergent structures. Second, most of the preceding work in design has used agent-based modeling without considering environmental parameters, like weather data or other data sets, resulting in agents operating in a generic three-dimensional (3D) box or on surface geometries.

Taking all of the above into account, the experiments begin by focusing on the way we can study emergent structures by empirically observing physical robots and using this data to inform abstract agents and digital swarm simulations. Another area the experiments focus on is developing agents that represent building components, whose environment is composed both from geometry (i.e., building envelope) and data (i.e., weather data) and address design problems that traditionally require the close collaboration of architects and engineers, such as facade and shell design.

The methodology used in the experimental designs was incremental to the overall objective, so as to be able to measure improvements in the design process and outcomes, both in formal terms and in design performance terms. Each case study is a methodologic decomposition of a typical design problem in different agencies, where a number of design parameters are coupled with structural and environmental performance targets, as well as process and constructability metrics to drive design exploration. Specifically, the design experiments include the following:

1. The exploration of emergent patterns and structures, by controlling the geometry of a very basic robotic system and by observing how changes in the geometry of the body can lead to emergent phenomena. Instead of generating a design and investigating how we can implement motion control mechanisms on a robot (i.e., inverse kinematics on a six-axis industrial robotic arm) to materialize it, this experiment looks into developing control mechanisms on a swarm of low-level robots by empirically analyzing the impact of the robot's body geometry on their motion. Via the implementation of agent-based modeling simulations, the objective is to find key design parameters that can lead to meaningful emergent swarm behaviors. This study develops an experimental testbed to investigate how real-life constraints can inform digital simulations of abstract agents and how such simulations can provide better insight into how to coordinate large swarms of robots in order to construct functional structures;

2. Facade design exploration, using agents that represent facade panels informed by environmental simulations. Facade panels for office buildings are probabilistically generated and their geometry and placement is informed by daylight and energy analysis.

https://doi.org/10.1515/9783110797435-004

3. Shell structure design exploration, using environmental parameters. Free-form shell structures with different topologies are generated, and their shape is augmented based on environmental and structural analysis.

Obviously, these experiments capture only a small portion of what architectural design problems entail; the reason for this selection is that such problems are more constrained and therefore more suitable to computational description and allow conclusions to be drawn. The experiments are developed in a way that allows the author to measure the capacity of the framework in the early stages in order to generate design alternatives by coupling parameters with functional requirements that relate to performance. Through the synthesis of the experiments, the research begins to apply the developed framework to different design problems and evaluate it; points to the successes and failures of the application of agent-based modeling approaches for early stage design, as well as its implications for autonomous construction; and allows one to start drawing conclusions on the affordances assumed through the behavioral modeling approach in combination with analytical methods, necessary refinements and future direction.

4.1 Experimental design 1: bridging digital agent-based modeling and simulation with physical robotic systems (agents)

The literature in Section 2.3.5 indicates that the most common paradigm of agent-based modeling in the field of architecture is that of swarm intelligence. Even though there have been several design projects using this agent-based modeling and simulation (ABMS) approach as a generative mechanism, in most cases, the agents are abstract and have no physical representation. A brief summary of a typical ABMS design process would involve designers implementing an existing agent-based (flocking) simulation and, in the majority of cases, providing a custom-geometric environment that triggers the bottom-up generation of a design. The designer then manipulates the agents' behavior based on the simulation and their design intention and, at some point, freezes the simulation and proceeds to the design materialization following a top-down approach. However, in complex adaptive systems, these behaviors are a direct result of the physical constraints and forces, as in the case of termites where the dynamic feedback between environment and behavior is what gives shape to the emergent structure of the mound. The first experiment addresses this lack of developing behaviors and simulations by transcribing empirical observations and analysis of a robotic system and investigates how to:

1. Develop a reciprocal relationship between digital swarm simulations and a swarm of low-level robots that operate in the physical world. We study how to best inform agent behaviors in digital simulations by developing an experimental setup that allows us to observe a swarm of robots in the physical world and use collected data to inform our digital simulation model, and

2. Explore whether and how the relationship between the environment and a swarm of active agents (in this case, low level) and passive agents (building blocks) can result in self-organizing structures that present desirable characteristics for architecture and construction.

The hypothesis of the experiment states that by observing how such a simple robotic system operates in physical space and how the locomotion (behavior) of robots is affected by the environment and the blocks, the author can develop agent-based simulations with behaviors informed by physical observations. The capacity to achieve collective behaviors without the need for top-down control is valuable in the context of robotic construction, where robustness, adaptation and scalability are of great importance. The objective of this experiment is to mechanically embed intelligence in the body of the robot–parts that allows the formation of emergent structures prone to exhibiting characteristics over time (i.e., stability, clustering and collective transportation) without extra software control.

4.1.1 Embodied swarm behavior

In this experiment, a simple type of walking robot, also known as a bristlebot, is used as a hardware platform to investigate how different behaviors can be mechanically encoded in the body of an agent by changing the geometry/shape of the robots and the environment they are operating in. This experimental design is based on an interdisciplinary workshop that included architects, computer engineers and roboticists working on collective construction [214]. The bristlebots (active agents) operate in a two-dimensional (2D) arena filled with differently shaped building blocks (passive agents–parts). The geometry of both the robots and the parts can be parametrically adjusted in order to test different basic behaviors (push brick).

4.1.2 Design process

We study the relationships between the properties of the emergent patterns (size, temporal stability) and the geometry of the robot/parts through physical experimentation and video analysis. This work combines our MAS framework with a robust robotic system and uses a set of simulation and analysis tools for generating and actualizing emergent 2D structures. By controlling the ratio of blocks (i.e., number of blocks × area of each block) in relation to the area of the arena (i.e., panel area), the level of occupancy in the arena can be defined, while the interaction of bristlebots with the blocks results in the formation of emergent structures. The design process consists of the following steps (Figures 4.1):

1. Designing custom bodies for a swarm of simple and robust robots, namely bristlebots

Flowchart of Experimental Design

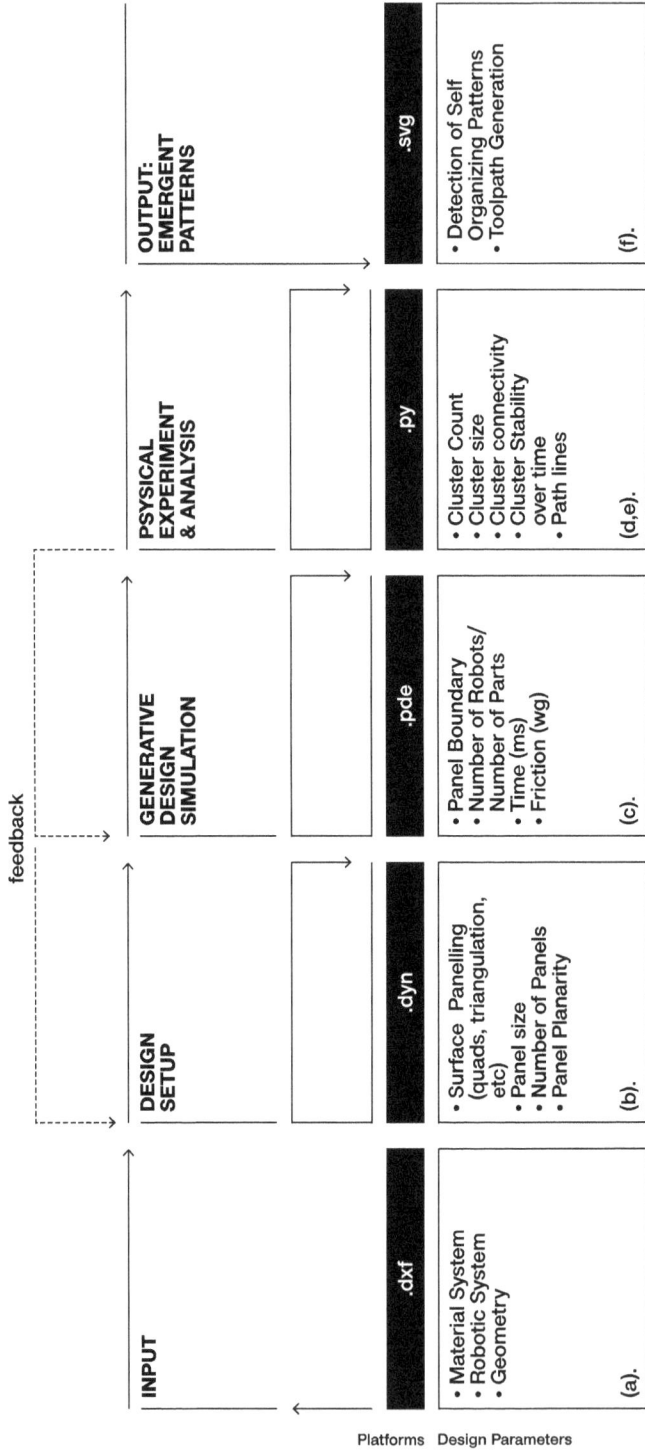

Figure 4.1: Framework diagram illustrating the design phases, parameters and tools developed.

2. Simulating their motion in a 2D environment filled with passive blocks of a certain shape
3. Investigating how the relationship of the robot and block geometry affects the formation of emergent structures
4. Performing experimental runs with 20–100 robots and 2 different types of block geometries in different types of environments (circular, triangular and rectangular)
5. Tracking the motion of the robots using computer vision algorithms and comparing the clustering with simulated results

As a first step, a small swarm of bristlebots is implemented, and the level of clustering that different geometries of the parts can achieve is investigated. Another investigation relates to how the robot's geometry (robot chassis) can affect the manipulation of the parts and the locomotion of the robot. Different shapes of robots and parts are studied along with different types of boundary geometries. Using Kalman filtering, the position and motion path of the bristlebots and the parts are identified, as well as their level of clustering [215].

By collecting data from multiple experimental runs, the relationship between the shape of the robots' bodies and the parts is analyzed with the objective of investigating how these geometric features can affect the formation of 2D structures with desirable characteristics, such as the area of clustering and cluster stability. In the second stage, to get more control over the robotic swarm, a custom bristlebot design (hprbot) was developed. The hprbot is equipped with two motors, a photoresistor, a hall sensor and a microcontroller that allows the bristlebot to be remotely controlled. The experimental setup and design process are described in the following section.

4.1.2.1 Experimental setup

The experimental setup is based on a workflow that was initially implemented at the Smart Geometry Conference 2016 [214]. It includes a reconfigurable 2D environment (arena) in which simple robots (agents) move and push around passive building blocks (parts) for specific time intervals (5–45 min) or until they reach a state of equilibrium (Figure 4.2). The shapes of the arena in which the robots operate can be considered as the boundary of an architectural panel to be designed. The design parameters studied at this stage include the shape of the boundary, the geometry and number of robots, the shape and number of blocks, the ratio of area to number of blocks (i.e., degree of transparency) and the runtime. The motion of both agents and blocks is tracked using a high-definition camera (GoPro) and analyzed using a custom-developed computer vision toolkit (Python, OpenCV, Matplotlib), which allows for the performative evaluation of the emerging structures and behaviors. The bristlebots used in the first stage are cheap and commercially available (www.hexbug.com/nano) and can move at an approximate speed of 11 mm/s by combining a simple vibrating motor and 6–14 angled soft legs. The way the bristlebots move is affected by the number and geometry of the bristles, the surface friction and the obstacles they encounter. Based on the variability of the ground surface and obstacles they encounter, the bristlebots can move in a relatively straight line or follow random trajectories. Since there is no microcontroller on

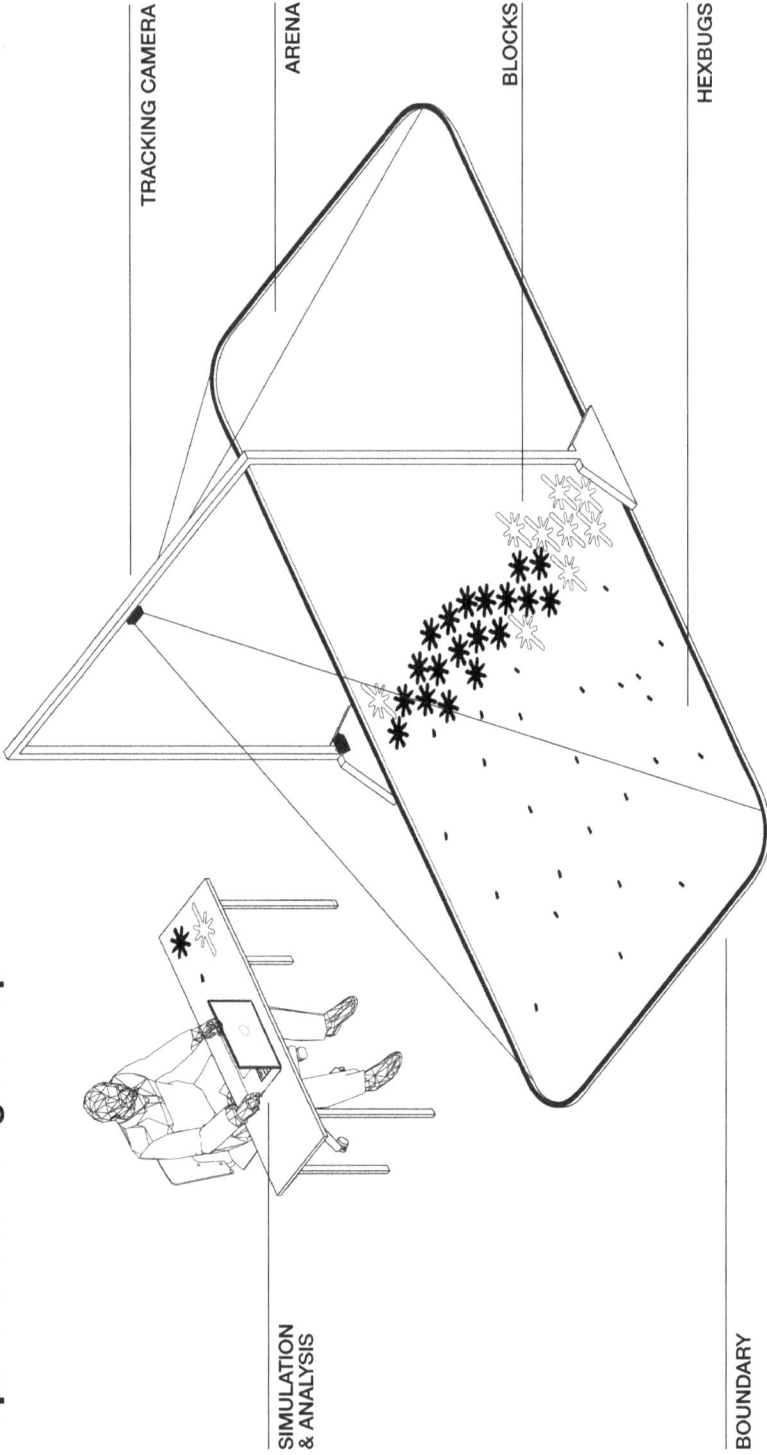

Experimental Design Setup

TRACKING CAMERA

ARENA

BLOCKS

HEXBUGS

SIMULATION & ANALYSIS

BOUNDARY

Figure 4.2: Illustration of the established experimental testbed.

the bristlebots, they are initially "programmed" by altering their body geometry and the environment where they operate. The body geometry of the bristlebots is altered by adding covers with variable geometry. Early experiments showed that big and/or front-heavy covers severely alter the locomotion of the bristlebots and consequently the patterns they generate [214]. Based on the first observation, different covers/bodies were designed to test how bristlebots grab and move blocks around the arena. In conjunction with the bristlebots' cover geometry, alternative geometric configurations were designed for the passive parts/blocks. For simplicity, the experiment starts by selecting two primitive shapes, the circle and the hexagon, and developing different types of blocks by topologically altering these basic shapes. The bristlebots operate in an environment (arena) that can be reconfigured into different shapes.

4.1.2.2 Simulation tool

An agent-based generative tool was developed in order to explore different design alternatives based on the physical setup described above. Unlike existing agent-based tools that implement swarm behaviors and are not connected to the physical world, this tool is modeled after my experimental setup and is used as a platform to fastly explore how different boundary conditions and agent geometries can lead to different self-organizing configurations. The tool was implemented in Processing [208], using Box2D as a physics engine and allows for fast iterations and design alterations, which can then be tested physically. Additionally, the tool enables the designer to test the scalability of different configurations. It also enables the observation of the global impact of local rules when a large number of agents interact in the environment. The aim is to use the data from the physical experiments to inform the simulation of the tool to match the behavior of the robots and their interactions in the physical world. By doing this, the global behavior of the system can easily be explored.

The tool consists of a 2D environment populated with active agents and building blocks, where the designer can parametrically alter the design of both the boundary and the robot/part geometry using the Dynamo visual scripting editor (DynamoBIM. org) and import them directly into Processing as a .json file.

To achieve a realistic simulation of the locomotion, the designer can, apart from using their geometry, control (a) ground friction, (b) object–object friction, (c) object density and (d) the coefficient of restitution. The vibrating motor that propels the robots forward is modeled as an applied vector force on active agents with some noise to account for the variable ground friction that affects their trajectory. The robots are placed randomly in the arena, and their position (x,y coordinates) is exported as a .csv file at specific time intervals for visualization purposes.

The video data is processed using blob detection algorithms implemented using Open CV, an open-source computer vision library for the C++ and Python languages with classes for analyzing the video in real time [216]. The aim of the analysis is to see how the geometry of passive blocks, robots and the shape of the boundary affects (a) the trajectory of the robots in the arena, (b) the formation of larger clusters of blocks through physical interlocking and (c) the collective transport of blocks by the robots.

Additionally, the video analysis is used to calibrate all of the above parameters (a–d) and improve the accuracy of the simulation. Changing the geometry of the robots affects how they interact with the passive blocks (i.e., grab and push one block at a time) and also with other robots (i.e., forming chains). By fine-tuning the geometric parameters of the blocks (parts), the formation, size and stability of the clusters can be controlled to a certain level.

4.1.2.3 Design of the robot (active agent)

Different speeds and levels of agility can be achieved by changing the shape and center of mass (body position) of the body fitted onto the bristlebot. The robot's motion is influenced by the weight and shape of its body, and the weight and total number of parts in the arena. The main design parameters that were found to affect the locomotion of the bristlebots are illustrated in Figure 4.3. Additionally, simple behaviors can be encoded mechanically by modifying the robot body. For example, a U shape (grabber) in the front enables the robot to easily grab and push a brick.

As shown in Figure 4.4, the author developed and tested different robot body designs in relation to the shape of the parts and created a taxonomy of robot bodies. Thin acrylic bodies with grabbers did not significantly reduce the robots' speed, provided directionality and enabled one-to-one engagement with the passive parts (push behavior). Additionally, by introducing a tail at the end with a shape complimentary to the front grabber, the bristlebots demonstrated a cohesive behavior, whereby they formed chains of three to four robots by pushing each other. Robot bodies without grabbers would not engage with the part for long and would demonstrate wandering behavior by diverting their trajectory, depending on which obstacle they encountered. Cardboard robot bodies were heavier and slowed down the bristlebots significantly, but they provided more stability and direction to the movement and the robots demonstrated a collective transportation behavior. An additional design parameter that emerged was the available energy in the system. When their batteries were charged, the robots could traverse the arena faster and engage with parts, but once the batteries discharged, the motor rotation (rpms) slowed down and the bristlebots would often move in circles without sufficient force to push anything.

4.1.2.4 Design of the part (passive agent)

As far as the parts are concerned, two different types of geometries have been tested along with their topological variations: (a) one set of designs based on a hexagon and four different types of block designs and (b) a second set of geometries based on circles, with four different types of blocks based on the topological variation of two, three and four circles (see Figure 4.4). The geometrical and topological variations consisted of making the parts bigger, the clusters more stable and creating connecting points with higher friction. The main design parameters are the parts' weight and contact surface with the ground, which affects friction. Moreover, the number and the length of the parts' edges affect their connectivity with other parts. Using computer vision, the size and number of parts in clusters is observed and analyzed over

Bristlebot's body part and the arena where they interact

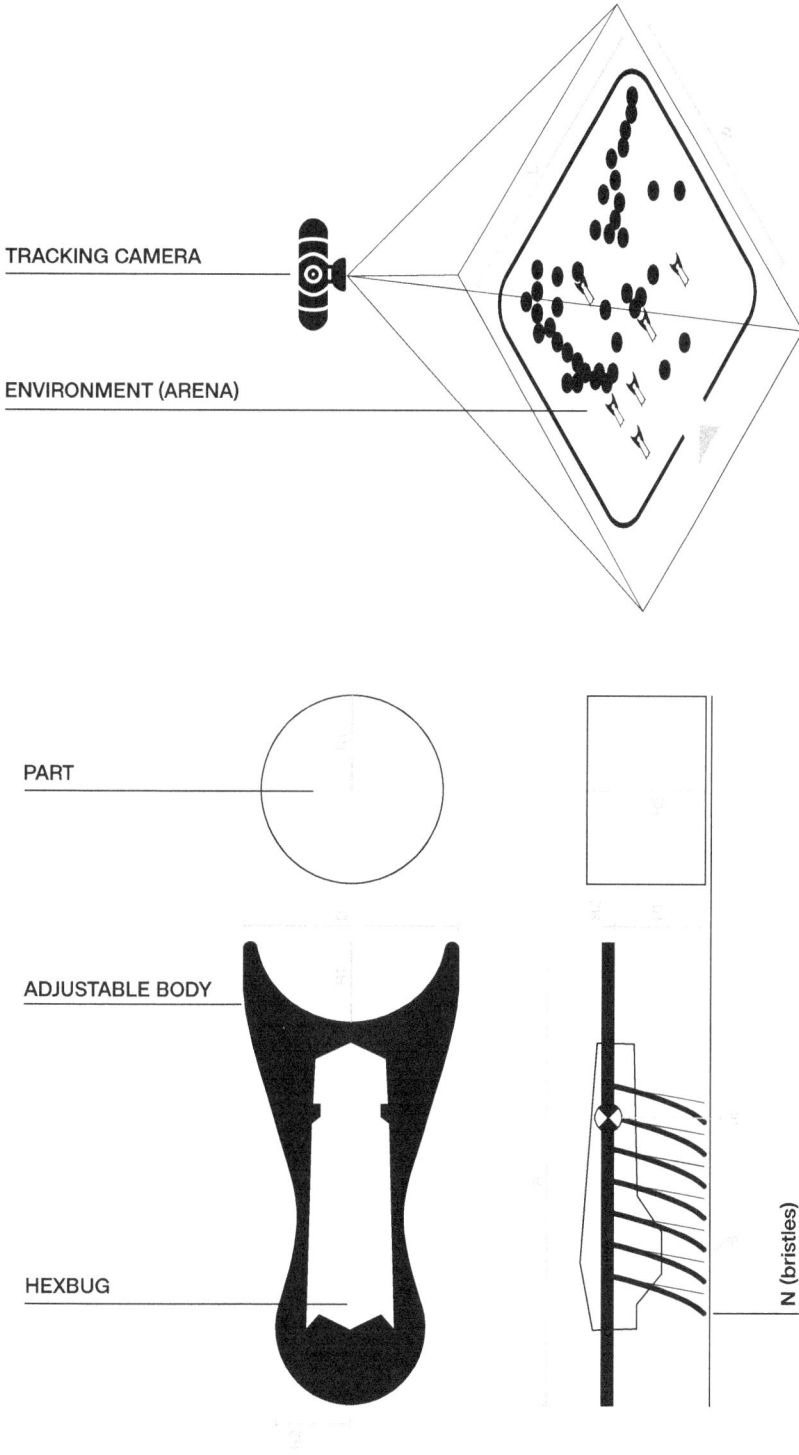

TRACKING CAMERA

ENVIRONMENT (ARENA)

PART

ADJUSTABLE BODY

HEXBUG

N (bristles)

Figure 4.3: The main design parameters of robots, parts and environment.

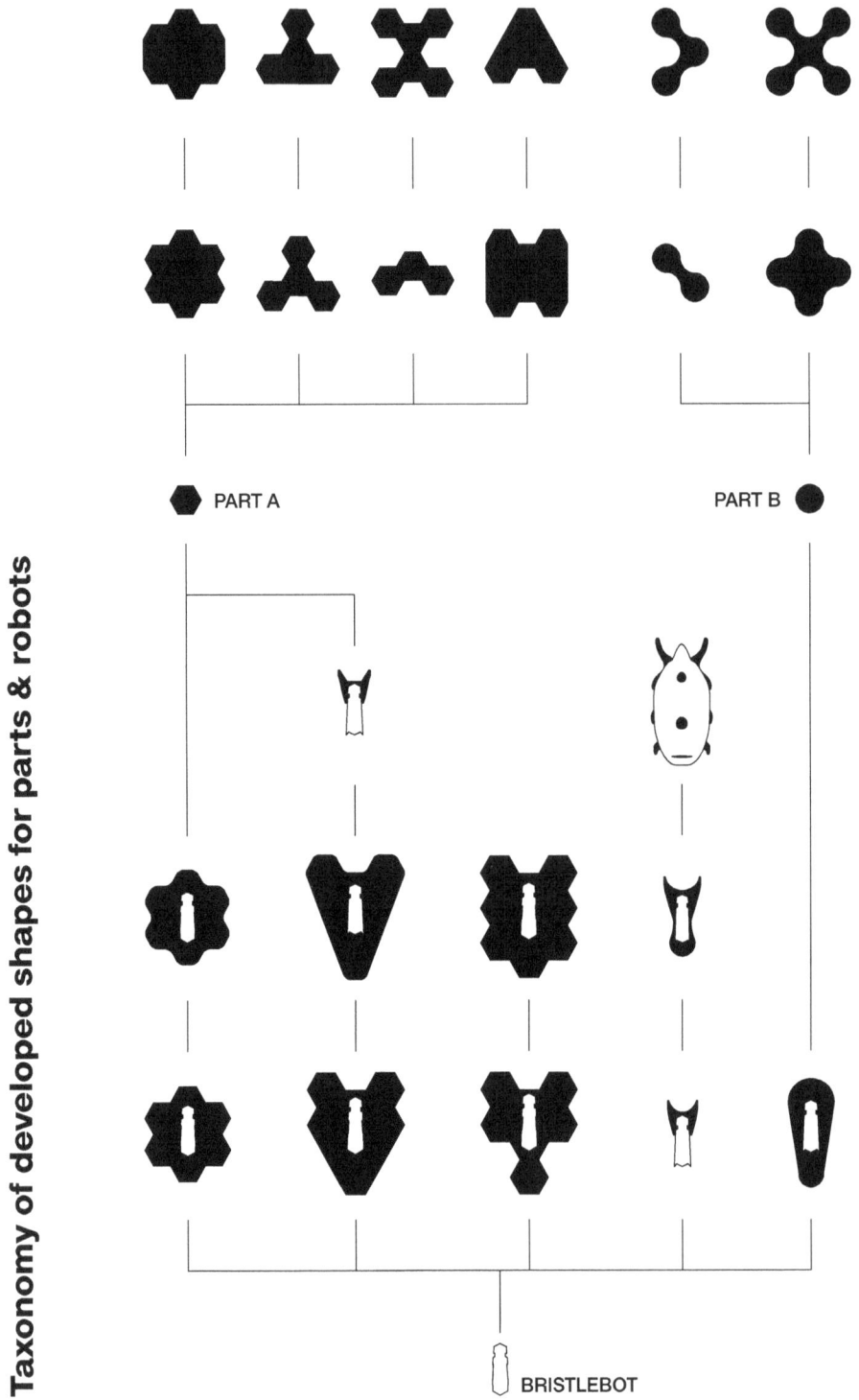

Figure 4.4: Taxonomy of robot and part designs.

time to see how the boundary of the arena and the robot–part geometry affect the formation and consistency of clusters (self-organizing patterns).

4.1.2.5 Design of the environment
Another major parameter that affects the behavior of the bristlebots is the shape and size of their environment. Three basic shapes were tested, namely a rectangular, a circular and a triangular form, to explore the impact of the global geometry on the behavior of bristlebots. Figure 4.5 shows the geometric configurations of the three different arenas whereas Figure 4.6 is a photo of the experimental design setup at Autodesk Build Space. Apart from the shape, the texture of both the sidewall and the floor of the environment influenced the locomotion of the robots significantly. For example, by adding more textured materials (i.e., fabric), the bristlebots could be directed through specific paths all the way to an endpoint without an extra control mechanism. Another parameter that was considered was the introduction of fixed parts (islands) in the arena; these reduced the variability of the clusters and, depending on their number (1–3), they constrained the motion of the swarm in specific areas. Lastly, although mostly flat 2D arenas were investigated, different sloping configurations were tested and experiments showed that if the slope was bigger than 8.5% the bristlebots could not move the parts, while the bristlebot itself was not able to move forward if the floor slope was above 15%.

4.1.3 Results and analysis

This first design experiment bridges the gap between agent-based swarm simulations in the digital realm and a physical manifestation of a (swam) robotic system. An experimental setup was chosen which allowed studying and developing agent behaviors based on the empirical observation of a robotic system. The initial experiment explored the generation of 2D self-organizing patterns using a swarm of basic, commercially available bristlebots (a.k.a. hexbugs).

Working with a simple and cheap robotic platform did not allow the robots to perform any sophisticated tasks, but it enabled me to explore the scalability of this approach and physically test the interactions of a large number of robots. Although more experiments need to be performed in order to reach statistical significance, the first indications show that the behaviors persist even when the number of robots was increased 10-fold (from 20 to 200 robots in the arena). Approximately 200 experimental runs were conducted in 3 different arenas, testing different geometries of both robots and parts.

The experimental runs were analyzed by measuring the number and size (area) of the clusters over time, using computer vision techniques mentioned in the previous section. The experimental results in Figure 4.7 show that there is a direct relationship between the velocity and the design parameters of the bristlebots. Additionally, the experimental runs show that hexagonal parts tend to form more stable structures due to the friction of the sharp edges, while circular parts do not cluster for long periods

Geometric configurations of experimental design setups (Arenas)

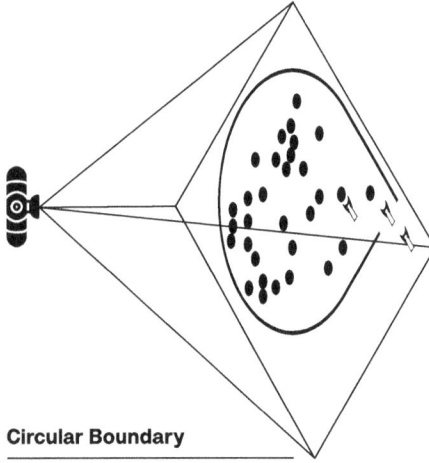

Circular Boundary

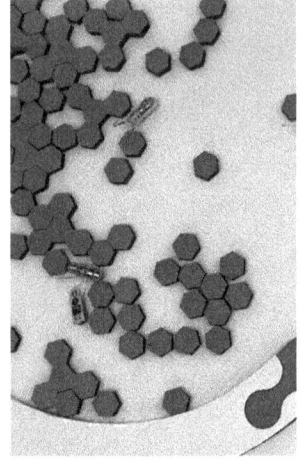

Ratio: Total Area/Parts Area: 30%

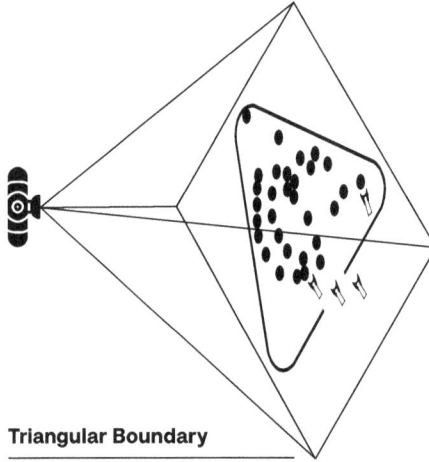

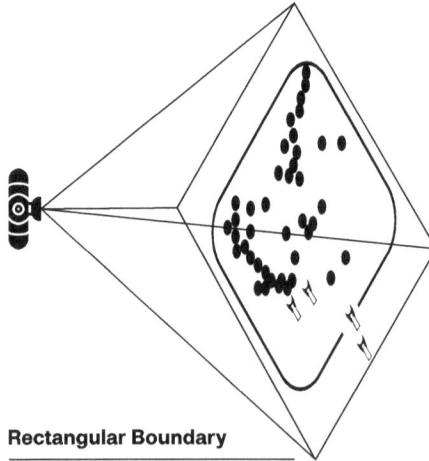

Triangular Boundary

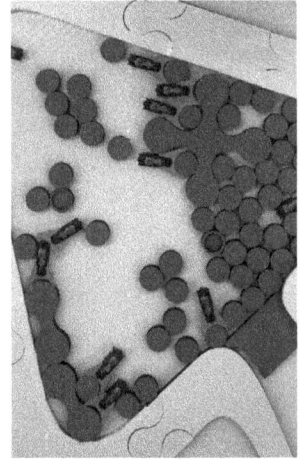

Ratio: Total Area/Parts Area: 30%

Rectangular Boundary

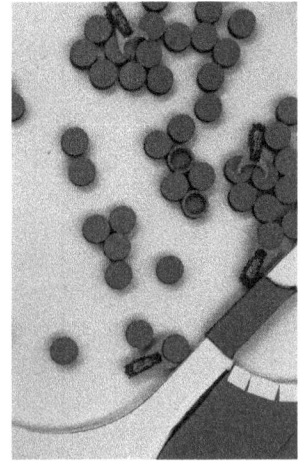

Ratio: Total Area/Parts Area: 30%

Figure 4.5: Configurations of different environments with active and passive agents.

of time. However, circular parts are easier for the robots to transport, as they provide multiple grabbing points. Lastly, the topological transformations of the shapes help the creation of bigger clusters but constrain the flexibility of the robots to move around the arena. These observations point to a direct correlation between the basic selected design parameters and the actual velocity and locomotion of the robots, which suggests that a more reliable simulation of the agents is attainable.

Figures 4.9 and 4.10 show plots of the experimental runs and calculate the size and number of part clusters over time to see how the boundary of the arena and the robot–part geometry affect the formation of clusters (self-organizing patterns). This analysis helps researchers develop an understanding of potential future applications of this system, such as the collective transportation of parts or assembly of 2D structures. By analyzing the graphs, we can observe that the hexagonal components form larger structures than the circular components and the emergent clusters stay together for longer durations. We can also observe that in the triangular arena, the robots formed bigger and more stable clusters (Figure 4.10d–f).

In terms of shape, robots with grabbers proved to be reliable in consistently engaging with one part at a time and pushing it forward. Robots that had the same geometry as the parts but did not have grabbers became part of the assembly and moved in unison with the cluster of parts. However, in more than 50% of the runs, the robots only clustered with other robots and the lack of directionality in their shape resulted in one robot counteracting the motion of another, causing them to go around in circles instead of moving forward.

Regarding the geometry of the environment, the initial experiments used one point of entry that was manually closed off once all the robots were in the arena. If the entry point was left open, bristlebots would eventually exit the environment. Different initial configurations for the parts were also tested, namely randomly placing parts in the whole arena, placing parts in one cluster in the middle of the arena and placing parts at the edge of the arena. The introduction of fixed parts (obstacles) in the arena was also tested and led to the quick formation of clusters. Depending on the number of obstacles (1–3), this significantly constrained the motion of the swarm. The robots tended to gather toward the boundaries and were able to draw constant trajectories following the boundary of the arena independent of their geometry. Sharp angles ($\theta < 90°$) tended to trap the robots.

By controlling a set of input parameters, specifically the population size, the shape of the part, the shape of the robot and the shape of arena (environment), I was able to test which parameters have a greater impact on the emergent structures. Control over the locomotion of the robots can be achieved without global coordination, either by developing specific body types or by introducing "soft" boundaries that have a higher level of friction (i.e., textile). It was demonstrated that by changing the agents' bodies, the shape of the parts, the ratio of the parts to robots inside the arena and the boundary condition of the environment where the agents interact, basic behaviors can be mechanically encoded in the robots (agents) without the need for extra programming. The agents' (robots) behaviors directly correlate to the geometry of the robot, and are as simple as engaging with a part, collectively pushing a part,

Figure 4.6: Photos of the experimental design setup at Autodesk's BUILD SPACE in Boston.

Experimental Results: Impact of design parameters on bristlebot's locomotion

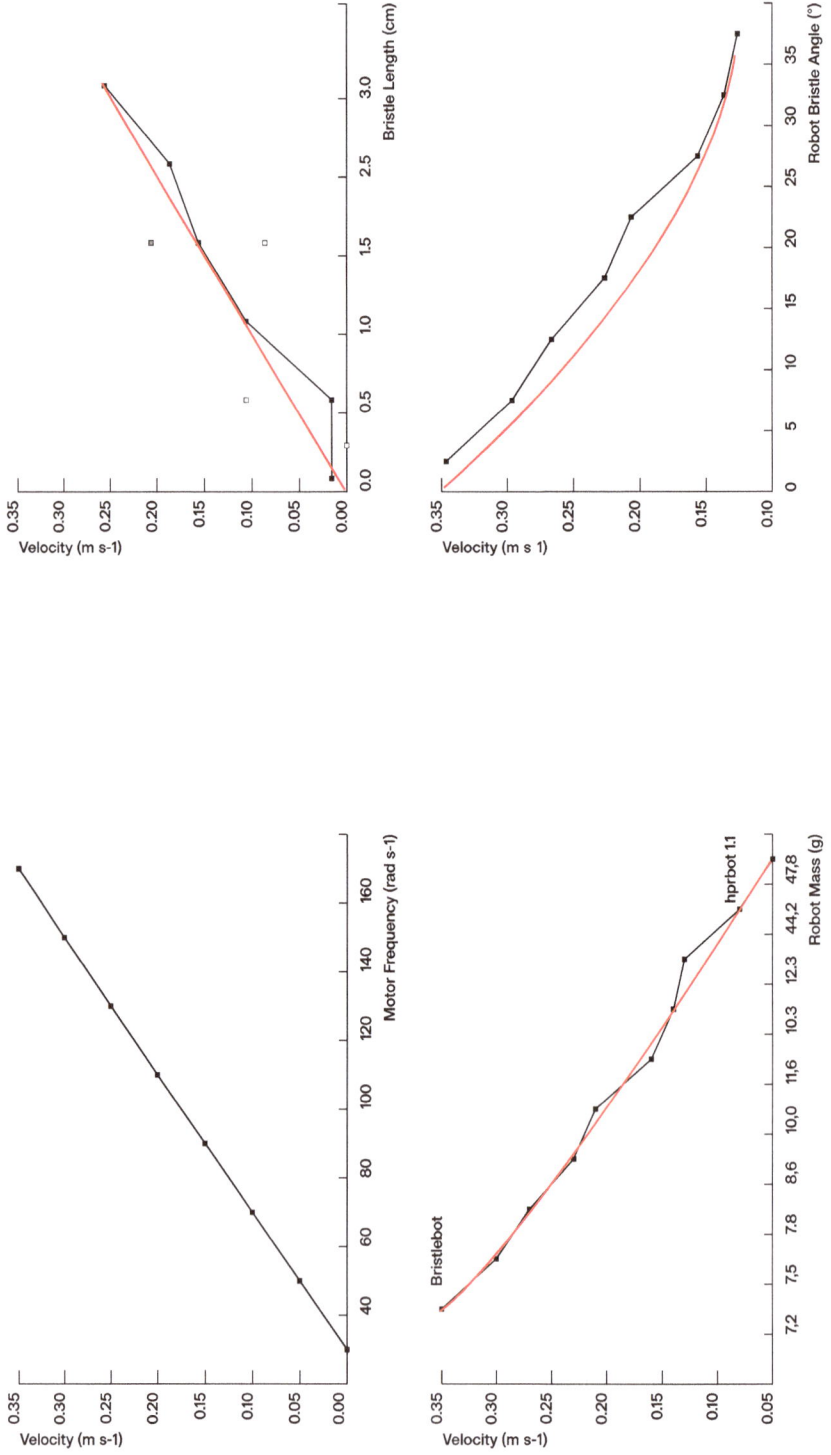

Figure 4.7: Plots showing the relationship of the bristlebots' design parameters with regard to velocity.

Functional Components of hprbot

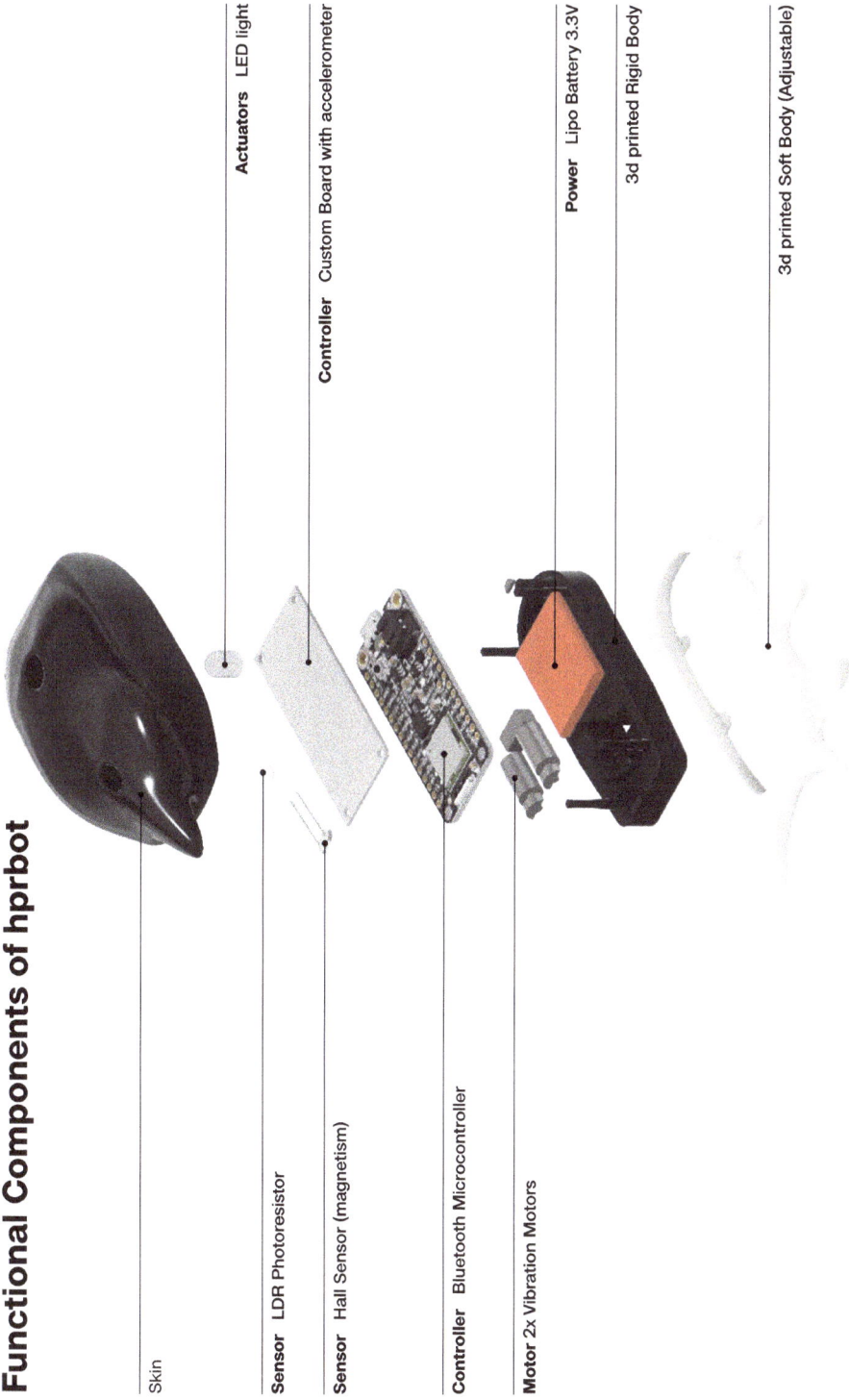

Skin

Sensor LDR Photoresistor

Sensor Hall Sensor (magnetism)

Controller Bluetooth Microcontroller

Motor 2x Vibration Motors

Actuators LED light

Controller Custom Board with accelerometer

Power Lipo Battery 3.3V

3d printed Rigid Body

3d printed Soft Body (Adjustable)

Figure 4.8: Exploded isometric view of the hprbot with all its components.

Experimental Results: Number and Size of bristlebots' clusters over time

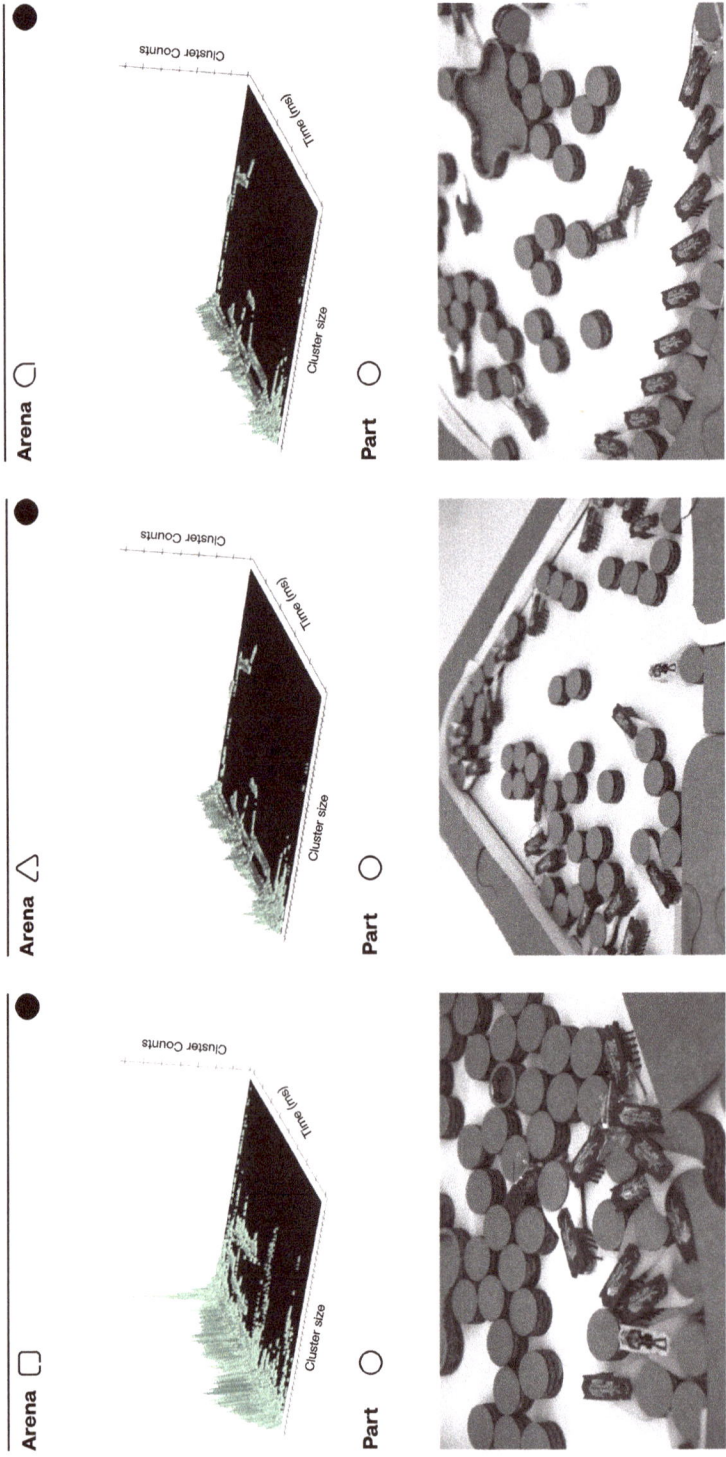

Figure 4.9: Plots showing the clustering of circular parts over time.

Experimental Results: Number and Size of bristlebots' clusters over time

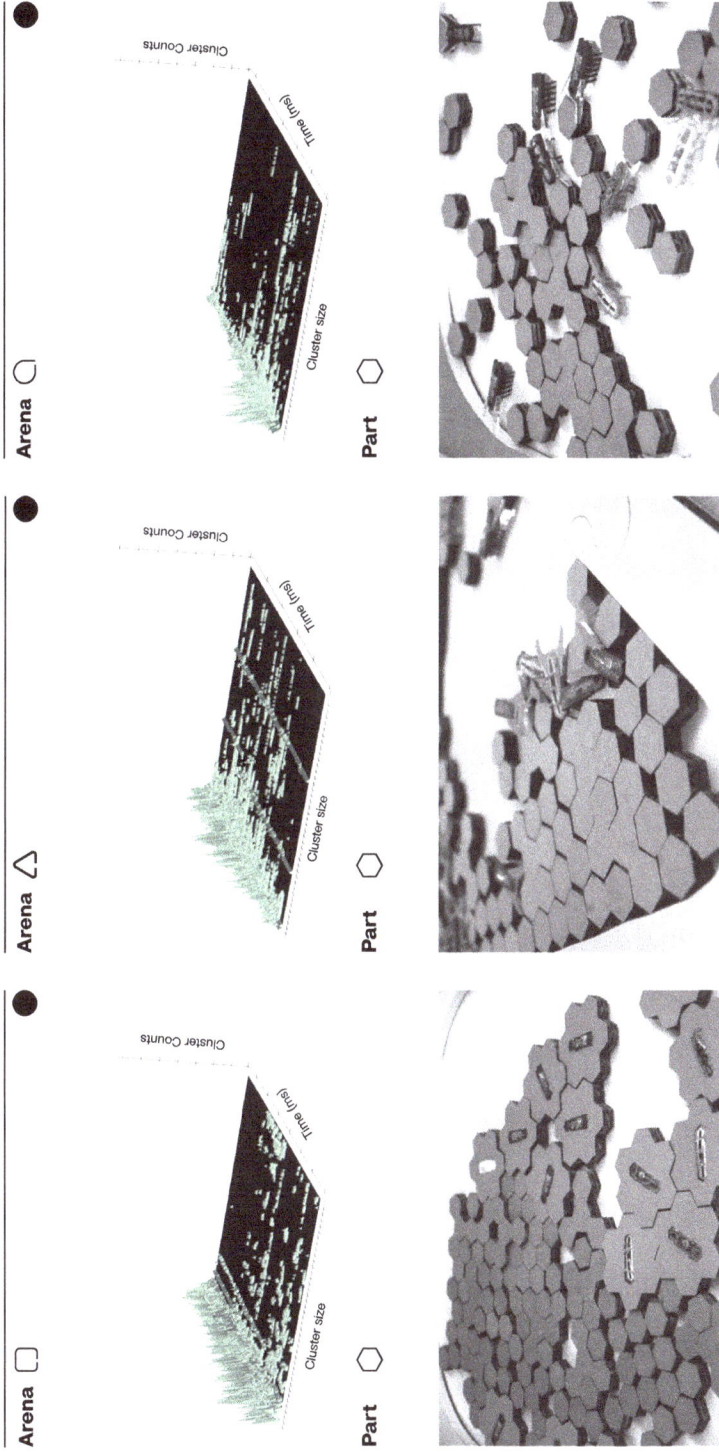

Figure 4.10: Plots showing the level of clustering of hexagonal parts over time.

Experimental Runs: Emergent Behaviors of bristlebots

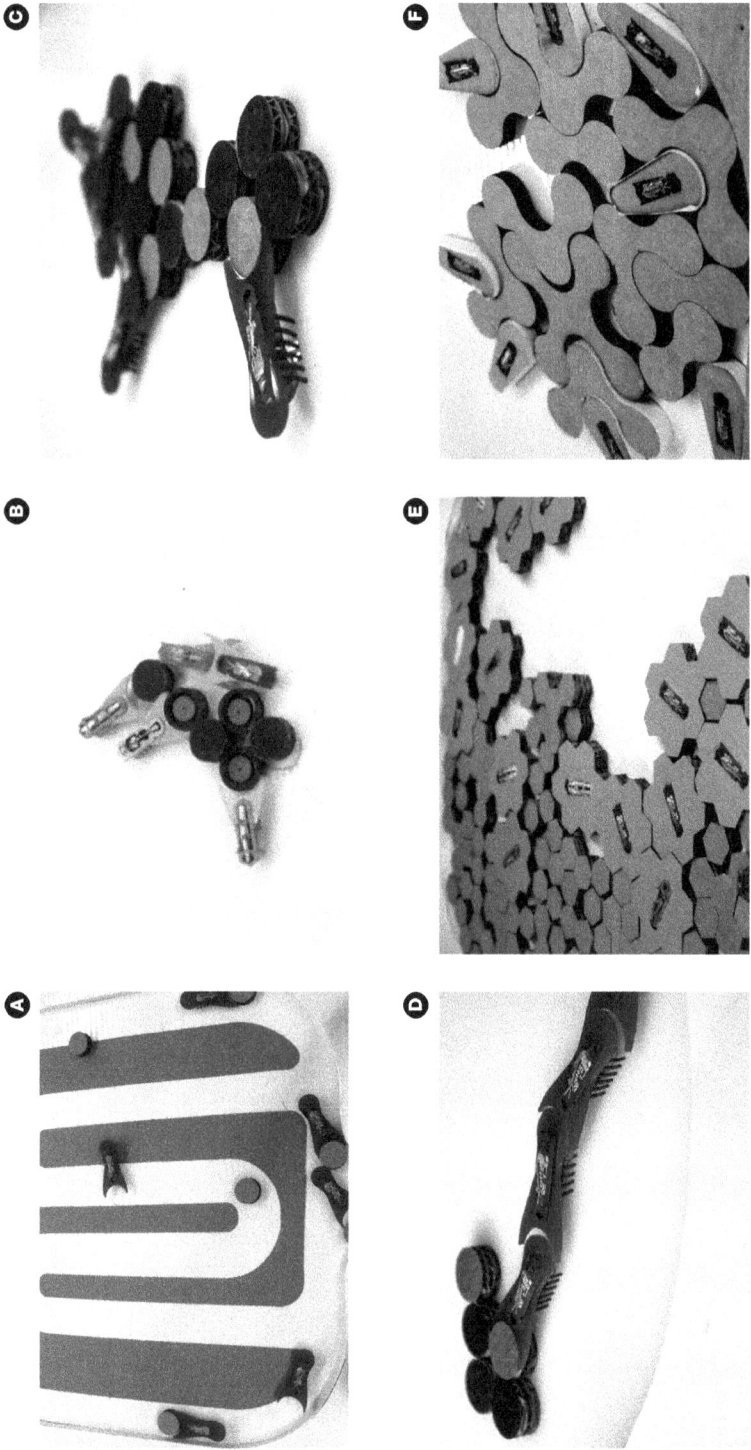

Figure 4.11: Images from experimental runs showing emergent robot behaviors.

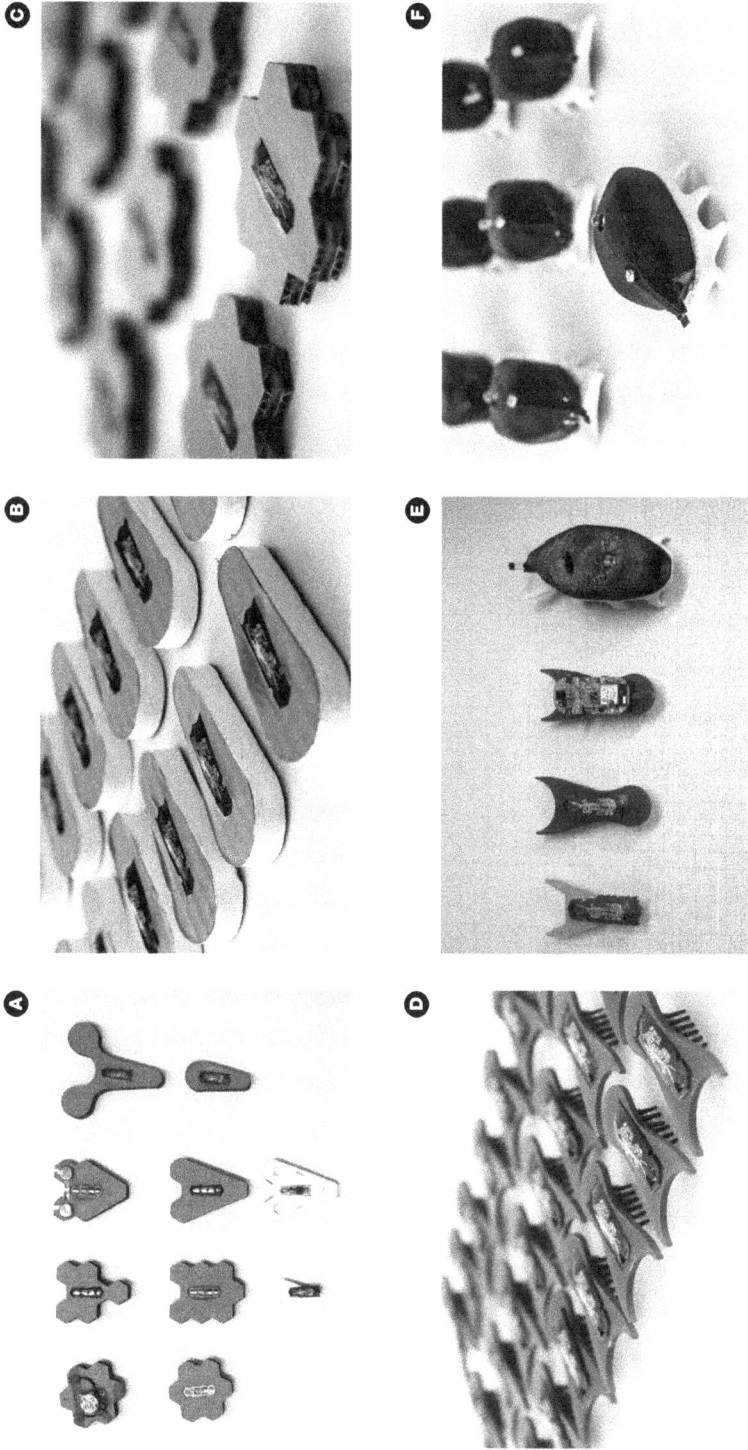

Prototypes

Figure 4.12: Photos of bristlebot prototypes. Image by author from author's archive.

Hprbots in arena

Figure 4.13: Photos of the hprbots interacting (clustering) within the arena by reacting to light intensity (if light is above threshold, then bristlebots start moving). Image by author from author's archive.

moving toward an attractor, forming a chain and creating a cluster. In Figure 4.11 a series of emergent behaviors are presented, namely: (a) bristlebots navigating a path defined by variation of friction, (b) bristlebots creating clusters of parts, (c) collective transport of circular parts by a group of bristlebots, (d) robots forming a chain by pushing parts in synchronous motion, (e) emergent clustering of bristlebots and hexagonal parts and (f) emergent clustering of bristlebots and circular parts. Figure 4.12 features photos from a series of robot prototypes, while in Figure 4.8 an exploded axonometric highlights the main components of "hprbot." Hprbot is the most evolved version of the bristlebot that was developed within this research project, featuring an adjustable 3D printed soft body with three pairs of bristles and two sensors, namely Photoresistor, Hall Sensor and a microcontroller equipped with Bluetooth that facilitates remote control. Lastly, Figure 4.13 shows a photo from hprbots autonomously interacting in the area.

4.2 Experimental design 2: agent-based facade design in office buildings

Facades are among the most complex architectural systems, combining aesthetic, structural, environmental and construction concerns [217]. In large part, the facade of a building determines the amount of available direct and indirect natural lighting [217, 218]. Since lighting accounts for 20–25% of the total electrical use in buildings and, more specifically, 30–50% in commercial buildings [219, 220], I am convinced that by reconsidering the performance of the facade, architects can improve the buildings' energy efficiency, as well as the occupants' comfort.

As a result, this experiment focuses on the generation of architectural facade designs. The experimental design investigates how to create facades that improve the design and performance of a building by combining (a) environmental analysis data – specifically solar radiation and luminance – with user preferences for light intensity within the office environment, and (b) robotic simulations. The key assumption is that, in the early design stage, an architectural design problem can be decomposed into the agents that cater to different aspects of the problem under consideration. In particular, the implementation of separate agent classes as presented in the previous section is proposed.

Five different classes of agents are implemented, with actions, properties, states and goals. These agent classes have goals related to different steps of the design process, namely (a) synthesis: generative agent, (b) analysis: specialist agent, (c) evaluation: evaluator agent, (d) simulation: construction simulation agent and (e) coordination: coordinator agent (Figure 4.14). Each type of agent relates to different design domains and communicates with a different type of software. By defining different agent classes and behaviors and selecting a design scheme, depending on a combination of the designer's experience and predictions about the performance goals of the design (i.e., based on previous analytical data), the solution space can be effectively reduced, allowing the exploration of multiple design solutions.

This experiment is developed in two stages; initially the experiment examines the generative aspect of design where the agents act autonomously; then, the experiment tests the optimization of generated outcomes where agents act collaboratively and negotiate to find optimal solutions. First, a set of generative agents and behaviors are modeled, based on a given design site's location and orientation, a building facade bounding context, and designer-defined parameters regarding the building's components (i.e., length, width, thickness and type).

Initially, the agents acted autonomously and developed design alternatives that satisfied the local rules and constraints from the geometric domain, avoiding specific areas reserved for window openings and views, and collision checking for constructability. During the second loop, the designs were analyzed by a set of specialist agents that communicated their data back to the generative agents to adjust the parameters and regenerate design alternatives, based on specific user preferences and performance goals. Two case studies are presented: in the first case, the specialist agent is using daylight simulations to evaluate each design; and in the second case, the specialist uses a robotic simulation to evaluate each design.

The objective of this experimental design is to demonstrate that (a) stochastic optimization can be combined with agent-based architectural modeling to enable design exploration, (b) analytical values and construction simulations can be used to inform and steer agent behavior toward the generation of higher performing design alternatives and (c) heuristic functions that are not combined with specific geometrical features (the shape of openings) but have more abstract relationships (glazing ratio and number of openings) can be developed and used in multiple cases in the early design stage.

4.2.1 Facade design exploration using heuristic algorithms

To drive the behavior of the generative agent toward producing alternatives that meet the design objectives, I correlated its input parameters with the analytical values obtained by a specialist agent. At each iteration, the specialist agent was responsible for performing a daylight analysis and communicating the analytical values to the generative agent (updating the values to the generative agent). In this case, there are multiple analyses, and each design parameter has a weighted value attached that indicates the impact of each parameter on the analysis.

Every time a parameter is updated, each agent gets a credit (e.g., +1) if the value obtained at the sensor point with the highest impact is close to the target value. If it is further from the target, the agent gets a penalty (e.g., −10). Based on this credit/debit system, the generative agent can decide how to update a design parameter. Initially, two basic local search algorithms were implemented: hill climbing and simulated annealing. These algorithms were chosen to test two fundamental heuristic search approaches that proved to be applicable on a wide range of design problems. For each of the approaches, a custom heuristic function was developed that associates a design feature (i.e., window position on a facade) with one or more analytical results (i.e., the amount of daylight entering a space). In order to draw conclusions, the performance of the different algorithms was evaluated vis-à-vis (a) generating unique solutions each time they run and (b) operating on small versus large sample size. To be able to validate the results, the problem was reduced to finding optimal opening positions in

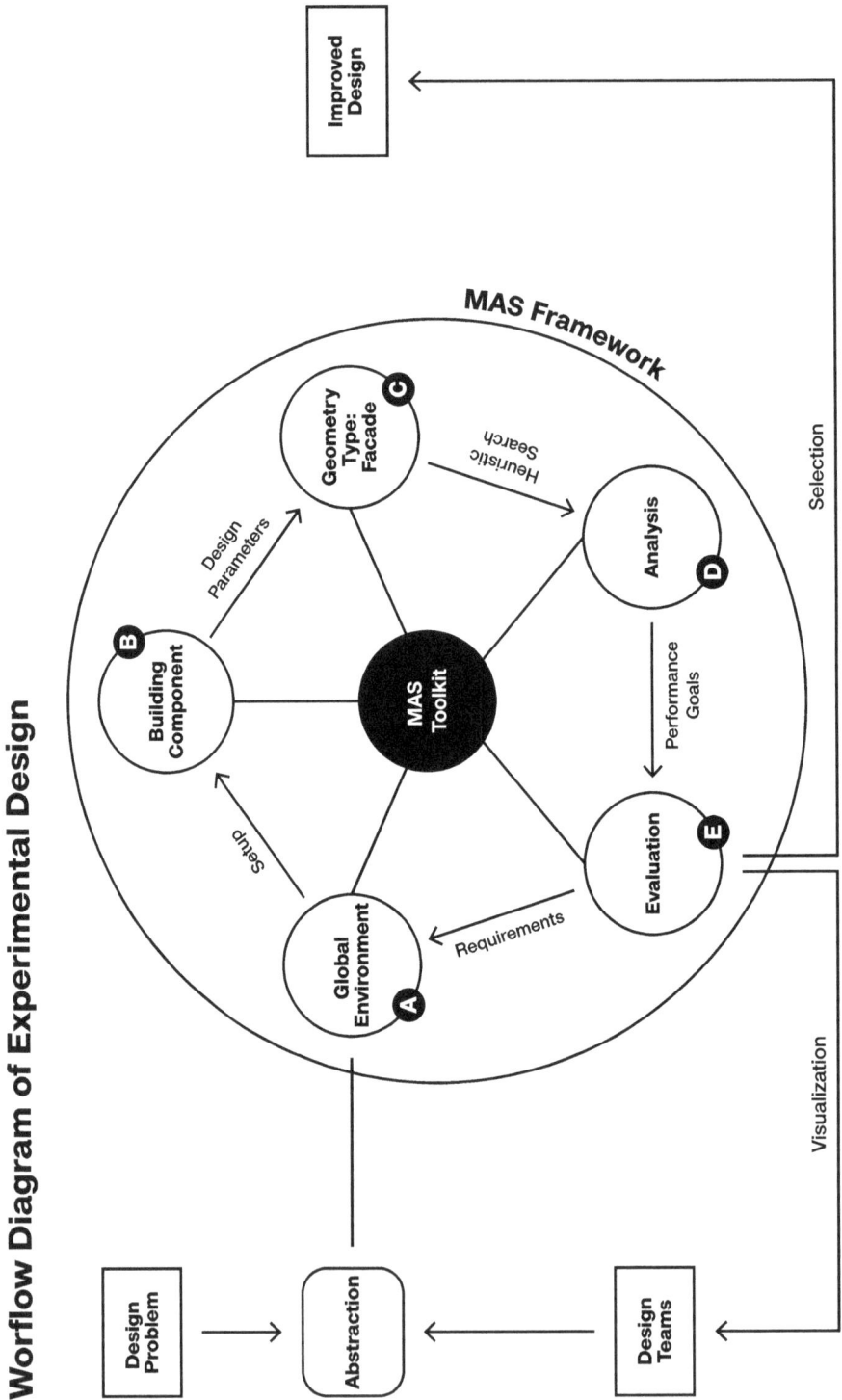

Figure 4.14: Decomposition of a problem design generation, simulation, analysis and evaluation.

a generic facade surface and searching linearly to map (brute force) the whole solution space. The relationship of the parametrization of the design features was observed in relation to the analytical results coming from environmental analysis. This information was used to guide the search more efficiently, based on this relationship and the agents' performance in each of the approaches discussed below.

4.2.2 Daylight metrics and design performance goals

Analytical values calculated by the specialist agents were used to steer the behavior of the generative agents. The fact that different disciplines are focused on different aspects of daylight use makes it difficult to evaluate a design strategy and therefore provide "goal values." To enable a comprehensive understanding of the trade-offs between daylighting performance and how different design features affect it, five different types of daylight metrics are considered. Of those, three are already established and relate to (a) illuminance levels, such as daylight factor analysis (DFA), (b) energy consumption with regard to artificial lighting such as continuous daylight autonomy (CDA) and (c) illuminance levels with regard to lighting preferences defined by the building's occupants, such as useful daylight illuminance (UDI). Reinhart provides an in-depth description of the calculation of the above metrics and their integration in sustainable building design [218]. Two additional metrics were introduced to calculate maximum light variance and the level of light diffusion within a space. The variance was calculated by selecting the minimum and maximum light values over the specified time across the office space, and the level of light distribution was measured by comparing the daylight values between neighboring sensor points. Last but not least, to help the designer build a holistic understanding of how specific design decisions about the amount of daylight affect the energy efficiency of the building, an energy analysis was performed on top of the daylight analysis to measure how the heating and cooling loads of the office space change based on the different facade alternatives presented by the agents (Figure 4.15).

All the metrics have been calculated using simulation software developed by the Department of Energy, as accessed via the Ladybug and Honeybee tools. The experiments required certain specifications: For the CDA, the designer had to define the minimum acceptable lux values and, for the UDI, an upper and a lower boundary goal. She or he also specified what time of day and year she or he was interested in analyzing in the generated design, that is, 9:00 am–12:00 pm, 9:00 am–5:00 pm, winter, summer, spring or autumn.

The designer had to define the size and height of the analysis plane and a specific resolution that was divided into a number of sensor points (in this case, 2,623 points). At each sensor point, the amount of incident daylight was calculated in lux for the DFA and a percentage based on whether the amount of daylight provided illuminance within the user-specified range for a specific time period (time of day and time of year) for the CDA and the UDI. To calculate the final goal based on the illuminance of a sensor plane, the values received for the CDA and the UDI were averaged at each sensor point. The heuristic function that describes this relationship is

Experimental Design Setup - Daylight Simulation

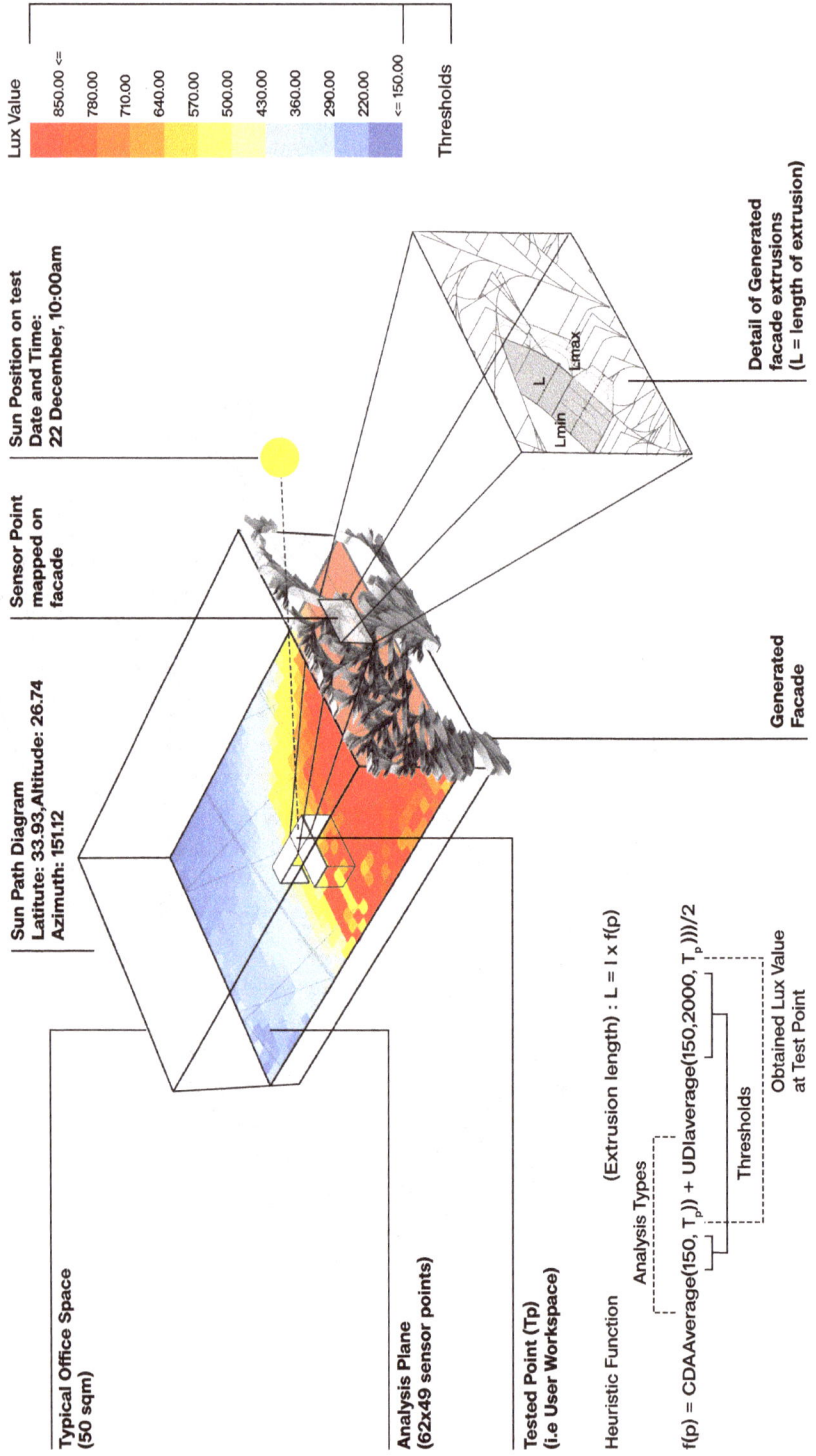

Lux Value

850.00 <=
780.00
710.00
640.00
570.00
500.00
430.00
360.00
290.00
220.00
<= 150.00

Thresholds

**Sun Position on test
Date and Time:
22 December, 10:00am**

**Detail of Generated
facade extrusions
(L = length of extrusion)**

Lmax
L
Lmin

**Sensor Point
mapped on
facade**

**Generated
Facade**

**Sun Path Diagram
Latitute: 33.93, Altitude: 26.74
Azimuth: 151.12**

**Typical Office Space
(50 sqm)**

**Analysis Plane
(62x49 sensor points)**

**Tested Point (Tp)
(i.e User Workspace)**

Heuristic Function   (Extrusion length) : L = l x f(p)

Analysis Types

$f(p) = CDAAverage(150, T_p)) + UDIaverage(150,2000, T_p)))/2$

Thresholds

Obtained Lux Value
at Test Point

Figure 4.15: Experimental design setup describing the environment, parameters and heuristic function.

$$\text{Extrusion}(L) = l \times f(p),$$

$$f(p) = \frac{[\text{CDA Average}(\text{Threshold1}, T) \pm \text{UDI average}(\text{Thres1}, \text{Thres2}, T)] \pm \text{credit}}{2}.$$

The office space was analyzed for each season by simulation, and the results were used to (a) inform the position of the opening, (b) change the probability of placing an agent type and (c) change the depth of the facade.

4.2.3 Facade design optimization using robotic simulations

Apart from the analytical values relating to daylight performance, and to add another layer of complexity and enable a comprehensive understanding of trade-offs between design generation, daylighting performance and constructability, I describe a class of construction simulation agents. This class of agents is tasked with simulating the robotic construction of the generated designs and evaluating them based on their constructability. To develop the robotic simulations, I assume an offsite robotic construction process using an industrial robotic arm. Based on that assumption, a number of constraints emerge that relate to the working volume of the robots, their positions, pickup locations and self-collisions. These constraints are used to develop a heuristic function used by the construction simulation agent to evaluate each design alternative.

Below, the list of measurements and definitions in the heuristic function is given. (i) Pos(x): the number of positions needed to assemble the whole facade and the maximum number of panels (Panels(n)) the robot can place from a given position. A position is considered optimal when it allows the robot to place the maximum number of panels possible. (ii) Col(j): the number of collisions between the robot and the panels already placed. (iii) Sing(k): singularity positions the robot can take while trying to reach a point in space. Singularity positions may result in self-collisions and therefore need to be avoided. (iv) $D(t)$: this calculates the sum of the travel distance from pickup location to panel placement (Figures 4.16 and 4.17). Finally, the constructability function is defined to enable the performance of the construction process to be measured and rank the designs that depend on the parameters described above (see equation):

$$\text{con}(\text{runID}) = \frac{\frac{\text{Panels}(n)}{\text{Pos}(x)}}{D(t) * \sum \text{Col}(j)} - \sum (\text{Sin}(k)).$$

The definition of the heuristic above was based on empirical knowledge gained via two robotic workshops held at ETH Zurich [221] and includes the following assumptions: more robot placement/calibration positions increase the construction time resulting in a lower score and singularity positions may damage the robot, further impacting the negative score. Based on a user-defined number of maximum possible collisions when placing the panels, the generative agent eliminates (deletes) the affected panels and updates the probability factor of adding a new panel in the next

Experimental Design Setup – Robotic Construction Parameters

R(d): Robot Reachability

F(x,y,z): Placement location (facade)

D(t) : Robot's Travel Distance

Loc(l) : Panel pick up location

Pos(x) : Robot positions

KUKA-KRL60: 6 Axis industrial robotic arm

Position II

Position I

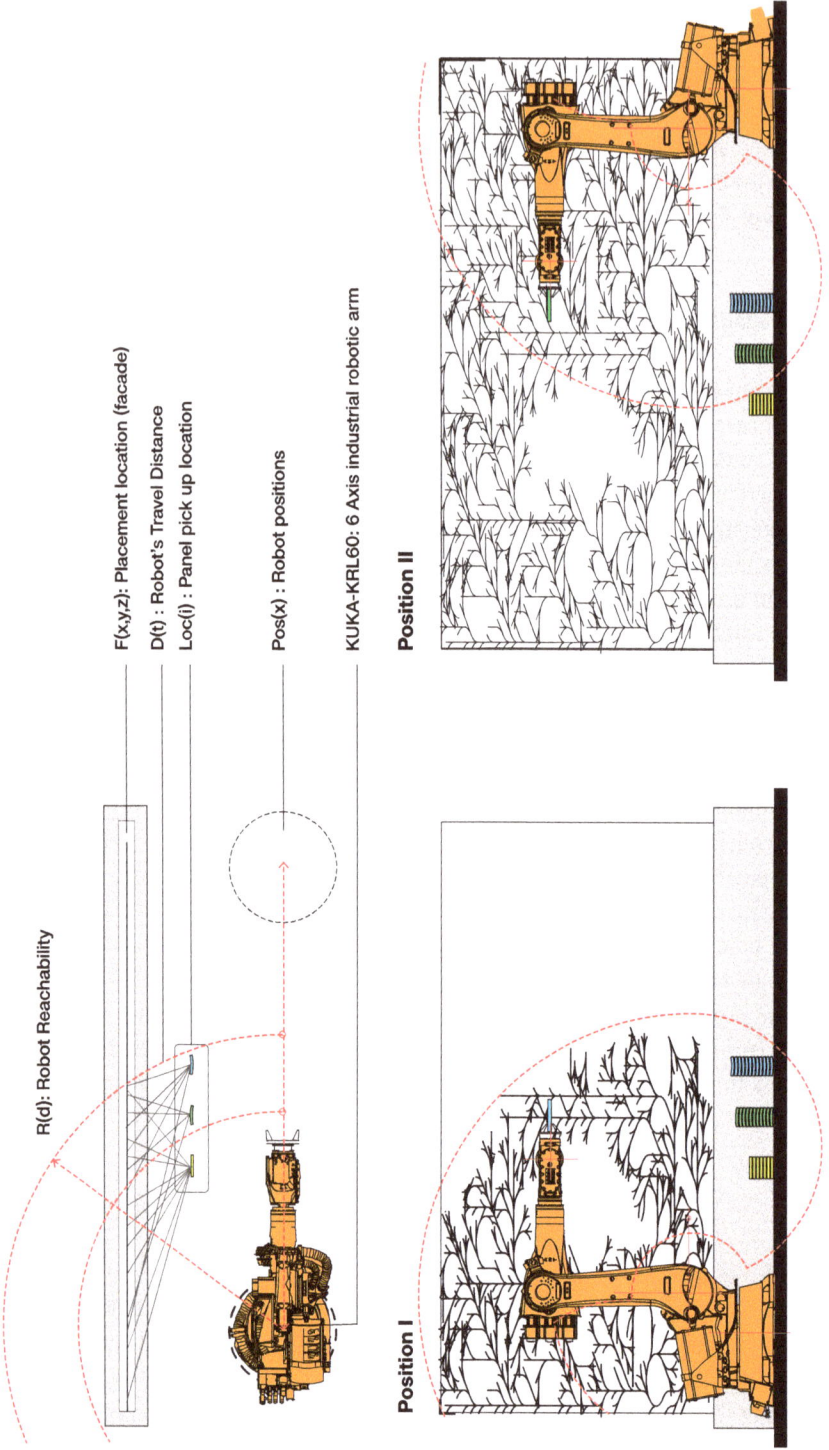

Figure 4.16: Diagram showing fabrication parameters of the construction agent.

Experimental Design Setup – Robotic Construction Parameters

Screenshot from Robotic Simuation in Rhino3d / Grasshopper

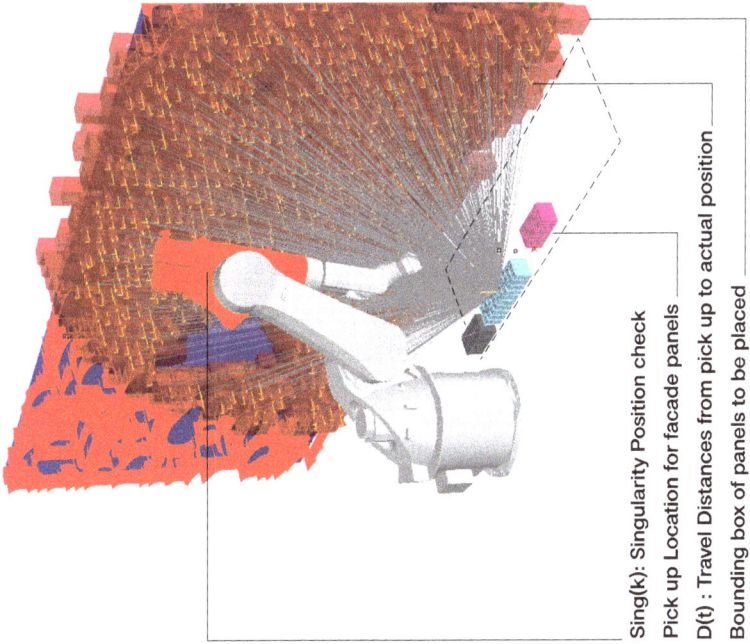

Facade Segmentation – Option A (rectilinear)

Facade Segmentation – Option B (circular)

Sing(k): Singularity Position check

Pick up Location for facade panels

D(t) : Travel Distances from pick up to actual position

Bounding box of panels to be placed

Figure 4.17: Robotic simulation of panel placement and two approaches for facade segmentation.

iteration. Each generated design is analyzed based on its construction simulation. The constructability score is used to (a) change the probability of placing an agent type to avoid collisions, (b) inform the position of the opening and (c) segment the design in ways that facilitate robotic building.

4.2.4 Case study A: facade generation for an office building in Los Angeles

Next, I considered the design of a facade for an office space located in Los Angeles, California. The 50 m^2 office space is to be located on the fourth floor of a 30-year-old commercial building and faces south. The MAS toolkit was used to generate alternative facade designs on a single bay (12.00 × 3.00 m) of a generic office building geometry using three different types of panels. The system is applied for the generation of whole facade geometry, and the objective was to determine the optimal positioning of windows with regard to performance goals, relating to daylight distribution and energy consumption. A basic design setup was defined with four global and eight local parameters. The global parameters were: (a) the location of the building, (b) the input design surface, (c) the orientation of the surface and (d) the number of openings. The global parameters defined the geometric (surface domain) and data (weather data and Sun's position) environment of the agent.

The local parameters are: (a) generative angle, (b) panel types, (c) panel length, (d) extrusion length, (e) extrusion type, (f) extrusion angle and (g) clearance between the panels. The hypothesis of the design experiment is that by running daylight simulations on a base case in a given location (i.e., Los Angeles), the design parameters can be coupled with the performance goals via agent behaviors. For instance, the position of the extrusion of the panels was coupled with the amount of daylight.

By observing the relationships between the design parameters and the goals, "interpolation" functions can be extracted which allow the system to predict the performance of the generated design solutions of any given design surface (in the same location). The case study is divided into two sections: (1) the development of a generative agent-based facade paneling system that is optimized based on environmental performance analysis and (2) the adaptation of designs based on specific user preferences, collected and profiled depending on how they relate to lighting conditions within the office space. The collected light preferences (input) are used as a means to formulate goals that drive the system toward optimality. The lighting preference data was collected from 89 participants. Details about the collection of user preferences are described in detail in a research paper developed in collaboration with professor Heydarian [222]. At this stage, to keep things simple and push the results toward more energy-efficient designs, the preferences of a single user group skewed toward more natural light (23% of all participants) were used. Although this group does not represent the majority light setting preference in the sample, it was selected to test the ability of the system to generate more energy-efficient solutions as one of the initial validation experiments.

The increase of natural light, along with its distribution inside the office and the satisfaction of user preferences, was defined as design requirements; the objective was to develop a design system that generated alternative facade designs that address the environmental performance criteria of program and location. The building facade described the design domain, while the number and size of the openings are described as areas where the components cannot be placed (Figure 4.19(B)). A construction sequence is then simulated by placing one component after another sequentially and implying fabrication and erection constraints. Each component type is based on a probability factor (explained in Section 3.6), where the components connect sequentially while ensuring that they do not self-intersect and the whole facade remains an interconnected structure covering the whole design surface (Figure 4.19(B)). The interior of the office is analyzed through 2,623 virtual sensor points (Figure 4.19(D)) that measure light intensity within the space at height levels that the designer can specify (e.g., floor level and table-top level). These virtual sensor points exist within the simulation environment of the analytical feedback loop. The designs generated are simulated and analyzed environmentally, and their performance is then combined with user preferences [223].

4.2.4.1 Design process
The designer must initially provide a geometry that describes the whole (massing model) or one part of a building envelope (one bay), as well as a file with the location and weather data. She or he then defines a number of desired openings, the building component and a basic generative mechanism (local design rules). A generative mechanism can include more or less complex design parameters, which in most cases pertain to the design problem. In this design experiment, the design parameters are: panel type, angle, length, extrusion length, extrusion uniformity and the maximum number of components.

Through a graphical user interface (GUI), the designer can test and visually evaluate different aggregations of components on the facade surface. When satisfied, the designer can save the configuration in an XML file. This file holds the core design information and is then run for a number of iterations (defined by the designer), and the values are updated and optimized based on the analysis performed. Each panel is defined as an agent and has three different states and eight design parameters (type, probability, angle, length, extrusion, extrusion type, extrusion angle and clearance), as listed in Figure 4.18. The panels populate the design surface in different configurations (based on probability), while trying to avoid areas reserved for the openings (window and clear glazing). The number, size and relative position of the openings and the extrusion length are updated by the stochastic algorithm at each iteration.

To be able to search for optimal alternatives, the designer defines (a) the type of analysis (i.e., DFA, CDA, UDI, etc.), (b) the resolution of the analysis and (c) the analysis period (e.g., daily and annual). She or he inputs user-collected data relating to user preferences and sets the targets for the heuristic function. The optimality can be adjusted by the designer and in this experiment, it relates to the following goals: (a) decreasing building energy use (annually) by increasing the amount of natural light available, (b) providing more distributed light during the day (daily cycle), (c) meeting the levels of light preferred by users. Finally, the designer runs the system and selects (a) the type of

Schematic of Building Data Model

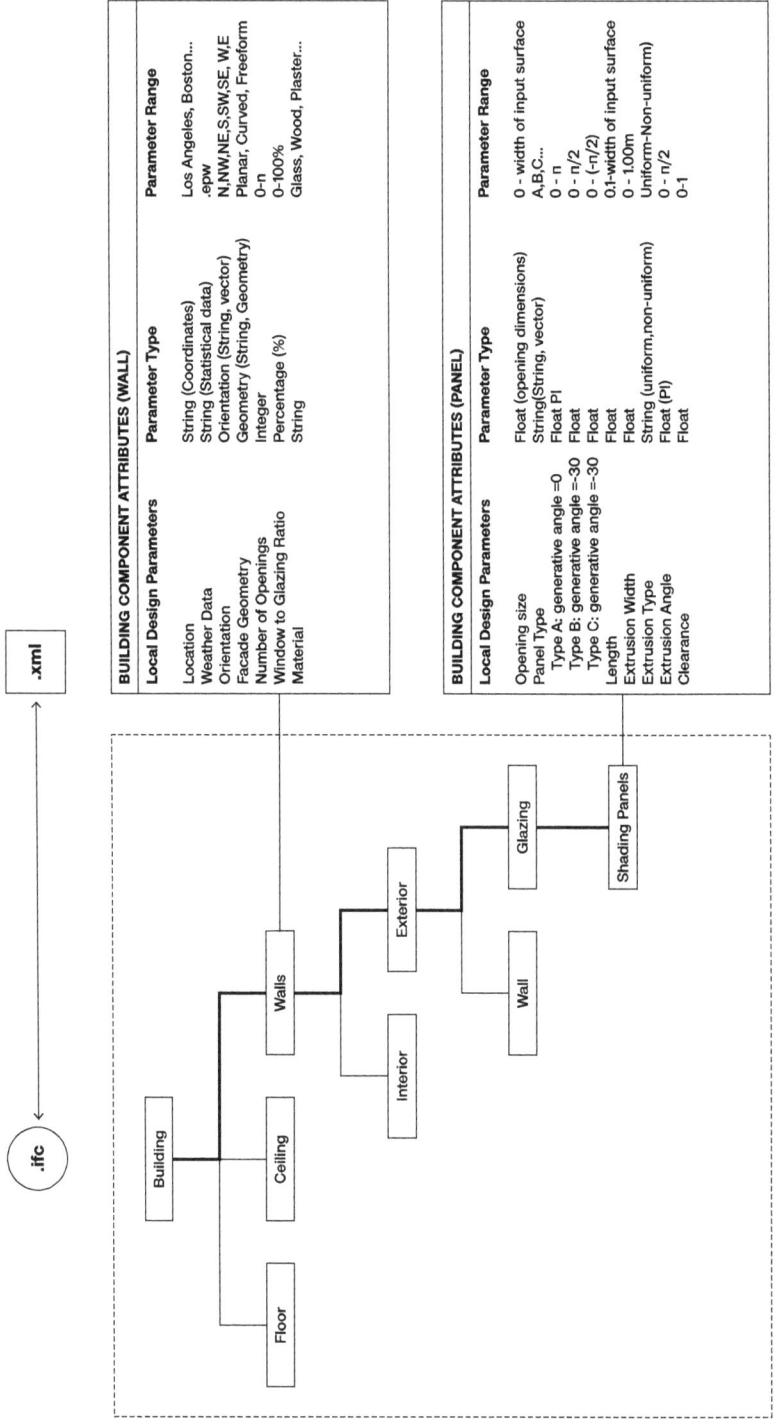

BUILDING COMPONENT ATTRIBUTES (WALL)

Local Design Parameters	Parameter Type	Parameter Range
Location	String (Coordinates)	Los Angeles, Boston...
Weather Data	String (Statistical data)	.epw
Orientation	Orientation (String, vector)	N,NW,NE,S,SW,SE, W,E
Facade Geometry	Geometry (String, Geometry)	Planar, Curved, Freeform
Number of Openings	Integer	0–n
Window to Glazing Ratio	Percentage (%)	0–100%
Material	String	Glass, Wood, Plaster...

BUILDING COMPONENT ATTRIBUTES (PANEL)

Local Design Parameters	Parameter Type	Parameter Range
Opening size	Float (opening dimensions)	0 – width of input surface
Panel Type	String(String, vector)	A,B,C...
Type A: generative angle =0	Float PI	0 – n
Type B: generative angle =-30	Float	0 – n/2
Type C: generative angle =-30	Float	0 – (-n/2)
Length	Float	0.1-width of input surface
Extrusion Width	Float	0 – 1.00m
Extrusion Type	String (uniform,non-uniform)	Uniform-Non-uniform
Extrusion Angle	Float (PI)	0 – n/2
Clearance	Float	0–1

.xml

.ifc

Building

Floor

Ceiling

Walls

Interior

Exterior

Wall

Glazing

Shading Panels

Figure 4.18: Schematic of simple building data model: relationship between components and object attributes (left) and table with local and global design parameters of the facade panel agent (right).

search method and (b) the type of analysis output. Once the system completes a cycle of iterations, it can suggest possible positions for openings based on the defined performance goals. Additionally, the designer can evaluate the results both aesthetically (geometry) and quantitatively (energy performance) through graphs that show the trade-offs between different goals (i.e., amount of daylight and total heating load).

The four phases of the design process that happen within the system are described in detail below. In the first phase, the generative agent iteratively grows 2D lines on the panel surface while trying to avoid areas reserved for specified window openings. In the second phase, the lines are transformed into 3D surfaces via extrusion, finalizing the window panel pattern. In the third phase, the position and size of the openings are changed, triggering the regeneration of panel pattern configurations. For each design generated, the design parameters that affect the environmental analyses the most, such as the depth of the panels, are kept constant. In the last phase, the aforementioned design parameters are altered in order to optimize the environmental analysis and converge toward the users' preferences. In the first phase, the parameters are: L, which defines the length of each line; $p1$, $p2$, $p3$, which are the probabilities of each agent's behavior; the connection angle between the lines; the maximum number of agents; and the number, size and position of the openings. In the second phase, the designer specifies d, the maximum extrusion length, and θ, the maximum offset in the vertical direction.

Hence, the lines are not only transformed into 3D surfaces according to the length but are also designed to be able to twist in order to better filter the light. In the third phase, the position and size of the openings are changed in order to find an optimal configuration for bringing more direct light into the space. In the fourth phase, a new type of specialist agent is created for each of the selected analyses. Three different types of environmental analysis are considered; therefore, three specialist agents were implemented: a DFA agent (DFAa), a CDA agent (CDAa), and a UDI agent (UDIa). The designer specifies a utility u for each of the analyses performed and the user preference profiles, thus defining the most important analysis, and adding a weighting factor to the agents' behaviors. The weighting factor is calculated as a percentage over all performed analysis types, depending on the performance target that the designer sets. All these aspects affect the amount (measured in lux) and the type of sunlight (direct or indirect) that enters the room, changing the illumination inside the space.

As shown in Figure 4.20, in the first phase of the algorithm the agent starts at a user-defined initial point on the panel and performs a series of iterations. In each iteration, the agent grows one line from its current position and moves to the end of that new line (Figure 4.20(G)). The agent can grow three different types of lines according to three different behaviors: straight, left-curved or right-curved, based on an angle (θ) as shown in Figure 4.20(G). In the beginning of each iteration, the agent chooses its next behavior according to the (user)-specified probabilities $p1$, $p2$ and $p3$.

However, the agent must also obey four constraints: (1) the new line must not intersect a previously constructed line, (2) the agent must not leave the boundaries of the given surface, (3) the agent must avoid specific areas reserved for openings and (4) the number of generated agents cannot exceed a maximum specified by the designer (Figure 4.20(I)). If the probabilistically chosen behavior violates these con-

Light Sensitive Facade Panel: Context

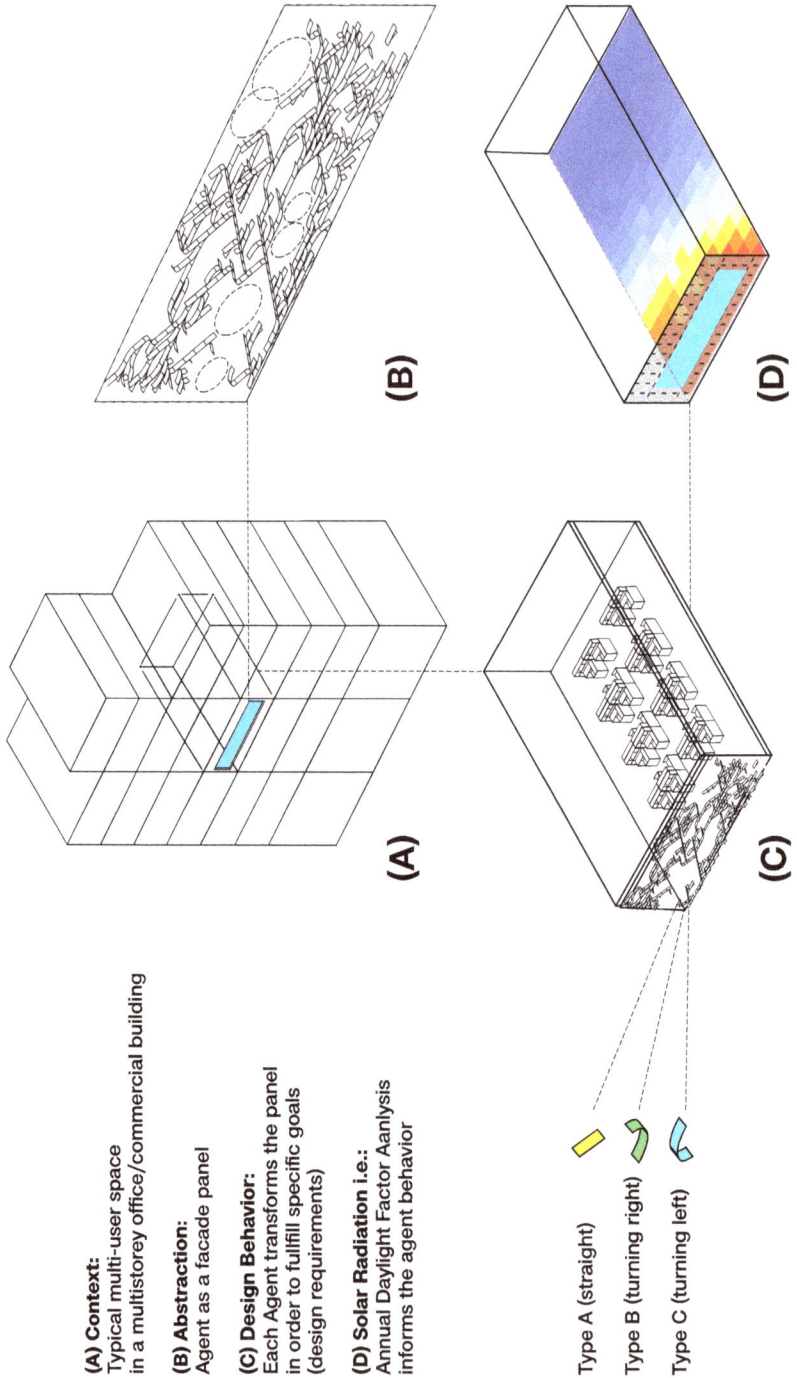

(A) Context:
Typical multi-user space
in a multistorey office/commercial building

(B) Abstraction:
Agent as a facade panel

(C) Design Behavior:
Each Agent transforms the panel
in order to fullfill specific goals
(design requirements)

(D) Solar Radiation i.e.:
Annual Daylight Factor Aanlysis
informs the agent behavior

Type A (straight)

Type B (turning right)

Type C (turning left)

Figure 4.19: (A) The specific design context of the building highlighting the design surface of a single bay; (B) the surface domain where the agent and specific behaviors are applied; (C) the panel types comprising the agent-based facade; and (D) the analysis surface with the virtual sensor points of this experimental case study.

straints, a new behavior of left, right or straight is selected until the behavior is valid. More specifically, the agent checks the history of all previous behaviors, as well as the life value of each agent; it then changes to the behavior that has the ratio furthest away from the one desired according to the given probabilities $p1$, $p2$ and $p3$, which naturally induce a ratio. The above probabilities are scaled based on the life state of the agent, which is influenced by whether the agent was created in a valid area (i.e., if an agent is created in a nonvalid area, its life state decreases significantly). This phase is terminated when a predefined number of iterations are reached or when the maximum number of agents has been created.

In the second phase, shown in Figure 4.20(E–H), the lines are extruded into 3D geometries. For each line, length and angulation are chosen according to the following equation: $d' = d \times w$; $\theta' = \theta \times w$, where $0 \leq w \leq 1$ is a weight given by the current Sun's radiation entering the panel in the position of the line. Hence, each line will have different d' and θ' bounded by the preference of the user. Depending on the user preference, the designer can specify two different types of extrusion: uniform or nonuniform (Figure 4.20(H)). The nonuniform case differs from the previous description in that the user can also specify a control point that affects the degree of the curves, generating the surface shown in Figure 4.20(F). Finally, these parameters define the aperture a' between surfaces (Figure 4.20 (E)), which in turn influences the amount and type of light that enters the space.

In the third phase, the designer assigns (1) a specific time of year (e.g., summer and winter solstice), (2) the minimum and maximum desired luminance obtained for the majority of the sensor points inside the office, (3) a specific time of day (e.g., 9:00 am–5:00 pm) as properties for the specialist agents (DFAa, CDAa and UDIa); the goal is to find the optimal combination of generated facade panels and (4) the positioning of n openings in order to provide luminance within the desired thresholds.

The specialist agents (DFAa, CDAa and UDIa) for each run collectively compare across all parameters and search the whole solution space for possible positioning (x, y) of a number of openings ($n = 2$ in this data set) on a given surface. The goal is to generate facade panel configurations that provide natural light availability above a designer-defined level (i.e., CDA > 150 lux) and closer to a user-defined level (i.e., 150 < UDI < 1,200). A run is then automated for an annual simulation, which calculates the amount of natural light available on the specific dates, as well as the average values throughout the year.

The optimal and suboptimal solutions from each run are selected and passed as inputs for the next phase along with the corresponding analyses to the CDAa and UDIa, whose goals are to alter specific design parameters (i.e., extrusion) in order to improve the environmental performance of the design. A heuristic function, as described above, is defined for changing the design parameter based on the analysis values. Finally, the User Preference agent (UPa) takes the best ranked designs in terms of the CDAa analysis as input and combines them with user-simulated data; it attributes a credit to each agent that is proportional to how much the analysis of each virtual sensor point deviates from the user preferences. At this stage, the user preferences reflect their preferred lux value. Based on this utility factor, all of the above parameters are recalibrated and the agents iteratively negotiate in order to better meet the performance values, while ensuring the minimum number of conflicts with the other specialist agents.

Light Sensitive Facade Panel: Design Parameters

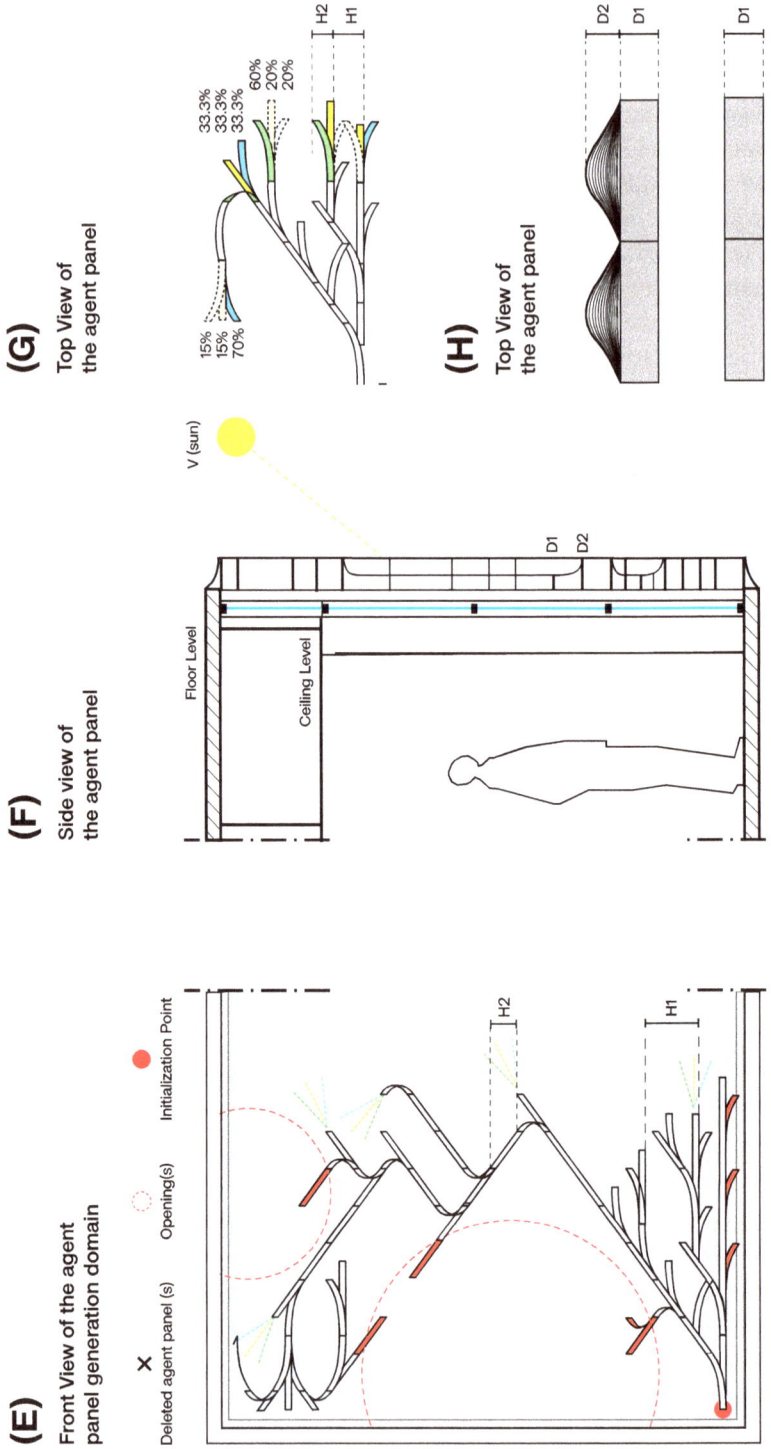

(E)

Front View of the agent
panel generation domain

✕ Deleted agent panel (s)   ⭕ Opening(s)   🔴 Initialization Point

H2

H1

(F)

Side view of
the agent panel

Floor Level

Ceiling Level

V (sun)

D1
D2

(G)

Top View of
the agent panel

33.3%
33.3%
33.3%

60%
20%
20%

15%
15%
70%

H2
H1

(H)

Top View of
the agent panel

D2
D1

D1

Figure 4.20: Rules and design parameters of the generative facade panel agent(s) (E–H).

4.2.4.2 Experimental runs and analysis

At first, the capacity of the system to generate and evolve designs was tested on different input geometries and different design parameters for the generative agent. To test the generative capacity of the system, multiple iterations were performed, and Figure 4.25 presents a selection of generated facade designs on different surface domains and with variable initial conditions and input parameters. Figures 4.23 and 4.24 show a set of design alternatives that were generated by the system and vary from the normative horizontal louvers to complex panel designs.

The diversity of the designs is either a product of their environment (the curvature of the design surface) or is achieved by encoding different agent behaviors (i.e., photophilic behavior: change the position of the opening to increase the amount of daylight available). Second, to compare the performance of generated designs and traverse the solution space more efficiently, the design parameters are coupled with the performance goals.

By doing this, the experiment explored the possibility of a relationship between the design parameters and the environmental performance that can be mathematically described in a heuristic function. Given the set of performance goals defined by the designer, two separate heuristic searches were implemented as described below. Initially, the system executed a linear search (brute force) to map the extent of the solution space in relation to the positioning of n number of openings ($n = 2$). Each design alternative was given a unique identity (hash ID), and at each step, the system checked whether the geometry existed to avoid generating duplicate geometries. Subsequently, a dictionary was created with all IDs for easy lookup. Once we had mapped the solution space, the system ran iteratively using two basic stochastic algorithms, namely simulated annealing and hill climbing.

Although these algorithms are not the most efficient in computational terms, they are suitable for open-ended problems [20] and form a good starting point for testing our framework. In Figure 4.26, the performance of each algorithm is graphed with regard to the DFA. It is evident that using either of the two algorithms will significantly cut down the search space (linear search) by a factor of 0.023 and 0.006 accordingly.

Additionally, the combined results from multiple analyses are plotted and are visually communicated to the designer using parallel line plots [224] (Figures 4.27 and 4.28). The plot includes all the design alternatives with their hash IDs, the environmental analysis performed, facade orientation, heuristic search algorithm used and simulation time. The designer can interactively adjust the boundaries of each analysis and filter out alternatives that do not meet their requirements. In this way, the designer can (a) gain insight about how design parameters affect environmental performance, (b) compare whether there is diversity among the solutions of each run and (c) evaluate how well each algorithm performs in a multiobjective context. In this case, diversity means design alternatives that have similar performance characteristics but distinct geometries.

Diagrammatic section of office space

Obstructed View
Unobstructed View

No Facade panels allow unobstructed views

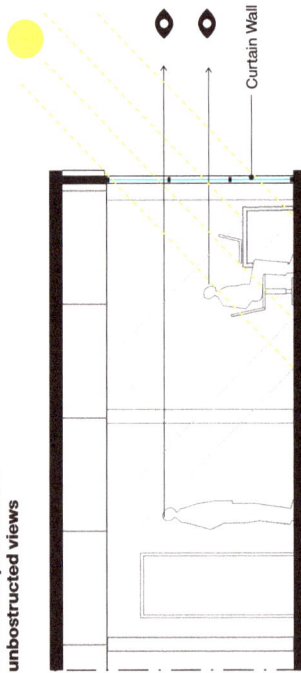

Curtain Wall

No Facade panels increase thermal gain and glare

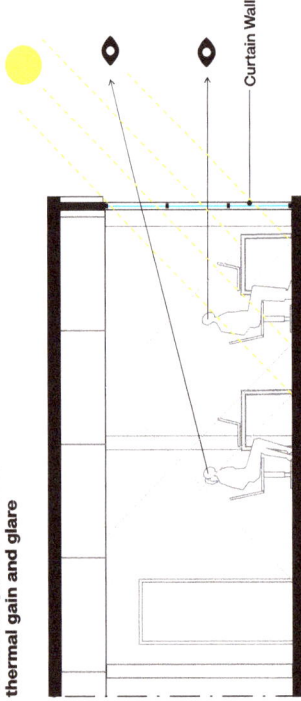

Curtain Wall

Shallow panels results in deeper light penetration and solar access

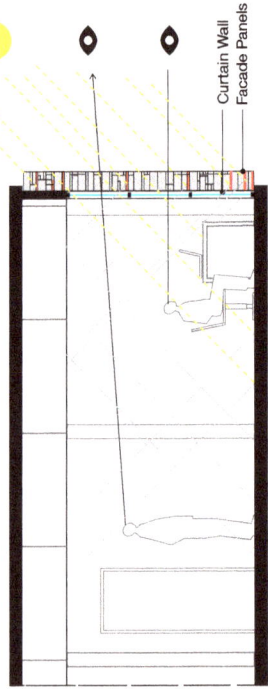

Curtain Wall
Facade Panels

Wide Panels resluts in high solar access, deep light penetration and low glare

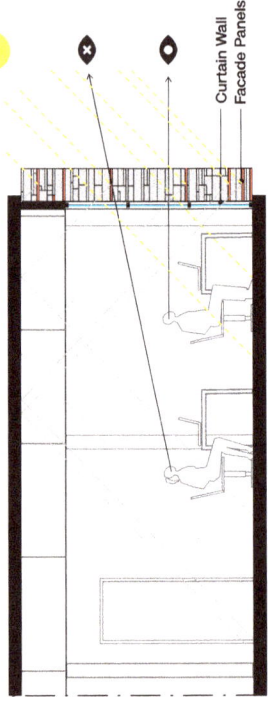

Curtain Wall
Facade Panels

Figure 4.21: Diagrammatic section of a generic office space showing the penetration of daylight and the obstruction/nonobstruction of view in relation to the facade panels.

Generation Process Diagram

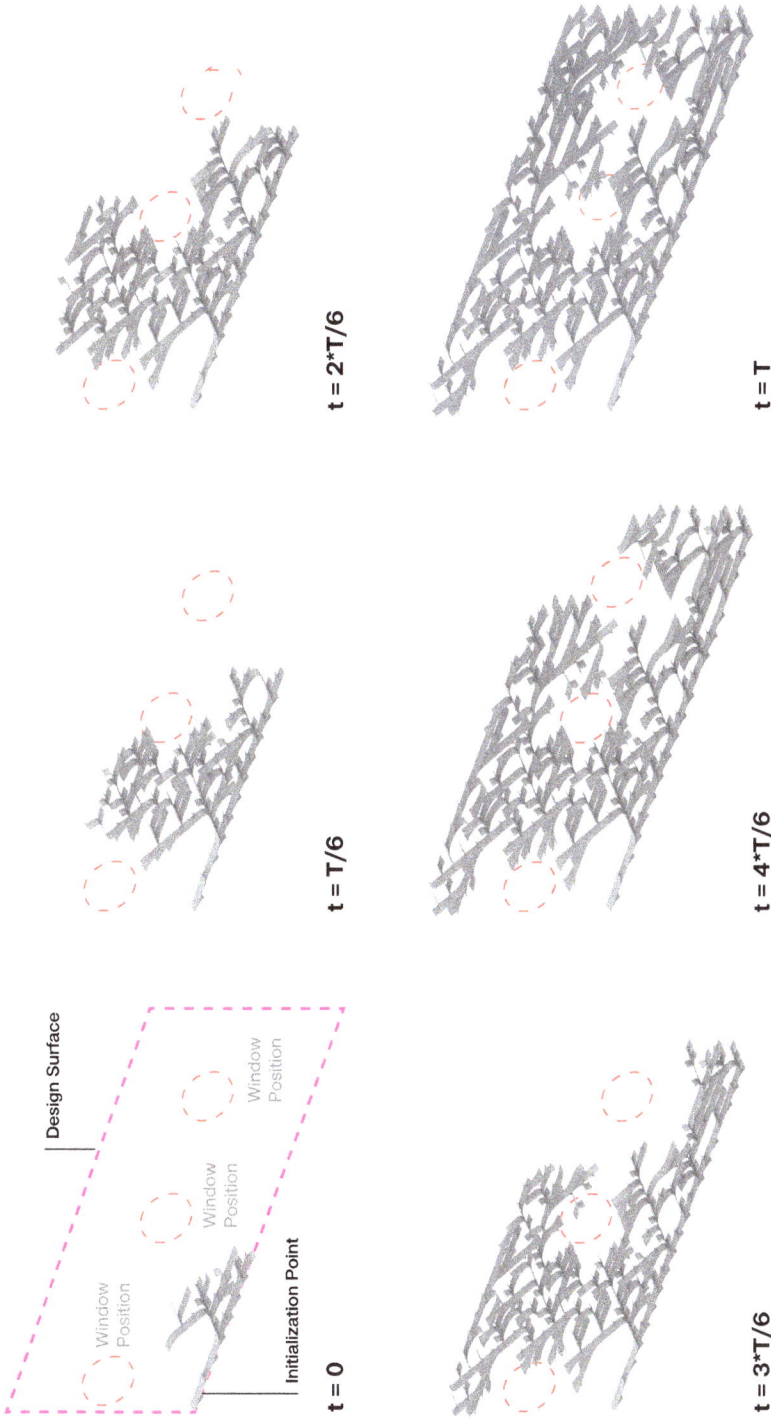

Figure 4.22: The generation process is shown in six sequential steps (from $t = 0$ to $t = T$, i.e., facade is fully covered or all max amount of panel is placed), including representative window openings.

Generated Facade designs with varying generation angle

F_Option_1:
0001_S100R00L100-hoR

F_Option_2:
0002_S100R00L100

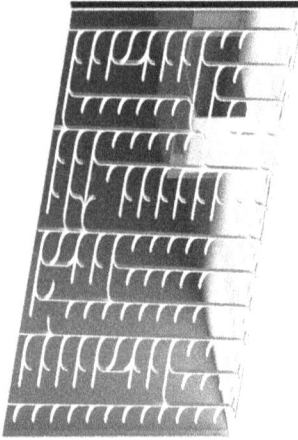

F_Option_3:
0003_S100R10L100

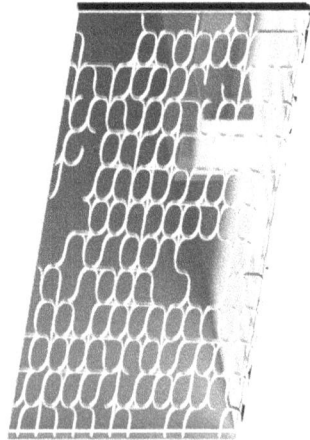

F_Option_4:
0004_S100R20L100

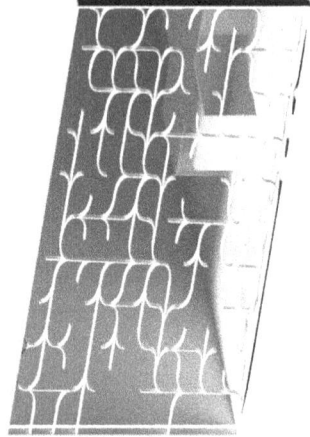

F_Option_5:
0005_S100R30L100

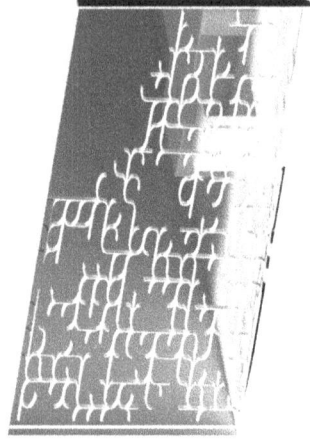

F_Option_6:
0006_S100R40L100

Figure 4.23: A set of generated panel designs on a planar facade surface.

Generated Facade designs with varying length and depth

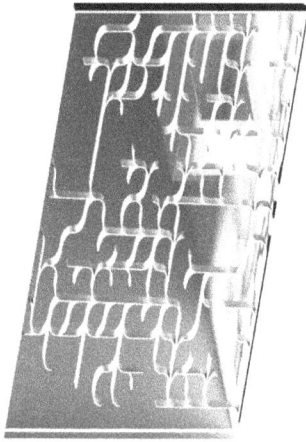

F_Option_9A:
0019_s_RE_0_2_0_299

F_Option_9D:
0022_s_RE_0_5_0_299

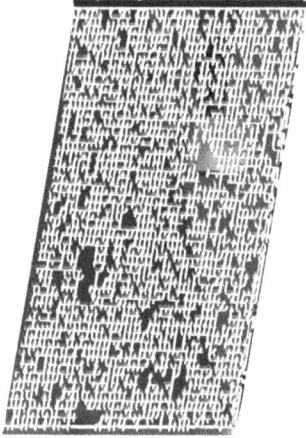

F_Option_8:
0018_s_RE_0_1_0_299

F_Option_9C:
0021_s_RE_0_4_0_299

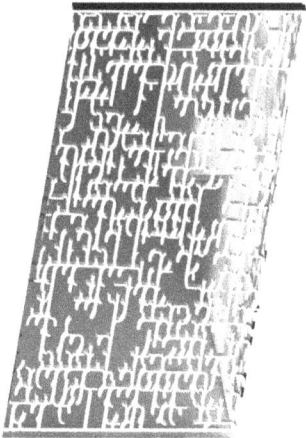

F_Option_7:
0007_S100R50L100

F_Option_9B:
0020_s_RE_0_3_0_299

Figure 4.24: A set of generated panel designs on a planar facade surface.

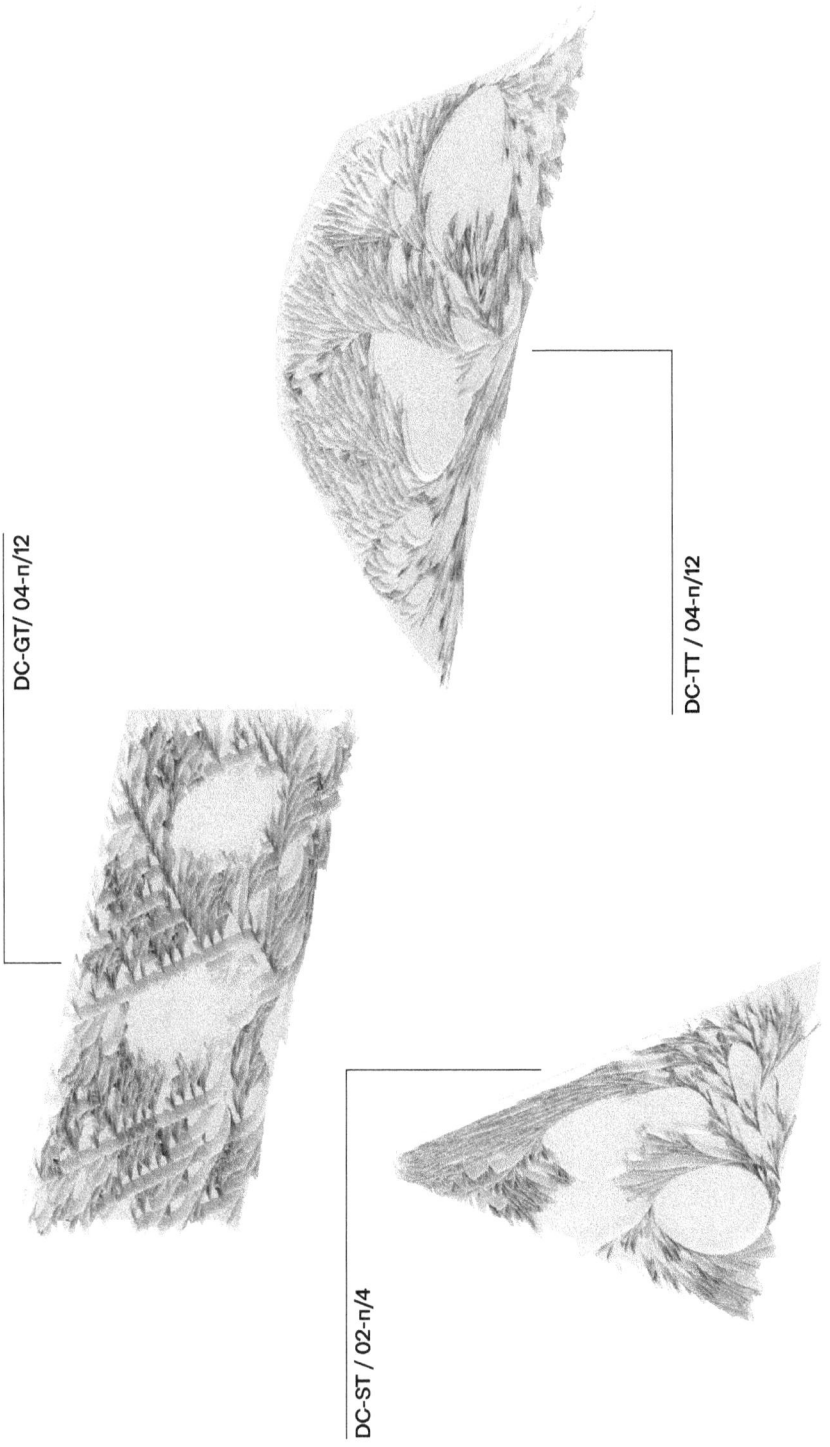

Generated Facade designs

DC-GT/ 04-п/12

DC-TT / 04-п/12

DC-ST / 02-п/4

Figure 4.25: A set of generated panel designs on a different facade surfaces.

Search for optimal window position with regards to dalyight factor analysis

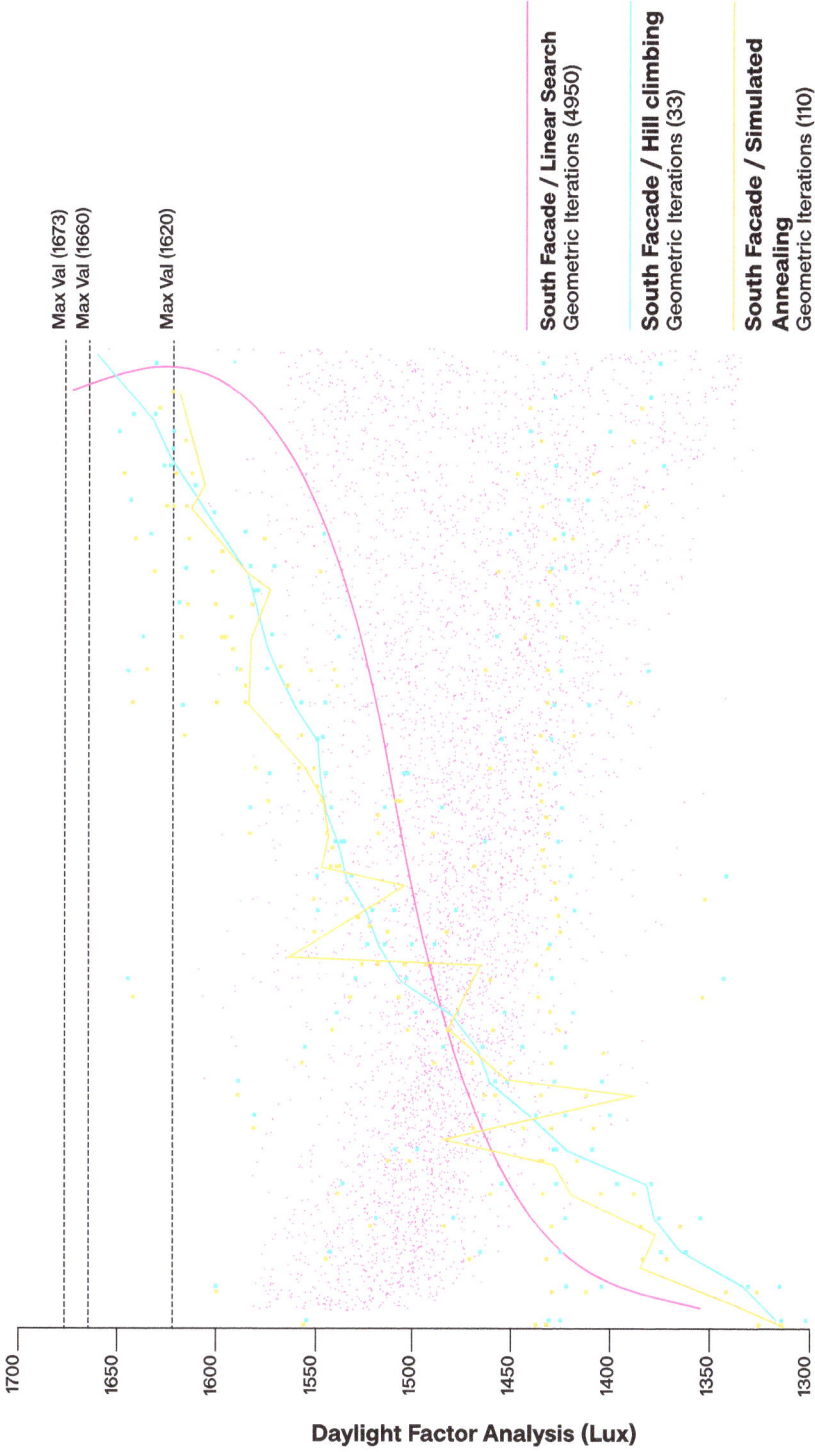

Max Val (1673)
Max Val (1660)
Max Val (1620)

South Facade / Linear Search
Geometric Iterations (4950)

South Facade / Hill climbing
Geometric Iterations (33)

South Facade / Simulated Annealing
Geometric Iterations (110)

Daylight Factor Analysis (Lux)

Figure 4.26: Comparative plot of simulation runs showing performance of different heuristic search methods.

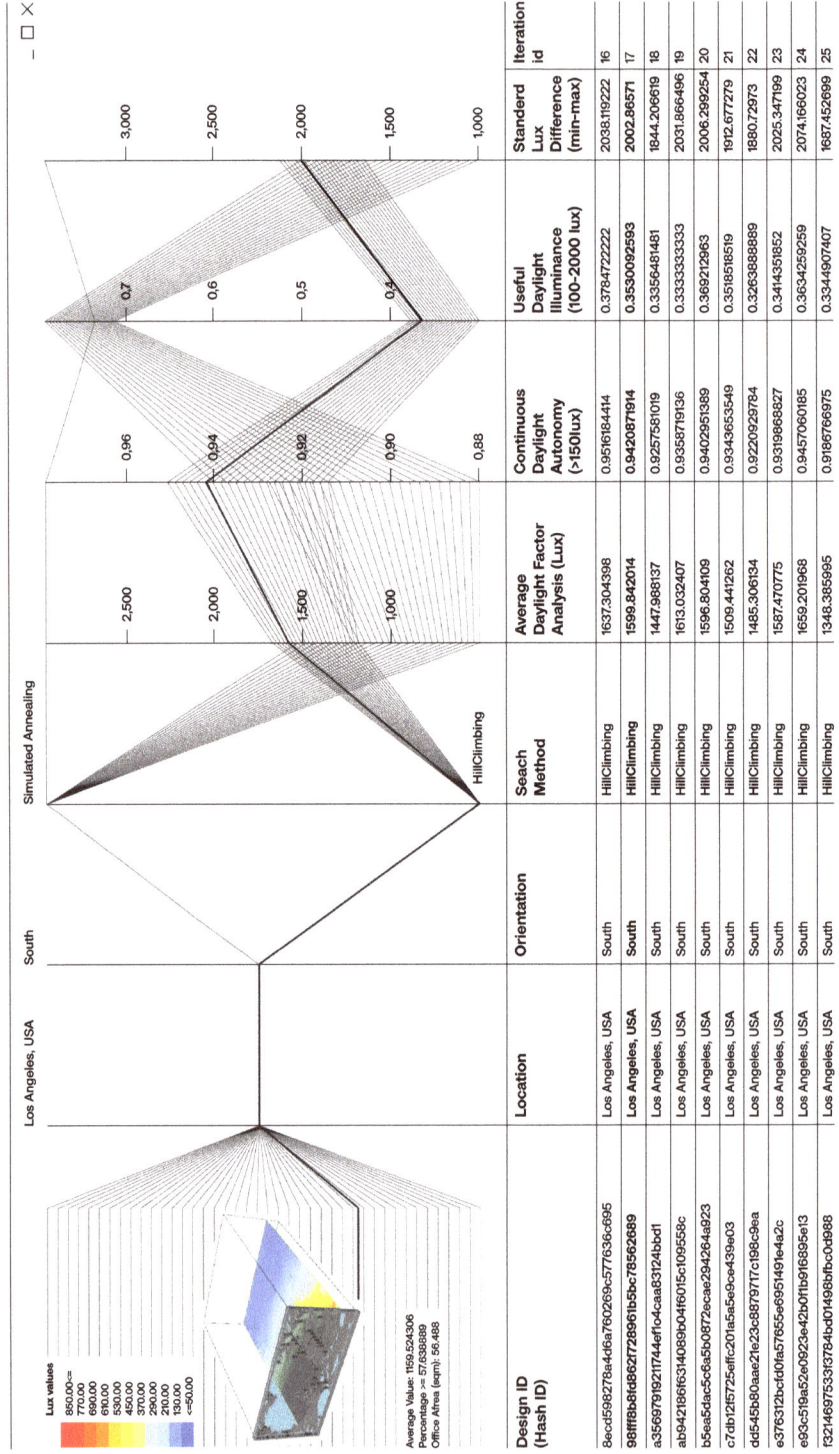

Figure 4.27: Visualisation of design alternatives using a parallel line plot.

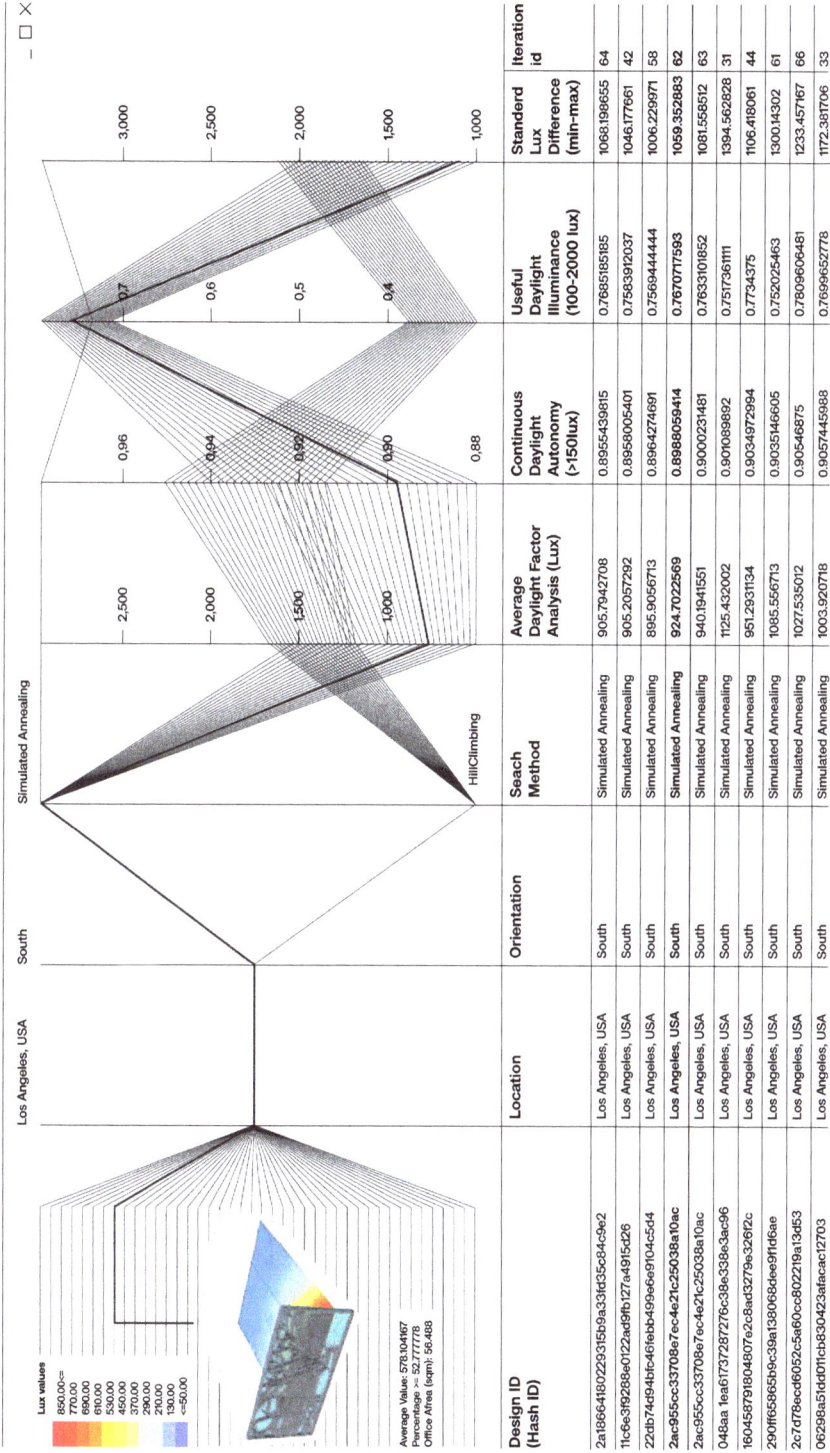

Figure 4.28: Visualisation of design alternatives using a parallel line plot.

4.2.5 Case study B: facade construction simulation of a bay on a generic office building

This case study considers the design and construction of a facade for another generic office space located in Los Angeles, California (see Figure 4.19). The framework was extended to explore how a robotic assembly simulation could become an integral part of the early design stage workflow and be integrated with daylight analysis in order to measure construction efficiencies. This is a progression of the previous case study, where the MAS approach is further extended to include the geometric constraints of the complex facade fenestration or louver pattern and to also include robotic simulations as a way of validating the constructability of the generated designs.

The objective of this case study was to initially search the design solution space for environmentally efficient facade alternatives based on different window configurations and variable design parameters, such as angulation. The most efficient results were passed to the construction simulation for testing and feasibility evaluation. Through this feedback loop, at each iteration the generative process is informed by both simulations and trade-offs between the agents, established through each agency's utility functions. In this way, the system iteratively limited the solution space of potentially higher performing designs (environmentally) by constraining the design parameters to those yielding combined constructible and environmentally efficient outcomes. The five agent classes mentioned in Section 3.2 were implemented after specifying the design, analysis and fabrication processes. These included: (a) a generative agent, (b) a specialist environmental agent, (c) a construction simulation agent and (d) a coordinator agent. The generative agent had the same set of design parameters described in the previous case study. This agent was tasked to create alternative geometries iteratively based on the input design parameters and context, as updated on each run. The design parameters of the generative agent were defined by the designer and were later conditioned by the constraints imposed by the construction simulation agent.

The specialist agent performed two types of environmental analysis on the generated designs and had the objectives of combining the analyses and the passing of the analysis data (lux values and CDA scores) to the coordinator agent, along with a set of messages in the form of a report. The goal for the specialist agent was to increase the amount of natural daylight entering a typical office space (i.e., the DLA), while keeping the lux values within a threshold (i.e., CDA) defined by the space typology and dimensions. The construction simulation agent received the generative designs as input, which were evaluated to determine their performance within a specified range in terms of environmental analysis. The system then performed the construction simulation and checked for constructability. Based on the constructability heuristic function described in Section 4.1.3, the designs were ranked when necessary, and the design parameters were updated and passed to the coordinator agent, forming one of the system's feedback loops. The goal of the implemented construction simulation agent was to decide how to segment the design to ease construction and to find optimal positions for the robots that speed up the construction time and eliminate robot self-collisions.

Finally, the coordination agent (also developed using Python) established communications between the generative process (generative agent), the analytical process (specialist agent) and the simulation process (construction agent) by passing the analysis data; it simulated the results and sent messages back to the generative process, thus forming the culminating feedback loop in the system. In terms of technical implementation, the Python and Grasshopper visual scripting editors are used to trigger the different agent processes that have been implemented in Java using Processing libraries [225]. For the creation and management of geometry generated by the agencies, the system uses the IGeo library (Figure 4.33) – an open-source NURBS-based library – in order to output geometries that can be further used for fabrication purposes [168]. For the construction agent, I implemented KUKA|prc, a simulation plugin for Grasshopper, to simulate the robotic construction sequence and the complex fenestration assembly process. The specialist agents were developed using Python and the Ladybug and Honeybee environmental simulation tools (Figure 4.34g,h) to combine three different environmental analyses: Energy Plus, Daysim and OpenStudio [35].

A series of simulation experiments were conducted to collect data and analyze how the results inform the generative process of an environmentally optimized shading system for a typical office environment. The workflow is applied on the south facade of three different generic office building typologies, whose geometry varies from orthogonal to free form. The geometry of a single bay of each tower type is shown in Figures 4.30–4.32.

4.2.5.1 Design process

Similar to the previous case study, the design process was divided into two phases. In the first phase, the designer developed (a) an initial design component of a facade and defined a subset of alternative panel types (three in total for this experiment); (b) provided the following data as inputs for the generative MAS for design system: length, angle, probability for each panel type and depth or extrusion of panel; (c) ran n number of iterations using a hill-climbing optimization algorithm; and (d) generated a solution space of design alternatives that were then evaluated for their performance across DLA and CDA metrics. In the second phase, the best performing designs were (a) passed to a robotic simulation software that explored different construction strategies; (b) collisions and errors were registered as negative scores in the ranking equation; (c) the designs were ranked based on their constructability in terms of time needed for construction; and, finally, (d) these scores were passed back to the generative system for further optimization.

The following steps were followed in the design of this experiment:

1. First, the designer set the input parameters and initialized a run of the generative system. The system used a hill climbing algorithm that searched for optimal window positions. In each iteration, the position of the windows was changed by adding a small increment to the surface domain of the facade. For each set of window positions, the system outputs one unique design with the position of the openings that provided the most daylight to the interior described as optimal. The data was collected by the system for six different generative angles: $\pi/2$, $\pi/4$, $\pi/6$,

Figure 4.29: A rendering showing the different designs generated for the southwest facade of the One Wilshire building in downtown Los Angeles.

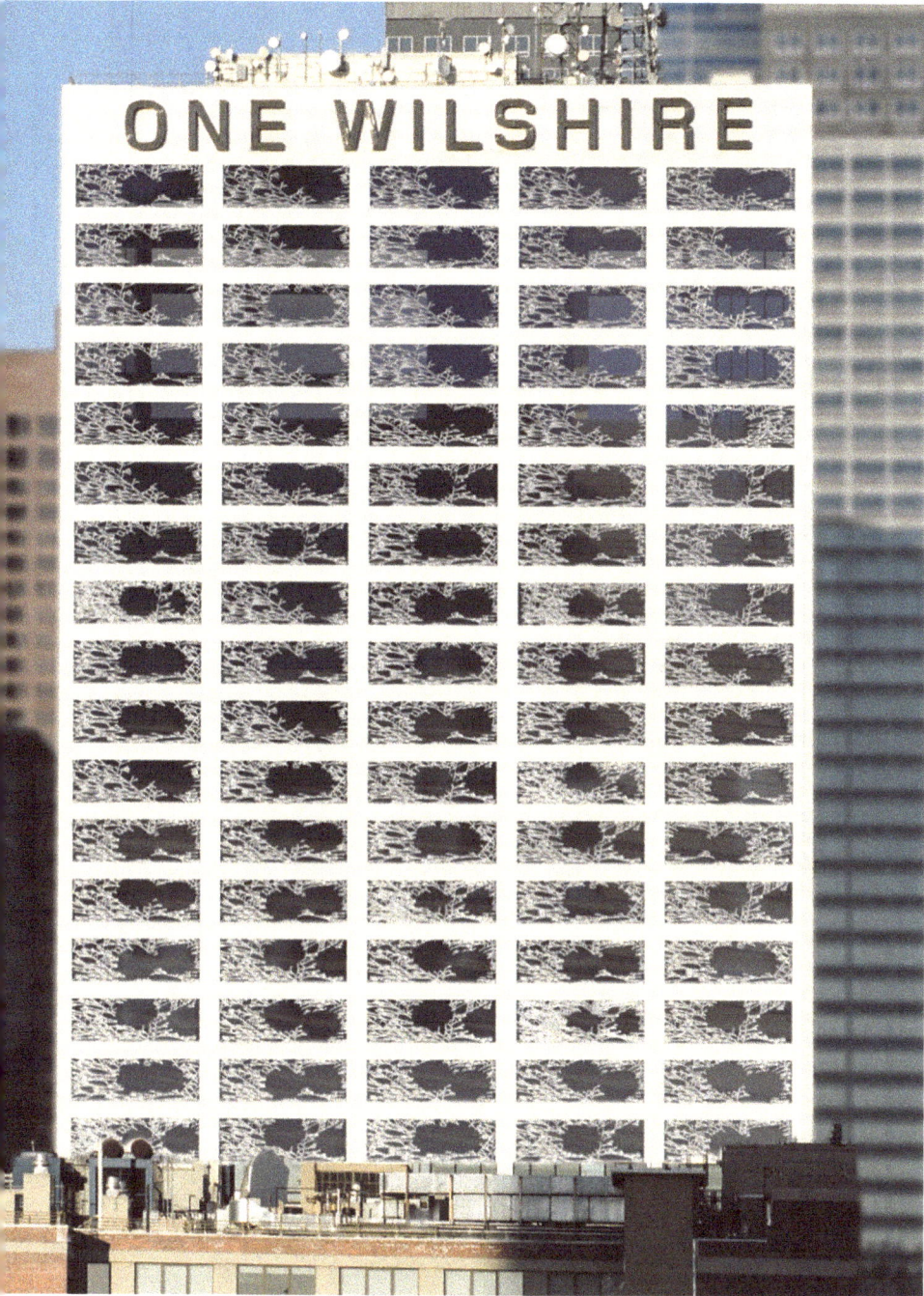

Design Case: One Willshire Tower

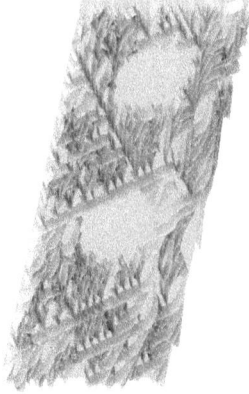

DC-ST / 01-π/2

DC-ST / 02-π/4

DC-ST / 03-π/8

DC-ST / 04-π/12

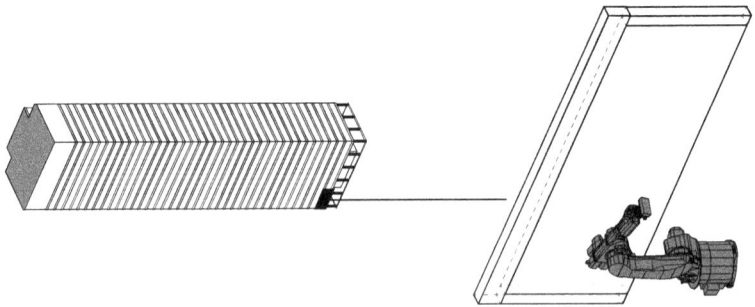

Figure 4.30: Design case A – Generic Tower: Generated Agent Panel Facade design of a single bay.

DC-GT/ 03-n/8

DC-GT/ 04-n/12

DC-GT/ 01-n/2

DC-GT/ 02-n/4

Design Case: Gherkin Tower

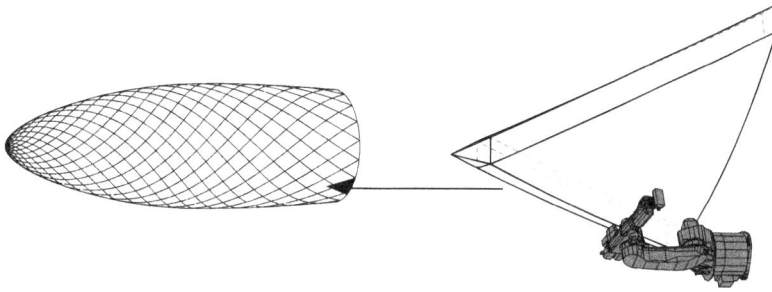

Figure 4.31: Design case B – Gherkin Tower: Generated Agent Panel Facade design of a single bay.

Design Case: Twisted Tower

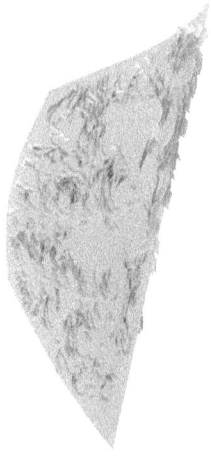

DC-TT / 02-n/4

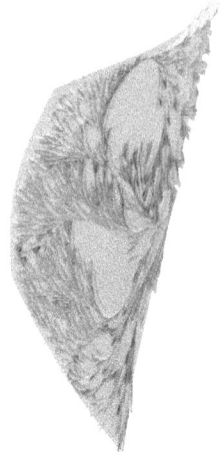

DC-TT / 04-n/12

DC-TT / 01-n/2

DC-TT / 03-n/8

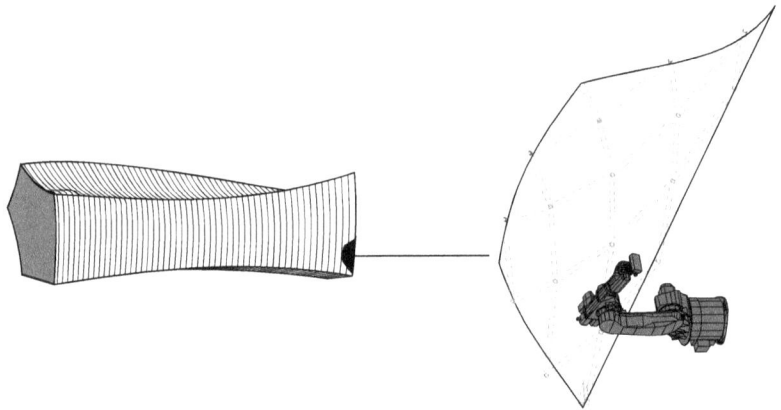

Figure 4.32: Design case C – Free-Form Tower: Generated Agent Panel Facade design of a single bay.

$\pi/8$, $\pi/10$ and $\pi/12$. The generated designs were automatically passed to the specialist agent for lighting analysis. The system then performed two kinds of analysis, a DLA and a CDA for the designs generated.

2. Once the analysis was completed, the designs with the highest scores were sorted and passed for robotic simulation (Figures 4.35 and 4.36).
3. The user selected the robot – in this case, the six-axis KUKA-KRL60 – and defined the pickup location and the robot movement axis – in this case, parallel to the design surface.
4. A planning process was selected, and the designs generated were segmented into groups based on the maximum reachability of the robotic arm. For this research, only two construction strategies were defined and thus divided the designs into rectangular and circular segmentations.
5. The robotic simulation was run, and the system measured collisions, singularities, panels and positions in order to develop the heuristic for the constructability agency.

4.2.5.2 Results and analysis

This section presents the initial results based on running the generative and analytical loops of the system for 450 cycles. The duration of each generation analysis cycle is approximately 5 min, including the geometry generation and the environmental analysis, while the robotic simulation cycle is approximately 3 min. Six different generative angles were tested to explore the system's capacity to create an expanded and highly varied solution space of 450 unique geometries. The 10 highest ranking designs from each cycle were selected and passed to the robotic construction agent. One pickup location for all three types of panels was set; the range of possible segmentation positions was set to 4 and the minimum was set to 2. The segmentation position refers to the way a generated design was divided into parts, based on the work volume of a specified robot. This study only considered circular design segmentation based on the maximum reach of the available industrial arm. Based on the simulation and the constructability heuristic function, the generated designs were improved for construction purposes, either by eliminating panels or by changing the sequence of panel types.

The updated parameters for each design were stored in XML files and were passed back to the generative agent. Figure 4.37 presents graphs of the constructability score of the initial and improved geometries for all six different generation angles. It is observed that the constructability score is highly variable with the initial conditions but becomes a line with zero or positive slope for the improved geometries across all the generation configurations.

The results show that geometries with a generation angle of $\pi/10$ performed better overall in terms of constructability score over time. At 6%, the deviation from the initial and final scores is the largest, compared to the rest of the runs. In the initial runs, the constructability score from one design iteration to the next ranges from 0.1 to 0.6, while in the optimized cases it drops below 0.15 and is constantly improved.

Pareto Chart of facade designs with regards to CDA and UDI Analysis

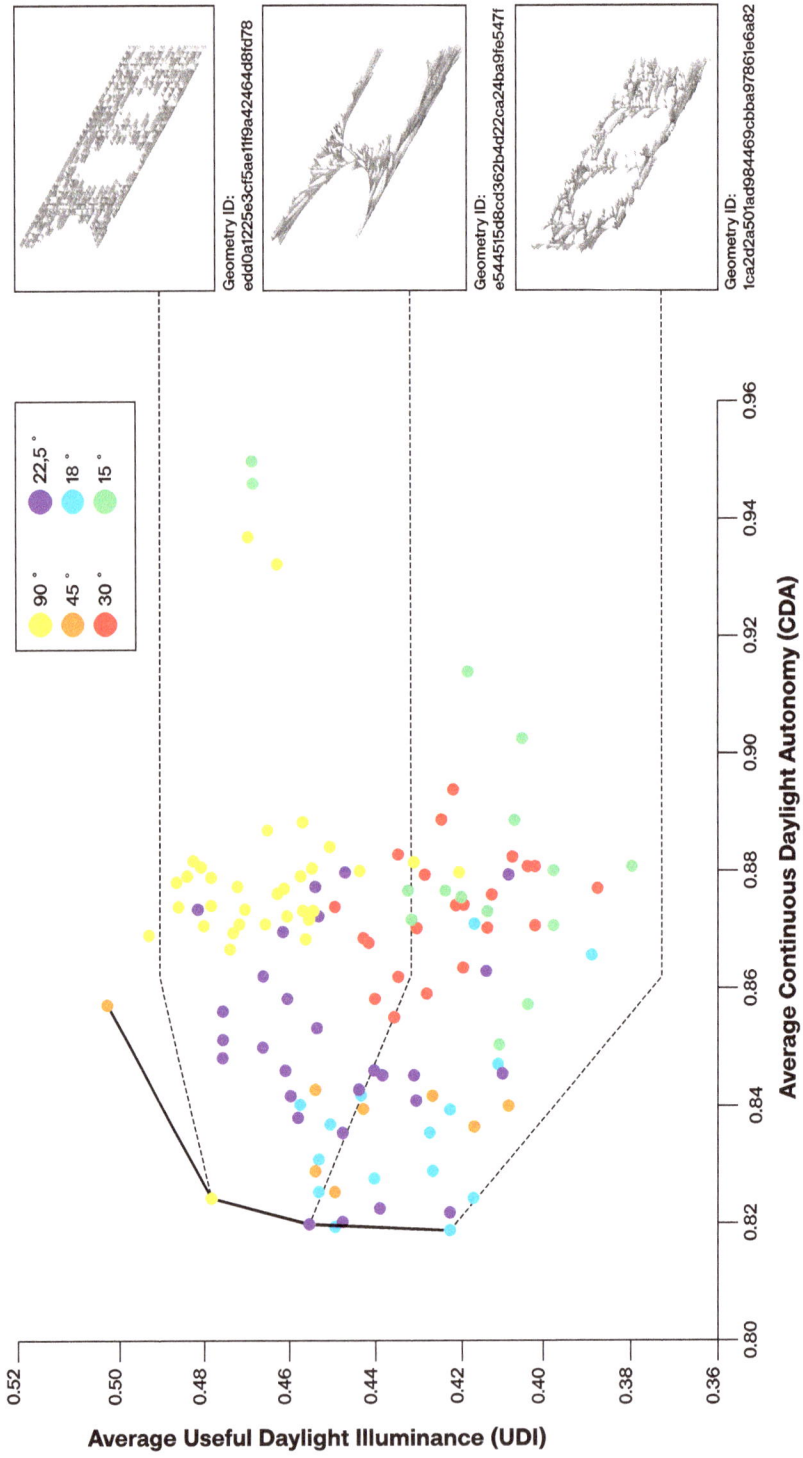

Geometry ID:
edd0a1225e3cf5ae1f9a42464d8fd78

Geometry ID:
e54451fd8cd362b4d22ca24ba9fe547f

Geometry ID:
1ca2d2a501ad984469cbba97861e6a82

Legend: 22,5° 18° 15° 90° 45° 30°

Average Continuous Daylight Autonomy (CDA)

Average Useful Daylight Illuminance (UDI)

Subset of generated geometries with minimum panel length of 30cm and six different angles that have been evaluated for both Continuous Daylight autonomy (CDA), Useful Daylight Illuminance (UDI)

Figure 4.33: Pareto chart of generated facade designs evaluated for CDA and UDI.

Experimental Design flowchart

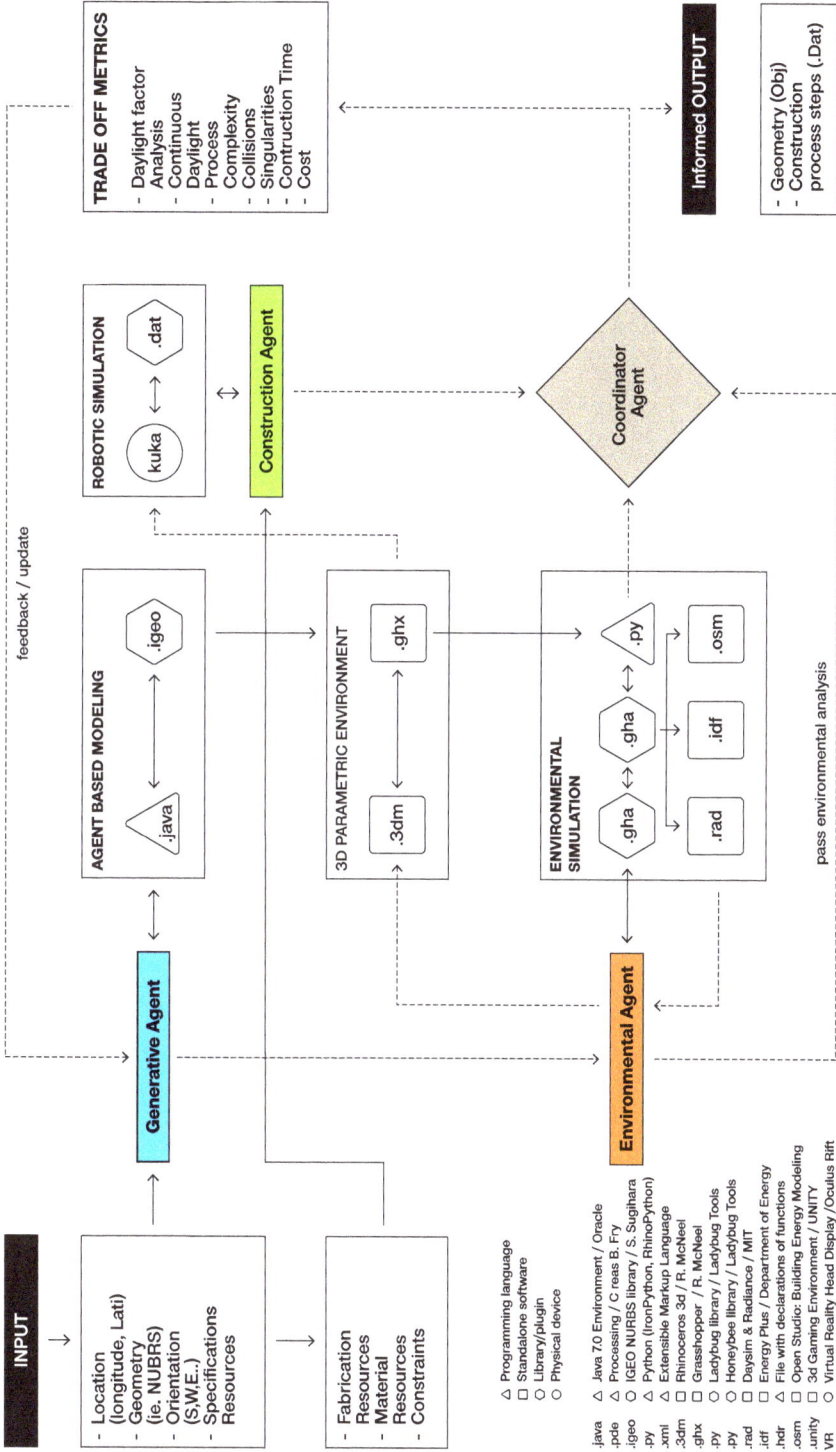

INPUT

- Location (longitude, Lati)
- Geometry (ie. NUBRS)
- Orientation (S,W,E..)
- Specifications
- Resources

- Fabrication Resources
- Material Resources
- Constraints

Generative Agent

AGENT BASED MODELING

.java → .igeo

3D PARAMETRIC ENVIRONMENT

.3dm → .ghx

Environmental Agent

ENVIRONMENTAL SIMULATION

.rad ← .gha → .gha ← .py → .osm / .idf

pass environmental analysis

ROBOTIC SIMULATION

kuka ↔ .dat

Construction Agent

Coordinator Agent

TRADE OFF METRICS

- Daylight factor Analysis
- Continuous Daylight
- Process
- Complexity
- Collisions
- Singularities
- Contruction Time
- Cost

Informed OUTPUT

- Geometry (Obj)
- Construction process steps (.Dat)

feedback / update

.java   △ Programming language   △ Java 7.0 Environment / Oracle
.pde   □ Standalone software   △ Processing / C reas B. Fry
.igeo   ○ Library/plugin   △ IGEO NURBS library / S. Sugihara
.py   ○ Physical device   △ Python (IronPython, RhinoPython)
.xml   △ Extensible Markup Language
.3dm   □ Rhinoceros 3d / R. McNeel
.ghx   □ Grasshopper / R. McNeel
.py   ○ Ladybug library / Ladybug Tools
.py   ○ Honeybee library / Ladybug Tools
.rad   □ Daysim & Radiance / MIT
.idf   □ Energy Plus / Department of Energy
.hdr   △ File with declarations of functions
.osm   □ Open Studio: Building Energy Modeling
.unity   □ 3d Gaming Environment / UNITY
VR   ○ Virtual Reality Head Display /Oculus Rift

Figure 4.34: Flowchart illustrating the inputs, types and relationships among the MAS agents.

Generated facade designs

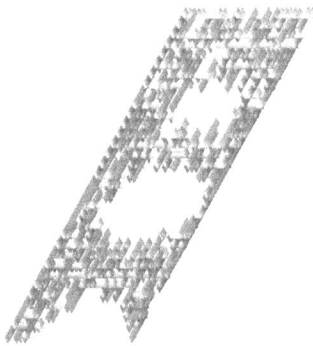

Geometry ID:
cbca40b3b761501c1774007e6ded998e

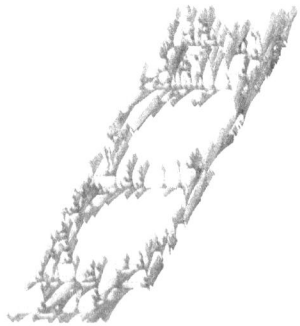

Geometry ID:
9cee02250cf5aa87edc3de8b802cd575

Geometry ID:
cf1f00ce61dffa5001057d5aca0bd72

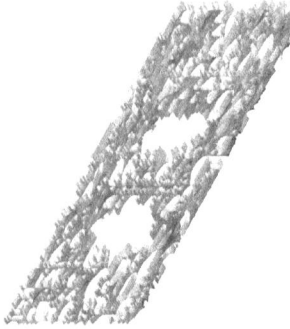

Geometry ID:
d53cdabfd877b1b299d90b592d24dcf

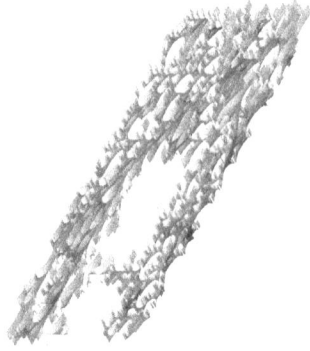

Geometry ID:
11453baa87c1461f174f609c674065246

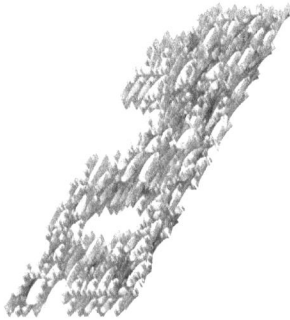

Geometry ID:
1ca2d2a501ad984469cbba97861e6a82

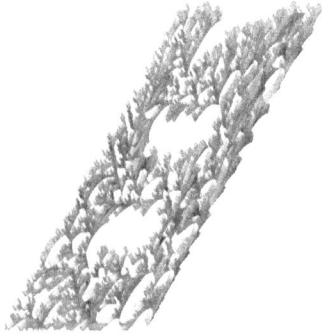

Figure 4.35: A subset of highly ranked facade designs with two or three openings.

Generated facade designs

Geometry ID:
e942bd8e7339ff194c7de56744898fde

Geometry ID:
23445767ea2650377c5793330dd27bc

Geometry ID:
e544515d8cd362b4d22ca24ba9fe547f

Geometry ID:
5ca1050df14977d3d52212217592deeb

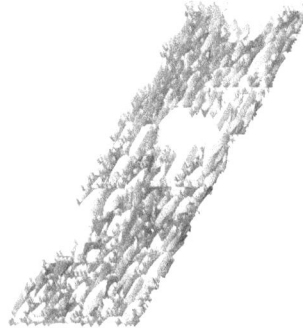

Geometry ID:
4dff8a51a1e882868ead5b2673013455

Geometry ID:
3b7c7326ebb9de4795567dc5e7a03474

Figure 4.36: A subset of highly ranked facade designs with one or two openings.

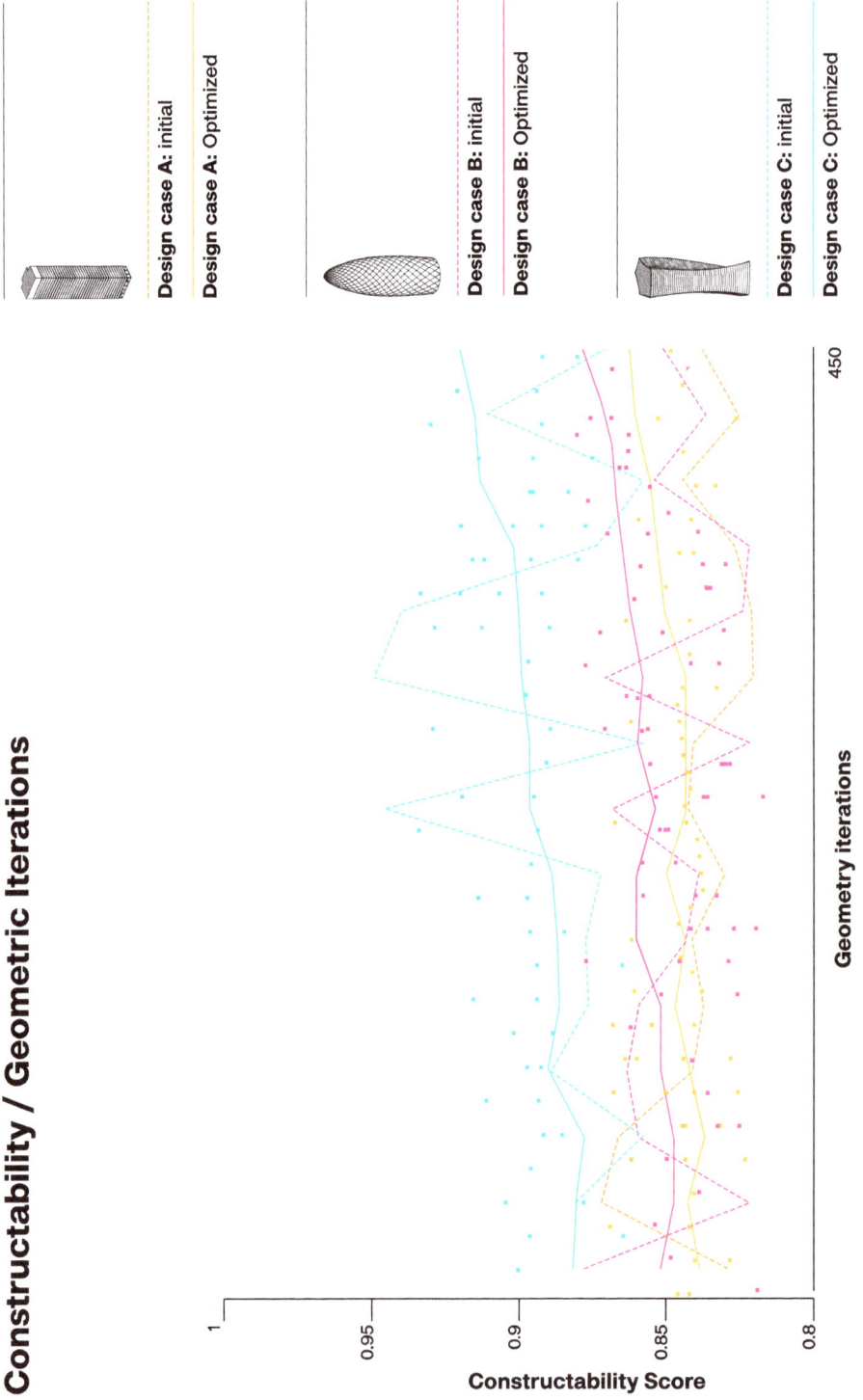

Figure 4.37: Plot showing the constructability score over the iteration cycles for three case studies.

The system continues to be tested, validated and further integrated. In this case study, a few limitations should be pointed out. First, for reasons of clarity, only one type of design segmentation was simulated and tested on a zero curvature (facade) surface domain. Moreover, at present, the system is not fully automated and still requires the designer to work across multiple interfaces. On occasion, during the generation of a design alternative, the criteria of creating a complete fenestration pattern across the domain of a surface, as well as being faced with a local minima problem within the hill climbing algorithm's manual deletion of collision points enabled the continuation of the pattern and design.

4.3 Experimental design 3: agent-based shell structure design

This experimental design tested the applicability of the framework on thin shell structures, a design problem that requires the close collaboration of architects and engineers to create articulated and efficient design solutions. The focus is on form-finding by describing agents as structural building components, where a number of design parameters are correlated to both the building's environmental and structural performance. The design parameters were derived from a vernacular structural system that analyzed the devised shell forms. The structural system, which is known as reciprocal frames, was selected due to its ability to span areas much larger than the length of its structural elements. Another reason for selecting this structural system is the fact that its inherent complexity was a hindrance for modeling it and using it in practice. The objective was twofold: (a) to extend our MAS in order to consider two types of analysis, the design generation and the environmental and structural analyses in particular; and (b) to compress the two steps of the design process used in the first experiment (design generation and design optimization) into one.

In an attempt to operate in an incremental fashion, the author initially conducted a pilot physical case study using a parametric form-finding approach to generate the shell geometries that take into account a vernacular structural system, as well as material and economic parameters. An agent-based simulation creates perforations and optimizes the effect of the solar radiation underneath the structure. The objective of the pilot study was to show the limitations of parametric modeling and demonstrate how a MAS approach could best be utilized in the context of form-finding for addressing environmental parameters. Based on the results of the pilot study, the author developed another case study in simulation, where the MAS framework was used for form-finding shell structures informed by environmental parameters, such as the position of the Sun and the structural parameters described in Section 4.2.2.

4.3.1 Structural form-finding

Structures can be categorized in many ways according to their shape, function and materiality. Shells are structures defined by a curved surface that is thin in the direction perpendicular to the surface, but there is no absolute rule as to how thin the surface has to be. This definition includes a wide variety of structures, from concrete shells to grid shells and bird's eggs. Shell structures can be further classified as form active when the form adapts to different loadings (i.e., a spider's web or a sail) and form passive if the shape does not change significantly (i.e., a dome) (Figures 4.38 and 4.39). The structurally efficient shell design challenge lies in determining the right structural shape that will resist loads within its surface without the need for extra structural systems [16]. Of all the traditional structural design parameters, such as material choice, section profiles, node type, global geometry and support conditions, the global geometry predominantly dictates whether a shell will be stable, safe and stiff.

The preceding work in the field includes the works of A. Gaudi, P.L. Nervi, F. Candela, E. Torroja, F. Otto, to name a few. Gaudi's genius was in his original approach to creating architecture by employing empirical methods and physical models (up until then, experimentation was not part of shape finding). He was inspired by the analysis of gothic architecture by Viollet-Le-Duc and the concept of structural rationality. As just one example from these precedents, Heinz Isler made extensive use and analysis of physical scale models that were cast in plaster upside down and then scaled to full size. Isler believed that physical models ensure a more holistic simulation of the problem, although they posed the ensuing challenges of accuracy and scalability of material and mass. Through this method, he was able to obtain different shapes that all had an even stress distribution by using physical model experimentation. This is a very rational way of thinking but it also involves huge costs, as the formwork must also be a free-form curved surface.

More recently, the contemporary works of J. Ochsendorf, A. Kilian, P. Block, C. Williams and M. Sasaki, among others, aim at replicating the shape creation process via the development of computer programs that can generate shapes [16, 226–228]. In part inspired by Gaudi's physical hanging chain models, A. Kilian at MIT introduced the use of particle spring systems for digitally simulating the behavior of hanging chain models in order to find structural forms solely composed of axial forces [229]. Another critical precedent is the work of Daniel Piker; he introduced an intuitive visual scripting tool, Kangaroo, that enables digital form-finding. Kangaroo, a nonlinear "physics-based" engine, is embedded directly within the Rhinoceros-Grasshopper computer-aided design environment, thus enabling geometric forms to be shaped by material properties, applied forces and interacted with in real time. By embedding rapid iteration and simulation in the early-stage design process, Kangaroo allows a faster feedback loop between the modification of a design and the engineering analyses [230]. This is particularly useful for the design of structures involving large deformations of material from their rest state, such as tensile membranes, bent-timber grid shells and inflatable structures. Kangaroo can also be applied to the interactive optimization of geometric and aesthetic qualities that may not themselves be intrinsically physical. Another research group led by Philippe Block has developed a structural form-finding software package, Rhinovault, that implements the thrust network approach (TNA) to create and explore compression-only structures. It uses projective geometry, duality theory and linear optimization, and provides a graphical and intuitive method that adopts the advantages of graphic statics for 3D problems [231]. Rhino-vault is based on relationships between form and forces expressed through diagrams that are linked through simple geometric constraints: a form diagram, representing the geometry of the structure, reaction forces and applied loads, and a force diagram representing both global and local equilibrium of forces acting on and in the structure [232]. Rhinovault takes advantage of the relationships between force equilibrium and 3D forms and explicitly represents them by geometrically linking form and force diagrams.

Drawing on these precedents, the author has observed an increasing interest in free-form shell design, partly promoted by the development of form-finding tools that implement contemporary mathematical models and digital simulations for computing and analyzing shell structures. In the case of TNA, linear optimization is used; in the case of Kangaroo, the geometry optimization is nonlinear. Moreover, TNA's reduction of the problem into 2D offers a more efficient computational model for calculating the force distribution. Although these tools and frameworks have greatly facilitated the design of shell structures using form-finding methods, they also tend to converge to-ward optimal shapes, which in architecture tend to be a rarity. What this means is unlike the design of a rocket where, based on a range of loads and objectives, you simulate an optimal shape and manufacture, in architecture a number of conflicting objectives are often the reason why suboptimal solutions are deemed more suitable.

Therefore, there is still a necessity for computational tools that enable designers to steer away from purely form-found geometries and help them explore alternative forms by (a) allowing them to introduce more architectural parameters in the early design stage and (b) integrating constraints related to the fabrication and construction processes. Behaviors are conceived as a way to express design intentions and/or con-straints in a more abstract way. This technique allows the utilization of basic struc-tural mechanics as an integrated design process (i.e., a free-form shell design is generated by following a set of structural rules), which is the opposite of the sequen-tial approach of conventional structural analysis (i.e., a free-form design is generated and is rationalized afterward based on material properties). The concept of behav-ioral form-finding is developed in the sections that follow.

4.3.2 Background and context of reciprocal frames

The principle of reciprocity in structural design and construction, that is, the use of load-bearing elements to compose a spatial configuration wherein they mutually sup-port one another, has been known since antiquity [233]. Etymologically, reciprocity de-rives from the Latin *reciprocus,* which is a compound of *recus,* meaning backward, and *procus,* meaning forward. The word "reciprocity" implies the practice of exchanging things with others for mutual benefit. Such a definition emphasizes the obliged stressed return of a certain action. The development of reciprocal frames has not had a linear history, and the evidence of its knowledge and applications around the world seems to be unrelated to one another. However, a common point in the use of this system in

A.

Historical Timeline of Long Span Shell Structures

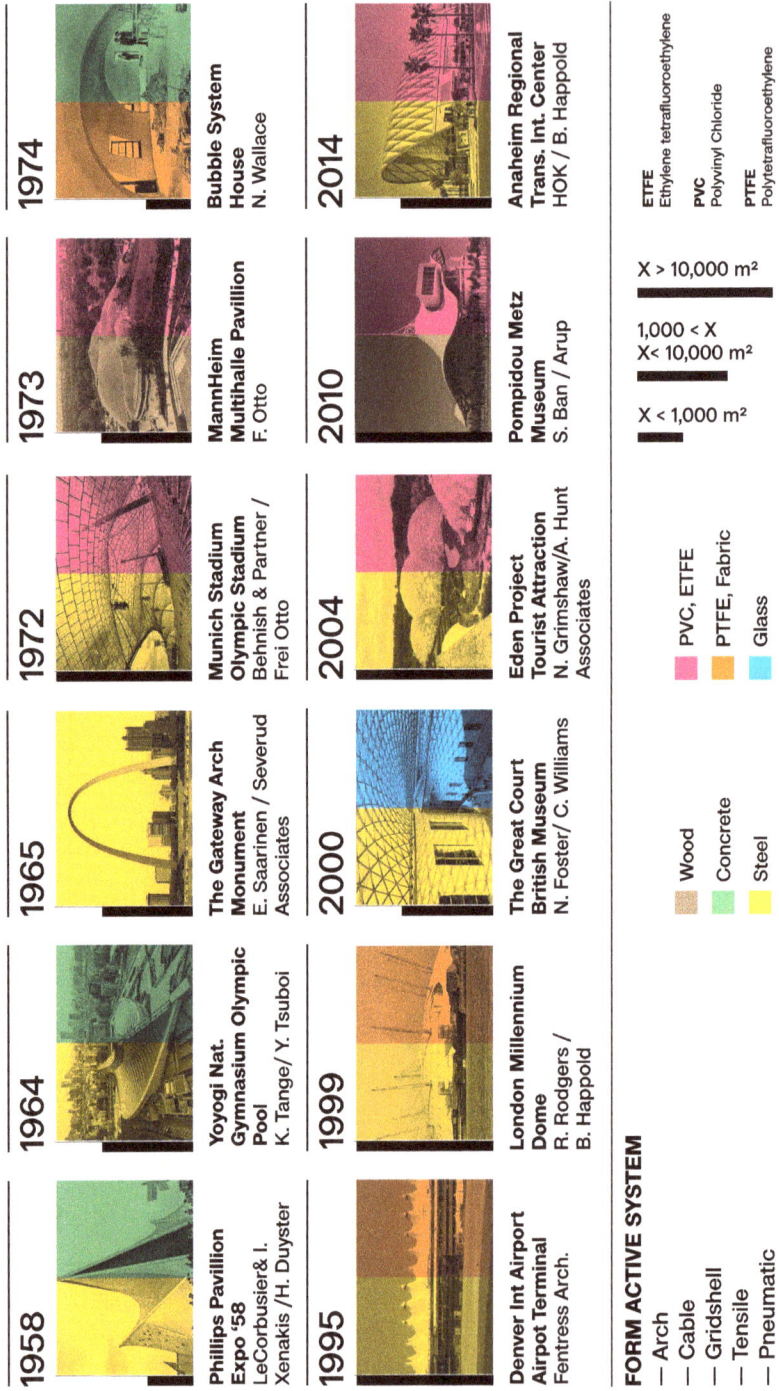

| 1958 | 1964 | 1965 | 1972 | 1973 | 1974 |

Phillips Pavillion Expo '58
LeCorbusier& I. Xenakis /H. Duyster

Yoyogi Nat. Gymnasium Olympic Pool
K. Tange/ Y. Tsuboi

The Gateway Arch Monument
E. Saarinen / Severud Associates

Munich Stadium Olympic Stadium
Behnish & Partner / Frei Otto

MannHeim Multihalle Pavillion
F. Otto

Bubble System House
N. Wallace

| 1995 | 1999 | 2000 | 2004 | 2010 | 2014 |

Denver Int Airport Airpot Terminal
Fentress Arch.

London Millennium Dome
R. Rodgers / B. Happold

The Great Court British Museum
N. Foster/ C. Williams

Eden Project Tourist Attraction
N. Grimshaw/A. Hunt Associates

Pompidou Metz Museum
S. Ban / Arup

Anaheim Regional Trans. Int. Center
HOK / B. Happold

FORM ACTIVE SYSTEM
— Arch
— Cable
— Gridshell
— Tensile
— Pneumatic

Wood
Concrete
Steel

PVC, ETFE
PTFE, Fabric
Glass

X > 10,000 m²

1,000 < X
X< 10,000 m²

X < 1,000 m²

ETFE
Ethylene tetrafluoroethylene

PVC
Polyvinyl Chloride

PTFE
Polytetrafluoroethylene

Figure 4.38: Historical timeline of selected form active shell structures from the 20th and 21st century. Images derived from photos by Wouter Hagens, Arne Mueseler, Buphoff, Frei Otto, N. Wallace, Soc23, Andrew Dunn, Grimschaw Arch, via Wikipedia.com, arch2o.com, ultraswank.com (CC License) (left top to bottom right).

B.

Historical Timeline of Long Span Shell Structures

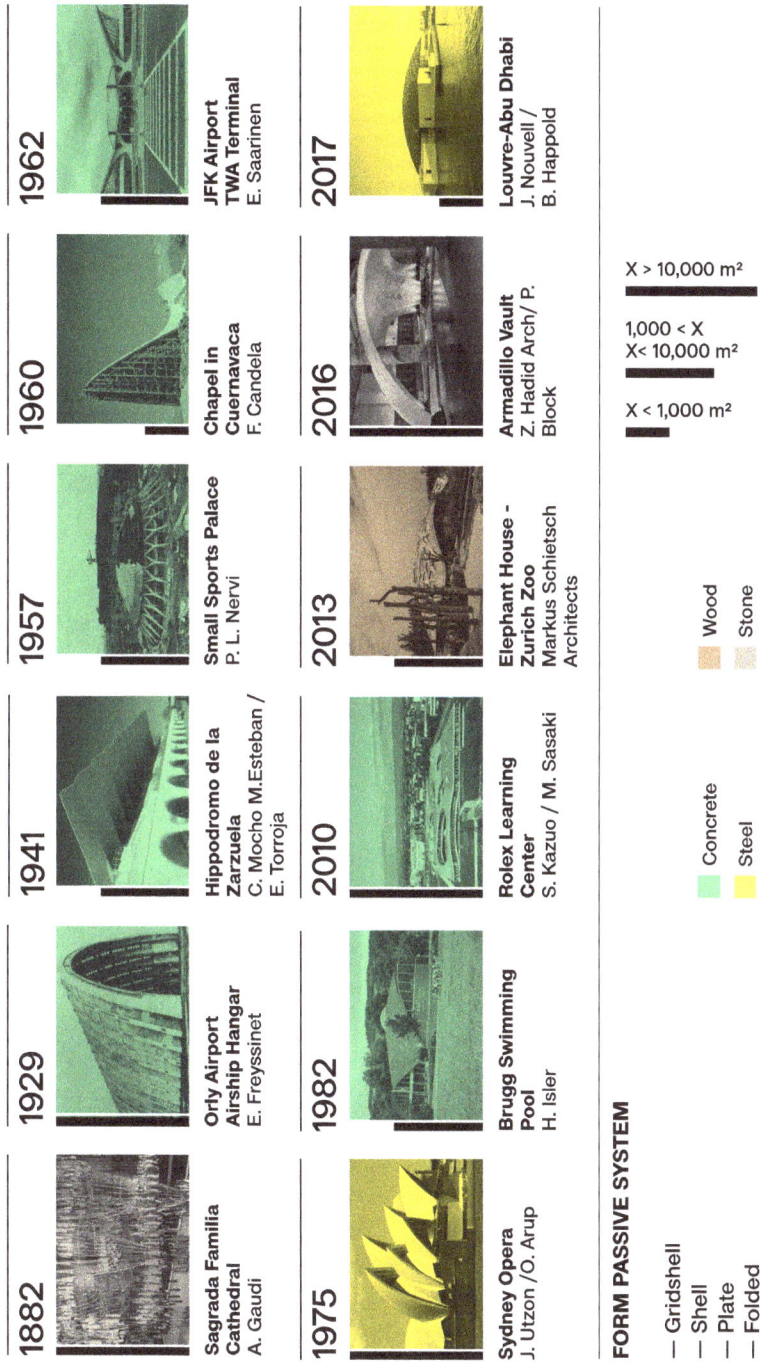

1882	1929	1941	1957	1960	1962

Sagrada Familia Cathedral
A. Gaudi

Orly Airport Airship Hangar
E. Freyssinet

Hippodromo de la Zarzuela
C. Mocho M.Esteban / E. Torroja

Small Sports Palace
P. L. Nervi

Chapel in Cuernavaca
F. Candela

JFK Airport TWA Terminal
E. Saarinen

1975	1982	2010	2013	2016	2017

Sydney Opera
J. Utzon /O. Arup

Brugg Swimming Pool
H. Isler

Rolex Learning Center
S. Kazuo / M. Sasaki

Elephant House - Zurich Zoo
Markus Schietsch Architects

Armadillo Vault
Z. Hadid Arch/ P. Block

Louvre-Abu Dhabi
J. Nouvell / B. Happold

X > 10,000 m²

1,000 < X
X< 10,000 m²

X < 1,000 m²

FORM PASSIVE SYSTEM

— Gridshell
— Shell
— Plate
— Folded

Concrete
Steel
Wood
Stone

Figure 4.39: Historical timeline of form passive shell structures. Images derived from photos by Outsin, Mister No, Gallery 400, Acoterion, B. Spagg, A. Buschmann, I. Baan, L. Boegly via Wikipedia.com, arch2o.com, dezeen.com, archdaily.com, ultraswank.com, efreyssinetassociation. com/ (CC License) (left top to bottom right).

both the Occidental and Oriental cultures is favoring timber as a construction material. It is worth noting that, in Europe, it was more of a practical and construction issue for the development of planar spanning configurations, while in Asia, it was mostly used for the ceremonial realization of 3D structures. Figure 4.40 presents an overview of reciprocal frame structures from antiquity until today, and Figure 4.41 presents contemporary experimental structures using the principle of reciprocity (April 2019).

The first reciprocal frame structures are traced back to Chinese and Japanese religious architecture in the twelfth century, as seen in the wood constructed roof support systems of the mandala roof. In Europe, the concept of spanning distances longer than the length of the available timber beams was the main reason for the use and development of the reciprocity principle [234]. During the thirteenth century, Villard de Honnecourt conceptualized roof support structures based on this principle in his sketches.

Later in the sixteenth century, Leonardo Da Vinci, who laid the foundation for a scientific study of reciprocal structures, explored at least five different spatial configurations based on the principle of reciprocity, experimenting with regular and nonregular 2D and 3D geometrical configurations. Sebastiano Serlio addressed the problem of planar roof construction with short beams in his book on architecture, dated 1556. A comparable structure system made of reciprocally supporting bar-shaped elements is the Zollinger system, which is mainly used in timber roof construction; Friedrich Zollinger obtained a patent for it in 1923 [235].

It has been the development of sophisticated timber products, such as glulam trusses and plywood that produce long-spanning structural elements through adhesive technology, which has led to the replacement of reciprocal frames and similar structural systems and has ultimately put a stop to any further improvement. Currently, the principle of reciprocity continues to stimulate the interest of designers and researchers, and it has resurfaced as a topic for academic research [236]. Architectural applications include the Mill Creek Public Housing project by Louis Kahn (1952–1953), the Bunraku Puppet Theater by Kazuhiro Ishii (1994), the Pompidou Metz museum by Shigeru Ban (2008) and the Apple Story at Fifth Avenue, New York City, USA, by B. C. Jackson and can be found around the world. Moreover, a set of experimental works related to structural, geometric and constructive issues of reciprocal structures are appearing, such as the Forest Park structure by Shigeru Bahn and ARUP AGU, the Serpentine pavilion by Cecil Balmond and Alvaro Siza (2005), the H-edge pavilion by Cecil Balmond and students from Penn University [237] and the research pavilions at EPFL Lausanne [238].

In these projects, the reciprocal principle has been explored by using different materials, element sections, joints and planar or 3D configurations, providing the fundamental evidence that adaptations of this typology should be further investigated for a diversity of architectural styles, patterns, performance characteristics and local sensibilities. Based on a specific type of reciprocal frame, the applicability of such a vernacular structural system was studied by analyzing its functional and material behavior. The research models a design system that fosters exploration by incorporating issues of recyclability and material efficiency coupled with design and spatial comfort performance objectives. This is partially achieved by implementing digital fabrication techniques and through the incorporation of agent-based design technologies enabling emergent and intrinsic performance.

4.3.2.1 Defining reciprocal frames

A reciprocal frame is a structural system formed by short bars that are connected using friction only. Most importantly, the reciprocal frame can span many times the length of the individual bars. "Reciprocal" refers to the fact that such structures are composed of a number of elements (also referred to as short beams) that structurally interact through simple support binding in order to create more complex structures of dimensions much greater than the individual elements they are composed of.

The application of the reciprocity principle requires (a) the presence of at least two elements allowing the generation of forced interactions; (b) that each element of the assembly must support and be supported by another one; and (c) that every supported element must meet its support along the span and never at the vertices in order to avoid the generation of a space grid with pin joints [234]. The space structures that conform to the above requirements are called reciprocal and consist of at least two interlaced linear elements where the final form is relative to a basic component type in its material, as well as the connection technology. The components can be identical or nonidentical but should follow a specific global tessellation pattern.

The joining of the components at the node points can generally be carried out without mechanical connections, solely by pressure and friction. Frictional forces can be supported with simple connection techniques, such as notching or tying elements together. In fact, in the context of wood processing techniques, it is recognized that the complexity of the connection technology becomes an important feature that distinguishes different structural propositions from a financial or a structural perspective [238]. From a structural point of view, each individual element in the system works as a single beam. Each edge of the beam either lies on another component or, when forming the edge of the structure, on the supports of the entire system. Each element bears the supporting force of one of the neighboring elements and optional dead loads or live loads. The interest of such a structural system lies in the fact that the "global" form is determined by the "local" condition of the building elements.

4.3.3 Case study A – new view pavilion: fabrication-aware form-finding

The first case study used the existing parametric form-finding methods to generate the shell geometries and analyze them in reciprocal frame. Then, an agent-based modeling was used to create perforations on the shell in order to enhance the solar radiation analysis underneath the structure. The emphasis of this case study was on using an integrated design-to-construction workflow and investigating the topic of material-aware form-finding through the perspective of a traditional structural system – that of reciprocal frames – as well as the affordances relating to file-to-factory processes, all while

A.

Historical Timeline of Reciprocal Structures

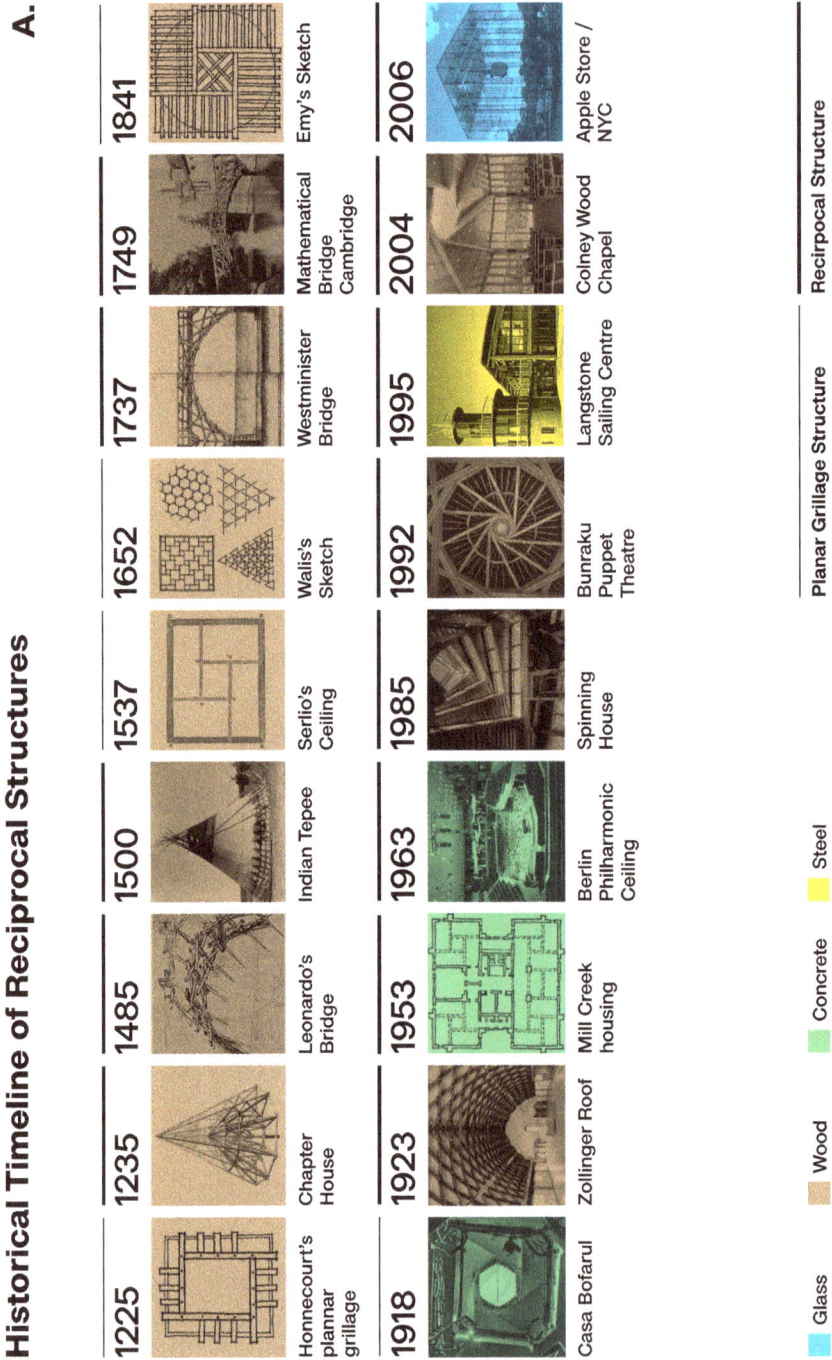

1225	1235	1485	1500	1537	1652	1737	1749	1841
Honnecourt's plannar grillage	Chapter House	Leonardo's Bridge	Indian Tepee	Serlio's Ceiling	Walis's Sketch	Westminister Bridge	Mathematical Bridge Cambridge	Emy's Sketch

1918	1923	1953	1963	1985	1992	1995	2004	2006
Casa Bofarul	Zollinger Roof	Mill Creek housing	Berlin Philharmonic Ceiling	Spinning House	Bunraku Puppet Theatre	Langstone Sailing Centre	Colney Wood Chapel	Apple Store / NYC

Planar Grillage Structure   Recirpocal Structure

Glass   Wood   Concrete   Steel

Figure 4.40: Timeline with examples of structures realized using reciprocal frames structural system dating from antiquity. Drawings and Images by author via northernarchitecture, parametrichouse, arup. com, wewanttolearn, Wikipedia.com, chusdoit.com, flickr.com (CC License).

B.

Timeline of Experimental Reciprocal Structures

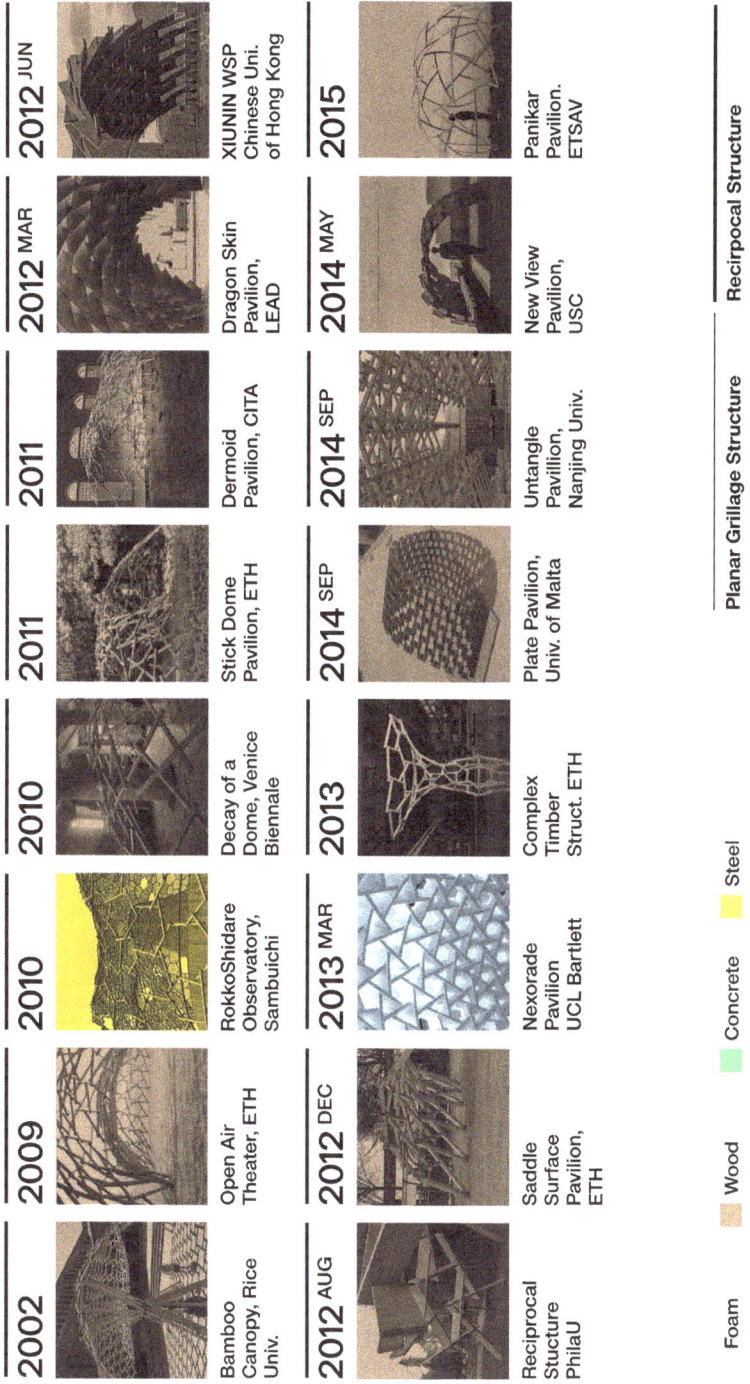

2002	2009	2010	2010	2011	2011	2012 MAR	2012 JUN
Bamboo Canopy, Rice Univ.	Open Air Theater, ETH	RokkoShidare Observatory, Sambuichi	Decay of a Dome, Venice Biennale	Stick Dome Pavilion, ETH	Dermoid Pavilion, CITA	Dragon Skin Pavilion, LEAD	XIUNIN WSP Chinese Uni. of Hong Kong

2012 AUG	2012 DEC	2013	2013 MAR	2014 SEP	2014 SEP	2014 MAY	2015
Reciprocal Stucture PhilaU	Saddle Surface Pavilion, ETH	Complex Timber Struct. ETH	Nexorade Pavilion UCL Bartlett	Plate Pavilion, Univ. of Malta	Untangle Pavilion, Nanjing Univ.	New View Pavilion, USC	Panikar Pavilion. ETSAV

Foam   Wood   Concrete   Steel   Planar Grillage Structure   Reciprocal Structure

Figure 4.41: Timeline with examples from contemporary experimental reciprocal structures. Images by author derived by own archive and via northernarchitecture, arup.com, wewanttolearn, flickr.com (Creative Commons License).

Figure 4.42: Photo of New View Pavilion in situ. The pavilion was designed and constructed based on the principle of reciprocity. Image by G. Sfakianakis via author's archive.

working with a real-world site, material, assembly and cost constraints. A self-standing shell structure at 1:1 scale was designed, fabricated and constructed on the rooftop of a cultural center in the center of Athens, Greece (Figure 4.42).

The MAS framework in this case study was not used for form-finding; instead, a parametric modeling approach and agent-based simulations were favored to create perforations. Kangaroo, a physics-based solver, was employed for form-finding the shell geometries by providing boundary conditions and material weight [230]. The form-found shells were analyzed parametrically into structural elements according to the reciprocity principle described in the section above. In Figure 4.43, the basic form-finding steps are shown, along with the stress analysis of generated geometry. In Figure 4.44, the reciprocal elements comprising the structure are arrayed in 2D. A solar radiation analysis was performed on the generated geometries to evaluate their environmental performance. The environmental data was then passed to a generative agent, tasked with creating perforations on the shell (in the Processing 2.0 environment).

The case study was realized following a design-to-production workflow that proceeds through the following steps: (1) form-finding of a canopy shell surface with two support conditions through the use of a mesh relaxation algorithm; (2) the discretization of the generated surface into its iso-curves that are populated with interlocking building components, where each component follows the principle of reciprocity and is modeled with an associative parametric geometry modeler; (3) informing the geometry of the basic element based on the selected material (thermoformed plywood) to optimize its transversal section and render it resistant to the forces under which the structure would be placed; (4) performing a solar radiation analysis, and using this data for informing the trajectory of the agents in order to control the permeability of the structure by introducing perforations probabilistically on the building components; (5) examining the fundamental mechanical properties of a single arch using the finite element analysis (FEA), while considering the nonlinear contact boundary condition; this is performed with the static behavior of the structure under the self-weight load case; and (6) constructing a prototype at 1:1 scale using a new type of thermoformable plywood panels, in which the material usage was optimized by opting for the minimum allowable thickness, and by grouping the components to be fabricated in panels with nonstandard dimensions that correspond to the available veneer width, rather than fixed plywood panel dimensions.

4.3.3.1 Design process

As the first step, the design of the experiment(s) began with setting the boundary and support conditions of the structure. By implementing a mesh relaxation algorithm, a series of surfaces was generated with the same support conditions. Based on a curvature analysis, a selection of these initially designed surfaces was parametrically discretized; their iso-curves were extracted and populated with a basic component type in an automated and parametric fashion (Figures 4.43 and 4.44). Interlocking components were placed sequentially in pairs on the division points of each iso-curve and were oriented parallel to the tangent vectors to create connected arches. The discretization of the arch in a different number of components and the thickness of each

Stress analysis of geometries with different support conditions

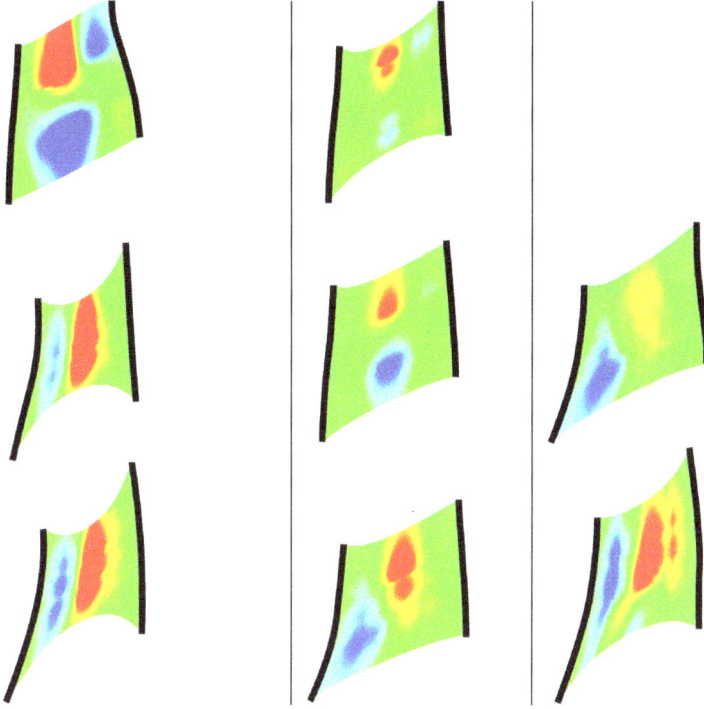

Form finding diagram

- Support conditions
- Covered area
- Loads

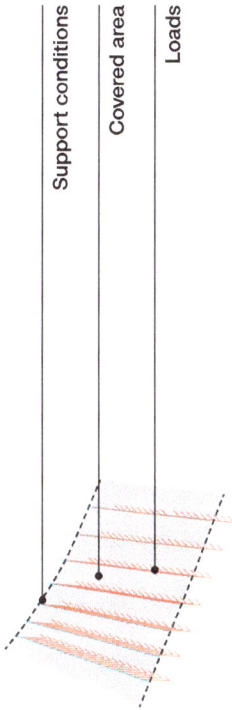

- Particle Spring System (Catenary Arches)
- Surface in compression (inverted)

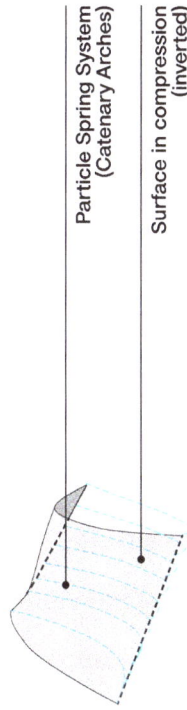

- Isocurves
- Form Found shell (Spring System in equilibrium)

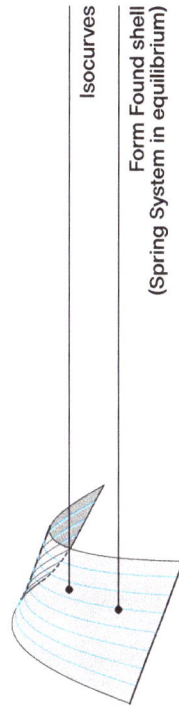

Figure 4.43: The form-finding process (left) and curvature analysis of alternative form-found shells (right).

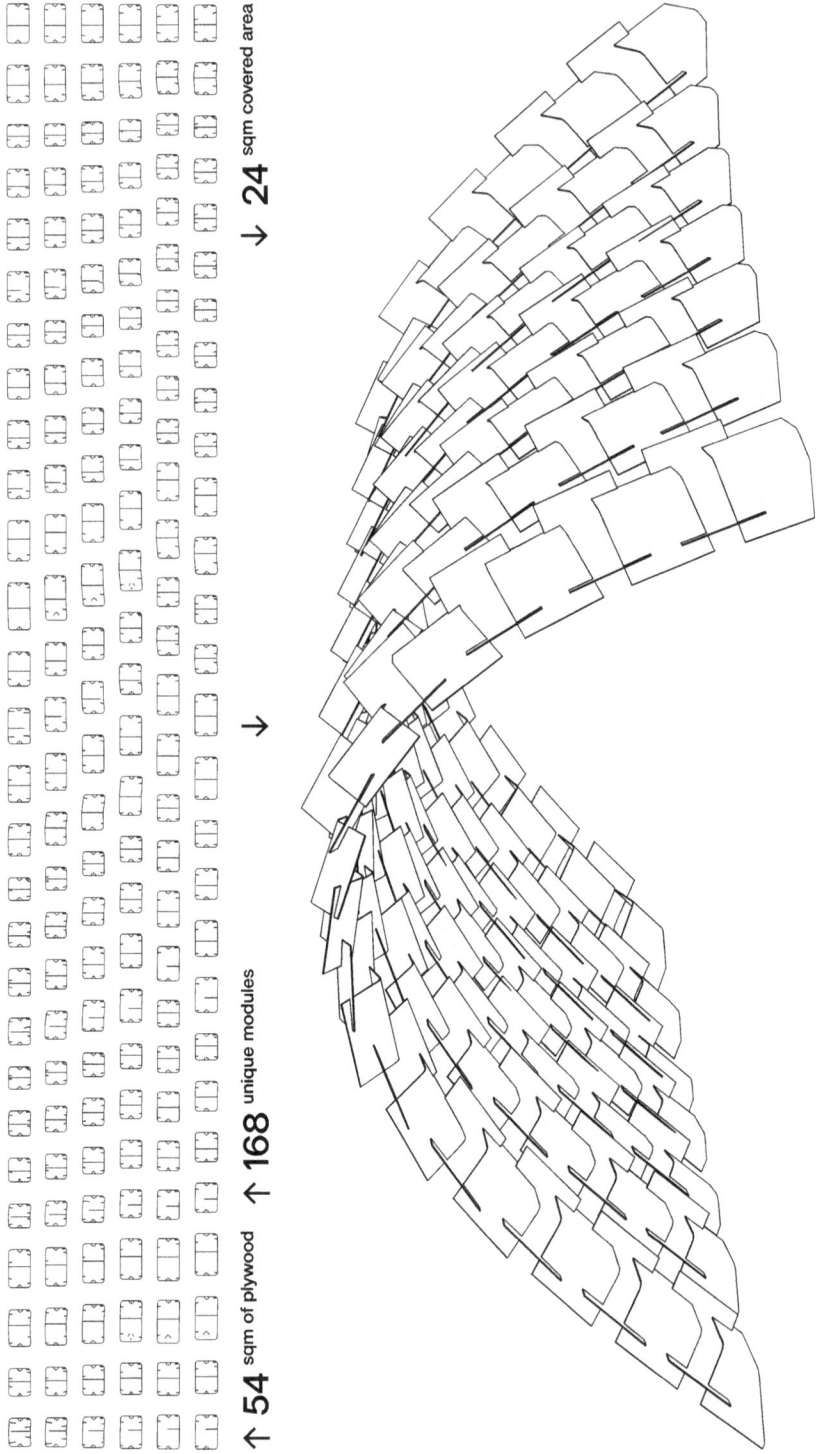

Figure 4.44: Panels developed in a plane (left) and assembled into the pavilion structure (right).

Structural analysis of a catenary arch with 3mm panels

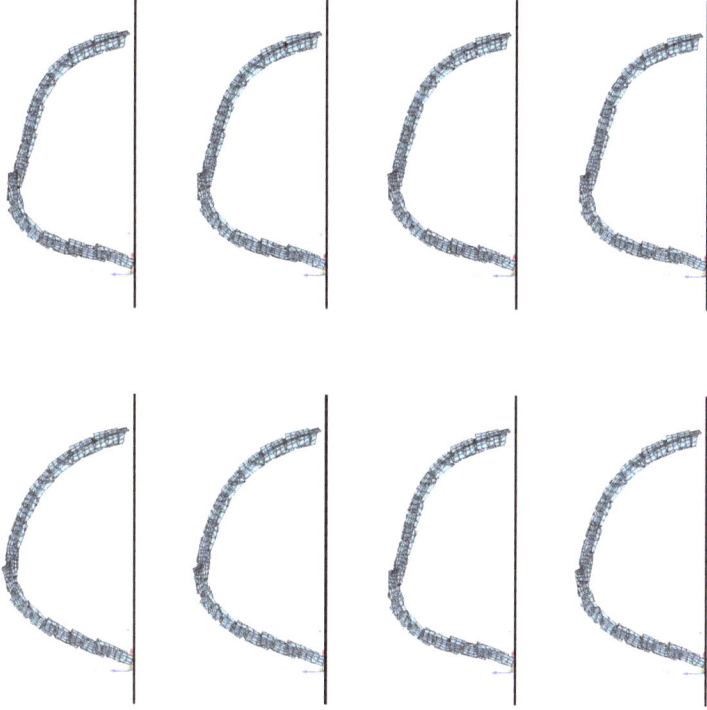

Part to Whole relationship of a catenary reciprocal arch

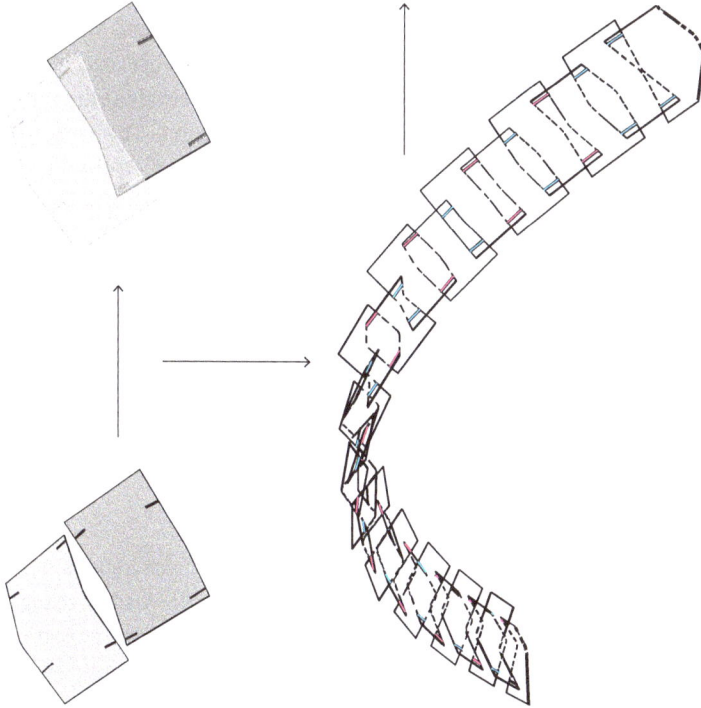

Figure 4.45: Interlocking logic of panels and structural analysis of an arch.

component were structurally evaluated by doing an FEA on a single arch (Figure 4.45). The generated structure was further analyzed with Ladybug, another Grasshopper plugin that runs a Radiance-based simulation (Figures 4.46 and 4.47). All the components were numbered and unrolled in flat panels with all the perforation and notch lines projected.

Thus, the material calculations were completed, relating the covered area of the shell surface to the simulation data that was exported as a text file. The text file was passed to the custom flocking algorithm in the Processing environment in 3D, where the agents read the vertices of the component surfaces and coupled data values. The agents were spawned and programmed to make movement and trajectory decisions based on the local information, including the intensity values from the Sun simulation, proximity to neighbors and trails left by other agents. Each agent had the capacity to read the data from the simulation, which was paired with its corresponding point in a mesh object, as well as data related to its neighbors and constraints as trajectories in time. Consequently, the agents' environment was a collection of points within which they were constrained, and each point was assigned an intensity score based on the data from the simulation. The agents' trajectories then became a generative geometry for material organization and reorganization, which happened in a collective recursion. The agent-based trails were exported again as a text file into the Rhinoceros/Grasshopper scripts to be incorporated as another layer of information that controls the permeability of the panels through a process of CNC-driven material erosion through milling.

The third step was to inform the component design based on the selected material and the condition of the neighboring elements. Furthermore, each component was analyzed to determine the amount of solar radiation it receives during a given period, and a value (kwh/sqm) was given for each mesh point. The weighted values were passed on to the multi-agent design system that controlled the permeability of the structure, by drawing the reaction of the agent swarm on the surface as paths for areas to protect or remove materials. Agents were created at the mesh points of each component and the associated value at each point influenced the agents' flocking behavior (Figure 4.46). The path of the agents was exported and translated in a perforation pattern via dashed slits for improved solar radiation protection and enhanced comfort beneath the structure (Figure 4.46 (c)). The last step included creating the prototype of the whole structure at 1:1 scale using curved thermoformable plywood panels that were CNC-milled flat and subsequently thermoformed.

4.3.3.2 Part-to-whole relationship

A basic module of a two-element reciprocal frame structure was parametrically defined and investigated as the basic structural system. Instead of using traditional linear or planar elements, mutually supported curved panels were used instead, because of formal design aspiration and in order to create a thickened pillow shell that provided a more complex shading condition. The self-standing reciprocal frame structure presented in this study shows an example of the design practice of a final form,

Agent Behavior Diagram

A. Agent motion towards a local maximum based on sun radiation

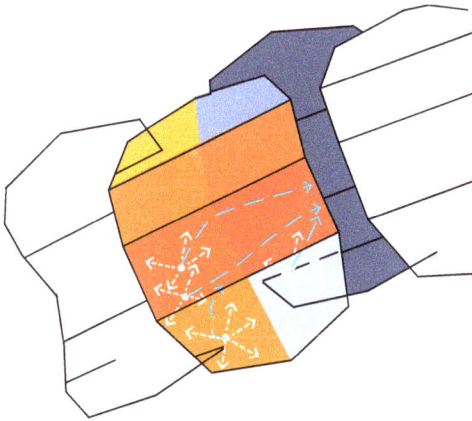

B. Motion Path / Trails of agents moving

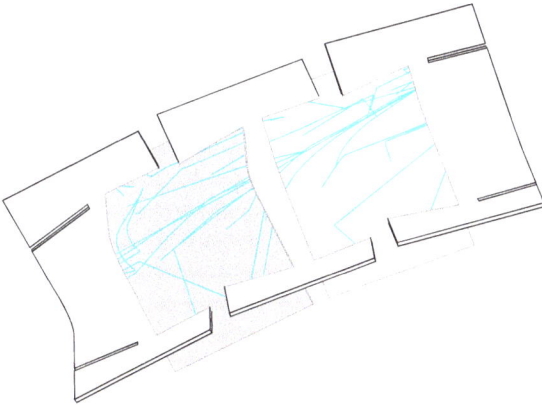

C. Module milling pattern based on agent's trails

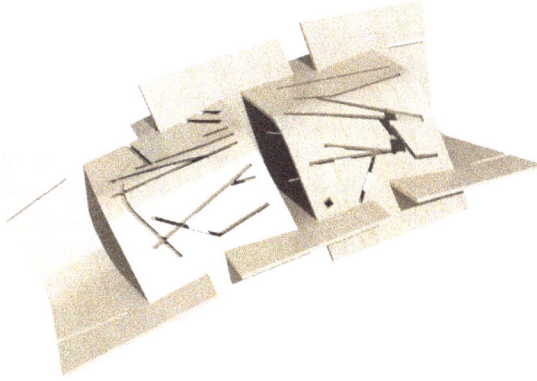

KWh/m2

62.92
125.84
188.96
251.68
314.59

RADIATION ANALYSIS
ATHENS_GRC
30 MAY 6:00 - 31 AUG 24:00

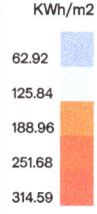

Figure 4.46: Diagram of the perforation system's agent behavior.

Radiation Analysis on Panels

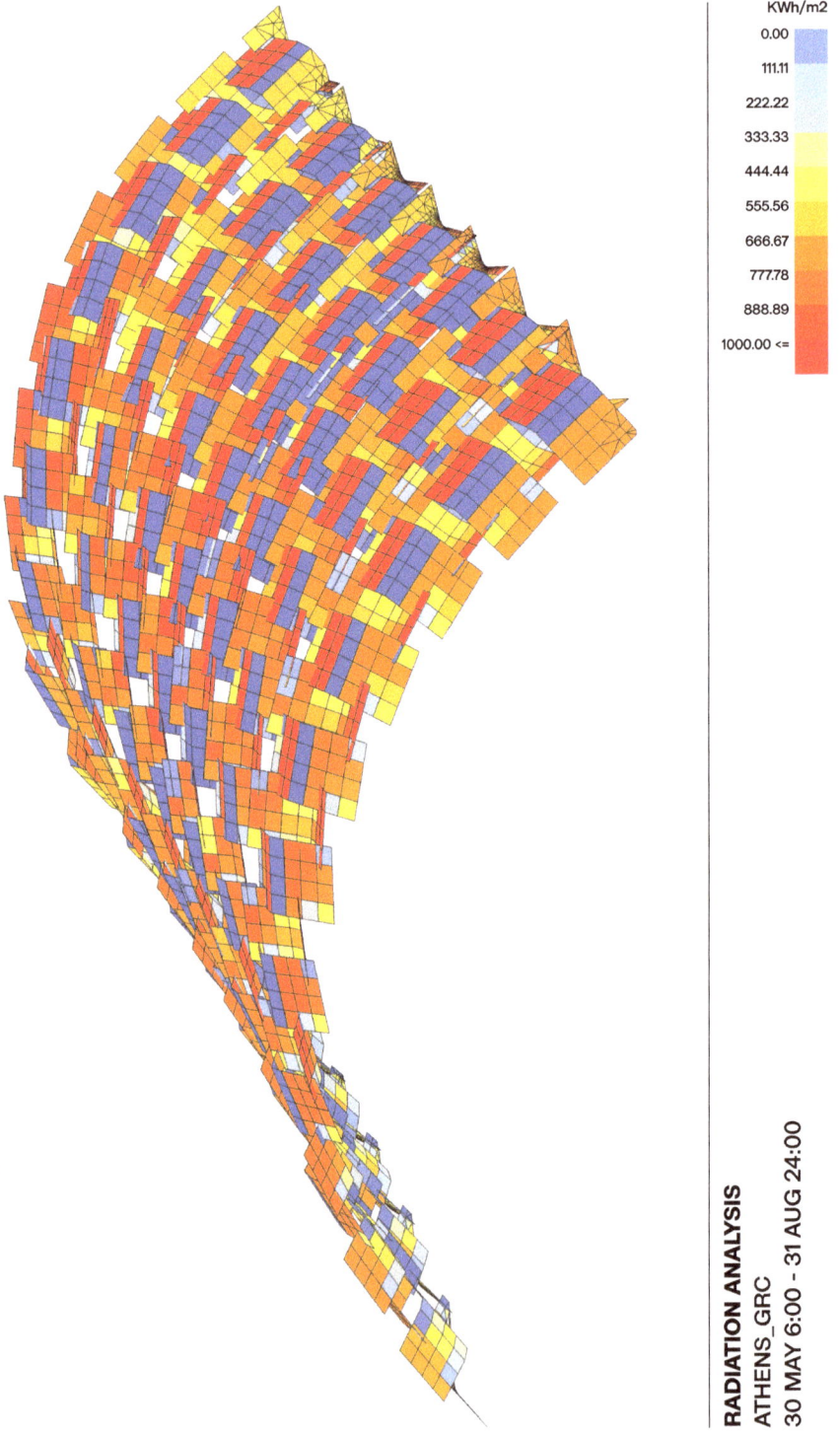

RADIATION ANALYSIS
ATHENS_GRC
30 MAY 6:00 – 31 AUG 24:00

Figure 4.47: Solar radiation analysis (6 months) on the reciprocal panels.

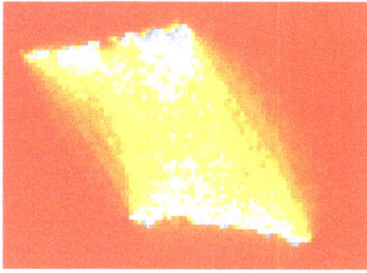

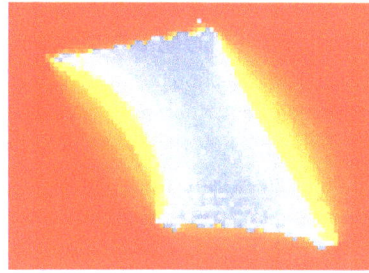

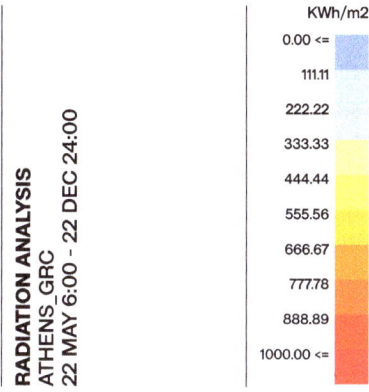

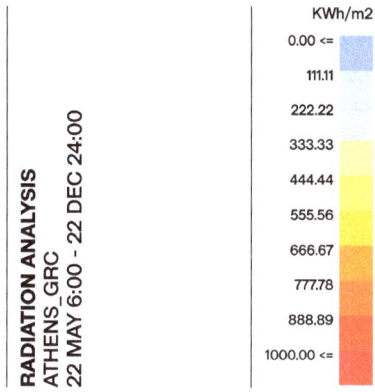

Radiation Analysis on ground surface

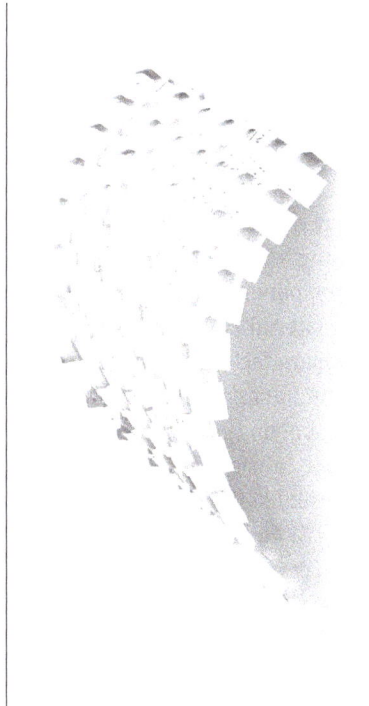

RADIATION ANALYSIS
ATHENS_GRC
22 MAY 6:00 - 22 DEC 24:00

KWh/m2
0.00 <=
111.11
222.22
333.33
444.44
555.56
666.67
777.78
888.89
1000.00 <=

RADIATION ANALYSIS
ATHENS_GRC
22 MAY 6:00 - 22 DEC 24:00

KWh/m2
0.00 <=
111.11
222.22
333.33
444.44
555.56
666.67
777.78
888.89
1000.00 <=

Figure 4.48: Generated geometries with agent-based perforations and solar radiation heatmaps.

driven by the connection technology in which a relaxed modular global form is discretized by means of mutually supported panels.

The proposed slide connection scheme inspires a new family of reciprocal frames, where folded or formed members are mutually supported instead of linear beams or bars. The connection between the building components is integrated as a notch with a specified angle within the geometry of its members, unlike the traditional reciprocal frame system in which the connector members are regular. A V-form module of a given angle is fabricated through the thermoforming of plywood panels and is then spatially multiplied using consecutive rotations and translations that follow the tangent vectors of a discretized catenary arch. The structure can be decomposed into three principal module types, each consisting of curved panels (convex and concave) with locally specific angles in relationship to their generatively form-found neighbors. These modules are then interlocked sequentially along their uniquely milled U shape cuts to form an arch (Figures 4.44–4.46). The inter-panel stability is provided by the rigidity of the slide connection and the axial contact of reciprocal panels. The structural performance of the whole structure improves when more than two arches are connected together. This is due to the fact that they can then act like a truss, with only axial compressive and tensile forces. Bending moments and shear forces are minimized in network arches [239].

4.3.3.2.1 From computational modeling to 1:1 physical prototyping

To test the feasibility of such a design approach, a prototype at 1:1 scale was produced. The design and programming of the structure was at first optimized for discretizing an irregular surface of curved panels using a single pressing mold. The size of the structure was parameterized based on a maximum amount (area) of material, that of 54 m^2. The whole production workflow included (1) material processing, (2) manufacturing of the pressing mold, (3) CNC cutting of the components, (4) thermoforming and postprocessing of the components and (5) the final assembly and erection on site (Figure 4.49).

Through this approach to the reciprocal frame, the potential of using curved plywood components was investigated instead of planar ones. For that reason, an innovative material called UPM Grada was selected. UPM Grada is specifically designed for the manufacturing of form-pressed plywood panels. UPM Grada uses the application of an adhesive film that allows the plywood to be thermo-formed after production by applying pressure through a custom-made mold. The UPM Grada technology allows the fabrication of custom panels, as the film can be cut according to the veneers available (Figure 4.50). A selection of photos illustrating the project is shown in Figures 4.51 and 4.52.

The creation of plywood panels in sizes that corresponded to the available veneer sizes instead of the standard dimensions was of great significance for material efficiency. Plies were used "as is" and reduced material processing time by not needing to be stitched together in larger panels before they were glued to form the final plywood panel that was CNC milled. Moreover, the module's geometry was optimized in order to be fabricated by a single mold to reduce production costs. First, the panels

Fabrication Workflow diagram

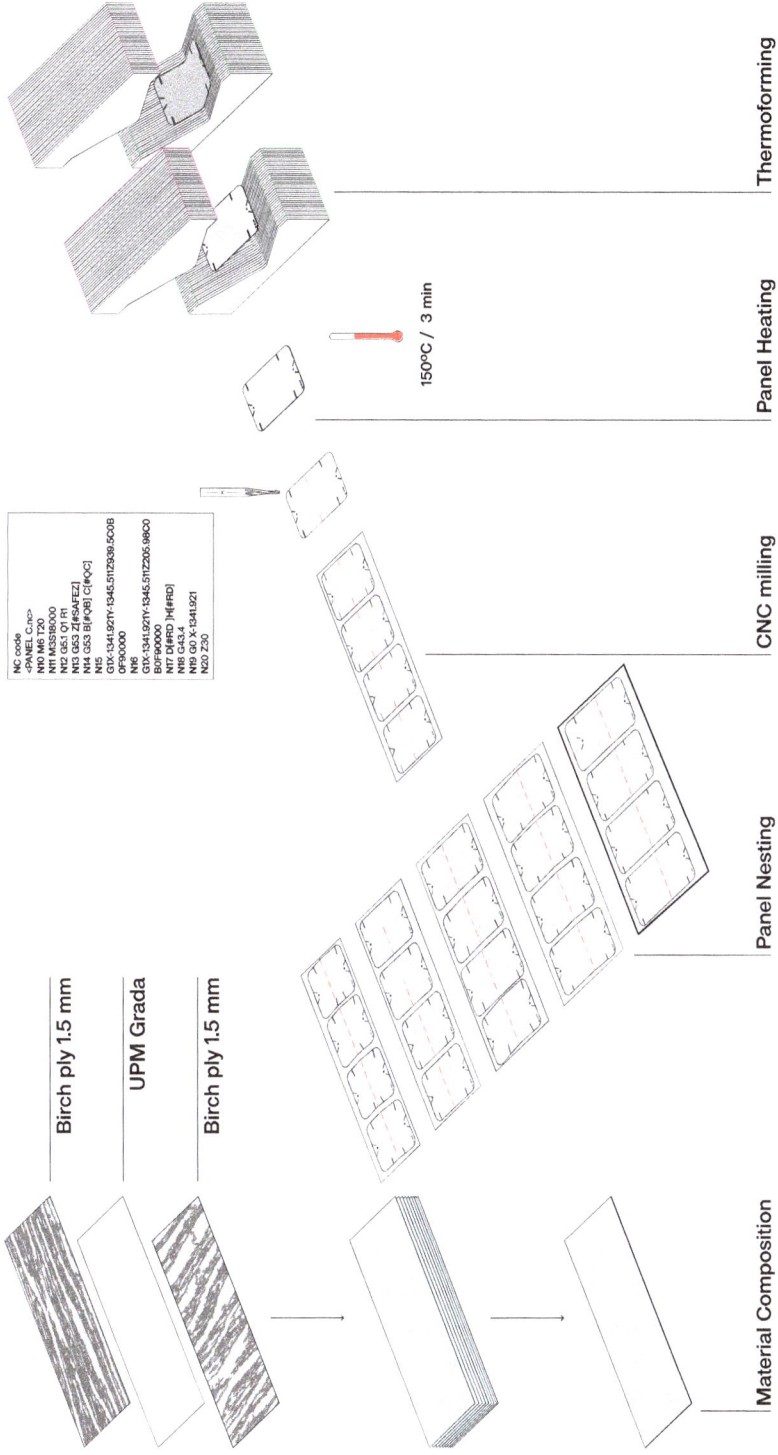

Figure 4.49: Fabrication process diagram of a curved panel.

Panel Fabrication Process

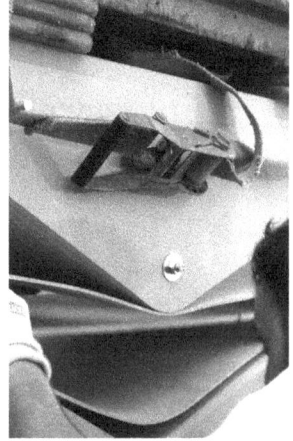

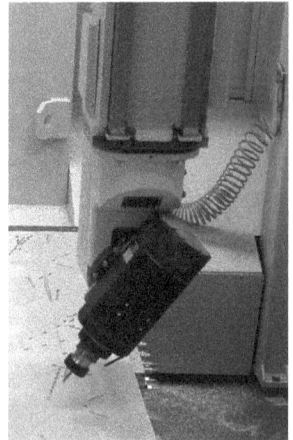

Figure 4.50: Steps of the fabrication process, namely material composition panels from plies of birch and UPM Grada, panel heating, thermoforming and milling. Images by E. Pantazis from author's archive.

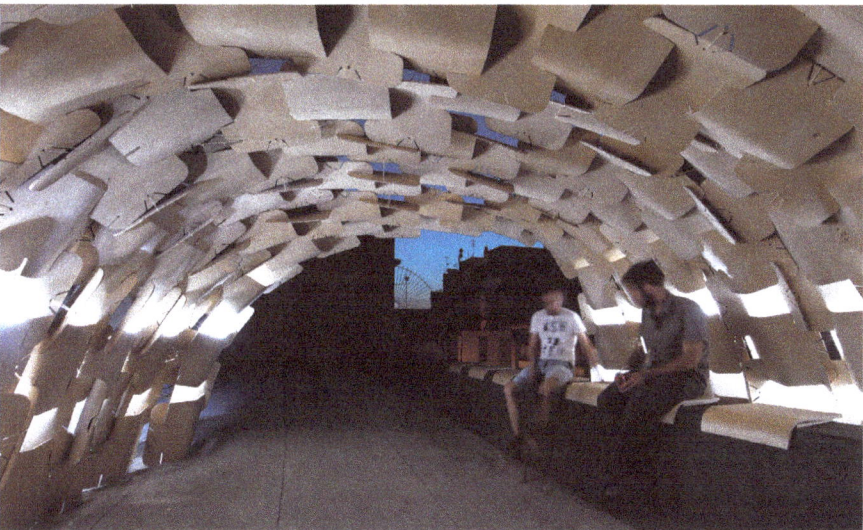

Figure 4.51: Photos of the assembled structure on the rooftop of Romantso Cultural Center in Athens. Image by E. Pantazis and G. Sfakianakis from author's archive.

Figure 4.52: Photo underneath the assembled structure. Image by G. Sfakianakis from author's archive.

were thermo-glued flat and then milled using a five-axis CNC machine. The cut pieces were reheated and pressed consecutively into the mold for final shaping.

A master parametric 3D model was developed in Rhinoceros 3D that generated all the cutting files for the 148 components in a file-to-factory process, where custom routines were developed to give each component their precisely calculated slots for the sliding joints, all in gradually shifting positions and variable angles in order for the pavilion to achieve its irregular funicular form. All components were uniquely labeled and numbered to facilitate assembling and dismantling the structure without the need of experts or detailed drawings. The plywood components were manufactured in Greece at a factory outside Athens and transported to the site.

4.3.3.3 Results and analysis

The thin self-standing shell structure with a footprint of 24 m^2 required 54 m^2 (0.4 m^3) of thermo-formed birch plywood. The on-site assembly of the pavilion was completed in 2 days by a group of five nonexpert workers. It was empirically discovered that the film adhesive and thermo-forming process produced panels with more elastic behavior than the traditional gluing technique using phenol–formaldehyde-based wood glue and high-voltage electricity for curing the glue while the panel was being pressed. This facilitated the assembly process as it allowed the panels to slide in place more easily. Given the common cost and local (and arguably global) material constraints for projects like these, this undertaking sought to be highly resourceful with material usage, aided by computational design tools and CNC manufacturing.

The total scrap wood did not exceed 5 m^2 and was used to create the seating below the structure. In further comparison to the preceding work [240], the author attempted to define benchmarks and measures for the affordances of the approach. By adapting the principle of reciprocity and a specific family of reciprocal frames with mutually supporting curved panels, the author benchmarked the construction of a maximum 600 cm span out of 45-cm-long, 7.5-mm-thick interlocking panels that required no additional joinery.

A previous structure following the same principle, realized by professors Y. Weinand and S. Nabaei and students in Lausanne, had achieved a span of 740 cm with 21-mm-thick panels [238]. While the project is clearly indicative and statistically descriptive of improved span-to-material depth, continued material investigations are expected for achieving further efficiencies of span, material usage and applicability to complex curvature and other design vernaculars. Another major thrust of future investigations is the impact of the swarm-generated pattering and its influence on the structural and environmental performance of the structure. Figure 4.48 shows a solar radiation analysis on the shell, a diagram of the perforation approach and its impact on the solar radiation analysis underneath the structure.

This case study is the very first built experiment that combines research toward the development of a holistic MAS framework using a basic agent model of flocking behavior with principles of parametric form-finding and material properties. The self-standing structure, from its inception through to its use, incorporates real-world material, assembly, cost and human constraints that have informed the future development of the MAS system. It is worth noting that if the internal forces are known, each element of the structure can be adjusted to its local stress, achieving optimized material consumption.

This pilot study explored the applicability of a bottom-up MAS approach for architecture through the combination of parametric form-finding techniques with an agent-based design system and digital fabrication processes. The aim of this design experiment was to test an agent-based workflow combined with a fabrication protocol that has the potential to integrate design intentions with a structural system and material constraints. The focus was on showing the disadvantage of using parametric modeling. The top-down nature of parametric modeling was restrictive in terms of adjusting the position and shape of the joints in order to facilitate easier assembly. The study emphasized how agent-based modeling can be used for encapsulating the intrinsic features of the material properties and assembly methods in the form of simple local rules. In such an approach, the design intentions can be represented by different agencies, and local constraints can be described as agent behaviors that lead to global characteristics instead of imposing "architectural gestures" in a top-down manner. The design optimization becomes an iterative process in which both the solution and the starting condition(s) are constantly oscillating toward an equilibrium defined by multiple performance criteria in response to the given topology and the user's design intentions. This experiment has presented an initial analysis of the potential generative and automation possibilities of integrating digital fabrication workflows driven by MAS design framework.

4.3.4 Case study B: environmentally aware reciprocal frames

This case study was an attempt to extend purely form-finding techniques through the integration of a generative (form-finding) agent with two specialist agents: one for structural design and one for environmental design. The design approach of this case study as well as its dissimilarity with traditional form-finding methods is graphically explained in Figures 4.53 and 4.54. The form found geometries are analyzed structurally and environmentally in terms of received solar radation (Figure 4.55) and then are rationalized geometrically into reciprocal frames.

As described in Section 4.2.2, the structural system relies on the principle of reciprocity, and its morphology is further analyzed in this case study. The advantage of this system is that it can cover large spans with short structural elements. The system is versatile and can be used either with planar or free-form geometries, but its complexity has restricted its popularity. In an attempt to work incrementally, the research proceeded from the simplest version of the system with two interlocking planar elements (as seen in the previous case study) to a reciprocal frame system with three elements and variable linear geometries. In Figure 4.59, the basic design parameters of the reciprocal system with three elements is illustrated, and the relationships of the design parameters are graphically explained. The objective was twofold: on a local level, to study the morphology of the system and extract rules to be used for informing the global con-

Form Finding Diagram

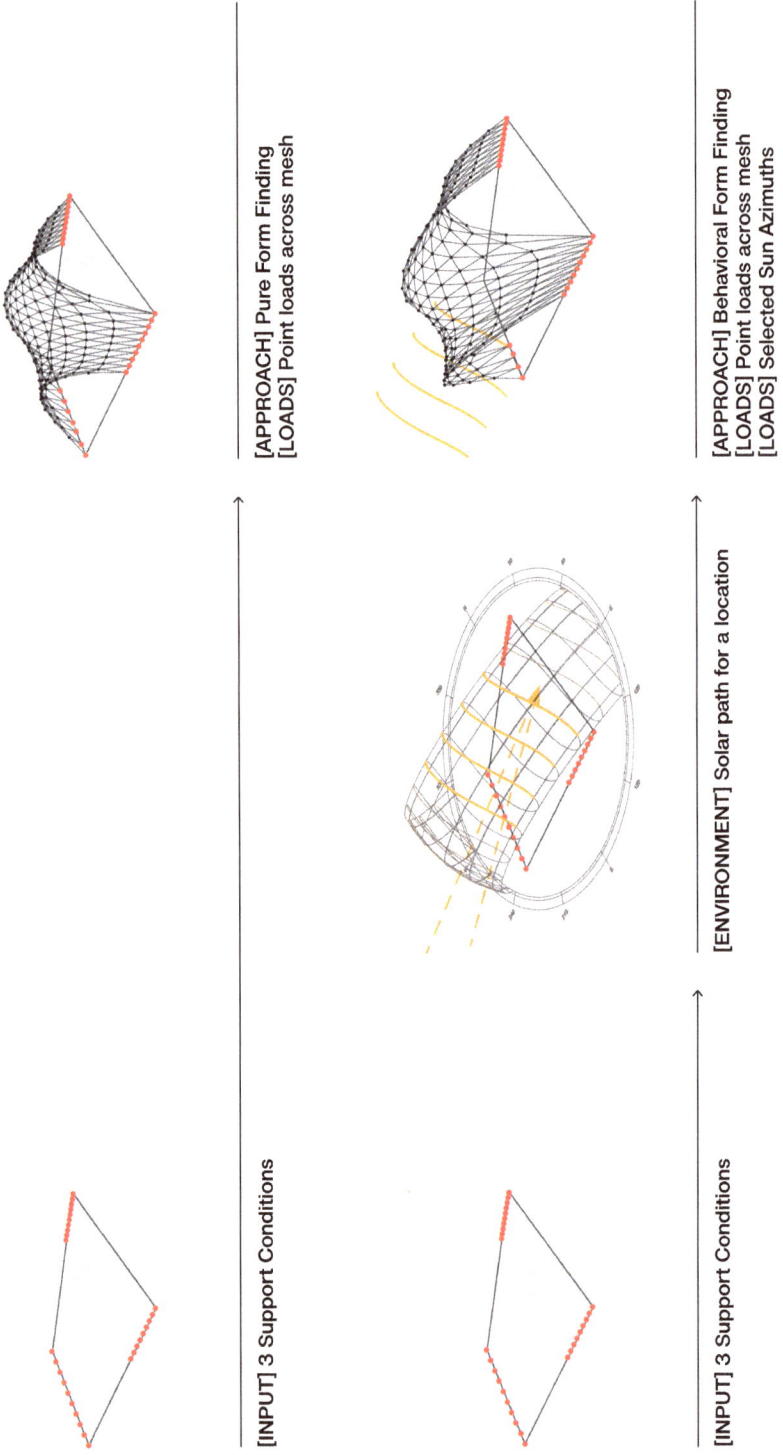

[INPUT] 3 Support Conditions

[APPROACH] Pure Form Finding
[LOADS] Point loads across mesh

[INPUT] 3 Support Conditions

[ENVIRONMENT] Solar path for a location

[APPROACH] Behavioral Form Finding
[LOADS] Point loads across mesh
[LOADS] Selected Sun Azimuths

Figure 4.53: Comparison of typical form-finding and agent based form-finding methods.

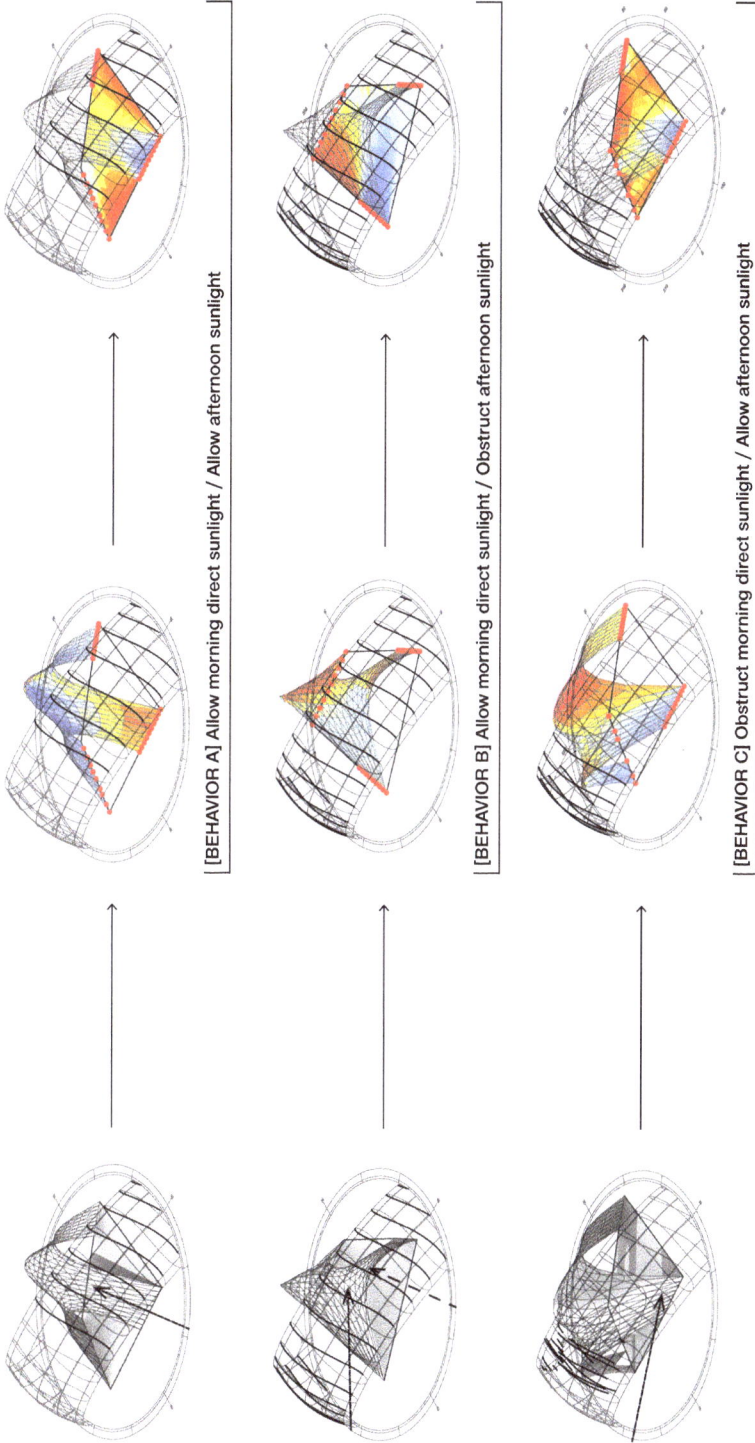

Figure 4.54: Inputs used for agent-based form-finding and examples of different behaviors.

Environmental Aware Form Finding: Context + Forces + Solar Path + Analysis

Case 1: Four peripheral support conditions (50m span)

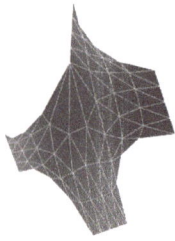

(i) Form Found Geometry

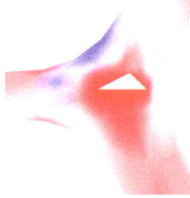

(ii) Stress Analysis

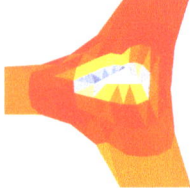

(iii) Daylight Analysis on Shell

(iv) Daylight Analysis Beneath Shell

Case 2: Four peripheral support conditions one hole (50m span)

(i) Form Found Geometry

(ii) Stress Analysis

(iii) Daylight Analysis on Shell

(iv) Daylight Analysis Beneath Shell

Case 3: Three peripheral & one central support (50m span)

(i) Form Found Geometry

(ii) Stress Analysis

(iii) Daylight Analysis on Shell

(iv) Daylight Analysis Beneath Shell

Figure 4.55: Different types of analysis used to inform the environmental behavior.

figurations, and on a global level, to use the MAS framework to explore geometries that were informed by the position of the Sun at specific times of the year.

4.3.4.1 Design process

The design process for this case study went through four phases of design (Figure 4.56). The first phase included the definition of typical parameters in structural design: F, the outline of the provided footprint in the form of a polyline; P, the topology of the network; $S(n,t)$ the number and type of support conditions; E, the material stiffness; G, the material weight; M, the maximum number of agents; and L, loads. Additionally, the designer provided: LL, the longitude and latitude of the site location; O, the orientation (N, S, E, W); weather data (.epw file); and EB, the environmental behavior (i.e., providing shadow beneath the structure in the morning). The steps are graphically illustrated in Figures 4.57 and 4.58.

In the second phase, based on the provided footprint and topology, generative agents were created, whose behavior was informed by a particle spring system. The behavior of the agents is further extended to account for environmental parameters (i.e., Sun's positions) extracted from the weather file input. At each time step, the position of the agents was updated based on the summation of forces that act upon it, including gravity, stiffness and Sun's attraction/repulsion. Introducing specific positions of the Sun as "virtual forces" directly linked the environmental parameters with form-finding. In this phase, the designer specified the duration of the form-finding process and the value force of the Sun, which relates to the environmental behavior. Once a global equilibrium was reached, a shell geometry was generated based on the optimal force distribution.

In the third phase, the generated shell geometry was exported and passed for environmental and structural analysis. The shell was analyzed both as a single mesh and a structural system with discrete elements – in this case, reciprocal frames. At each generation, the design parameters that affected the daylight analysis, such as the Sun force and position of supports, were updated based on the daylight analysis values. The designer was able to interactively update the parameters in order to steer the form based on her/his intentions.

In the last phase, once the desired daylight performance was reached, the topology and size of the structural members were updated in order to satisfy the structural requirements (Figure 4.57). Two types of specialist agents were created for each of the analyses performed; they communicate with the generative (form-finding) agent, namely the structural analysis agent (SAa) that accounts for displacement, and a DFAa that accounts for sunlight on and beneath the shell structure. The local analysis of different types of reciprocal frames provided the designer with insight about how the design parameters of the selected structural system (i.e., the number, size and geometry of the structural member) affected its structural behavior. The designer specified a utility, u, for each of the analyses performed; this defined the most important agent type, and the designer also added a weighting factor to the agents' behaviors. The weighting factor was calculated as a percentage over all the analysis types performed and depended on the performance target that the designer set. All the above aspects affected the amount

Worflow Diagram of Experimental Design

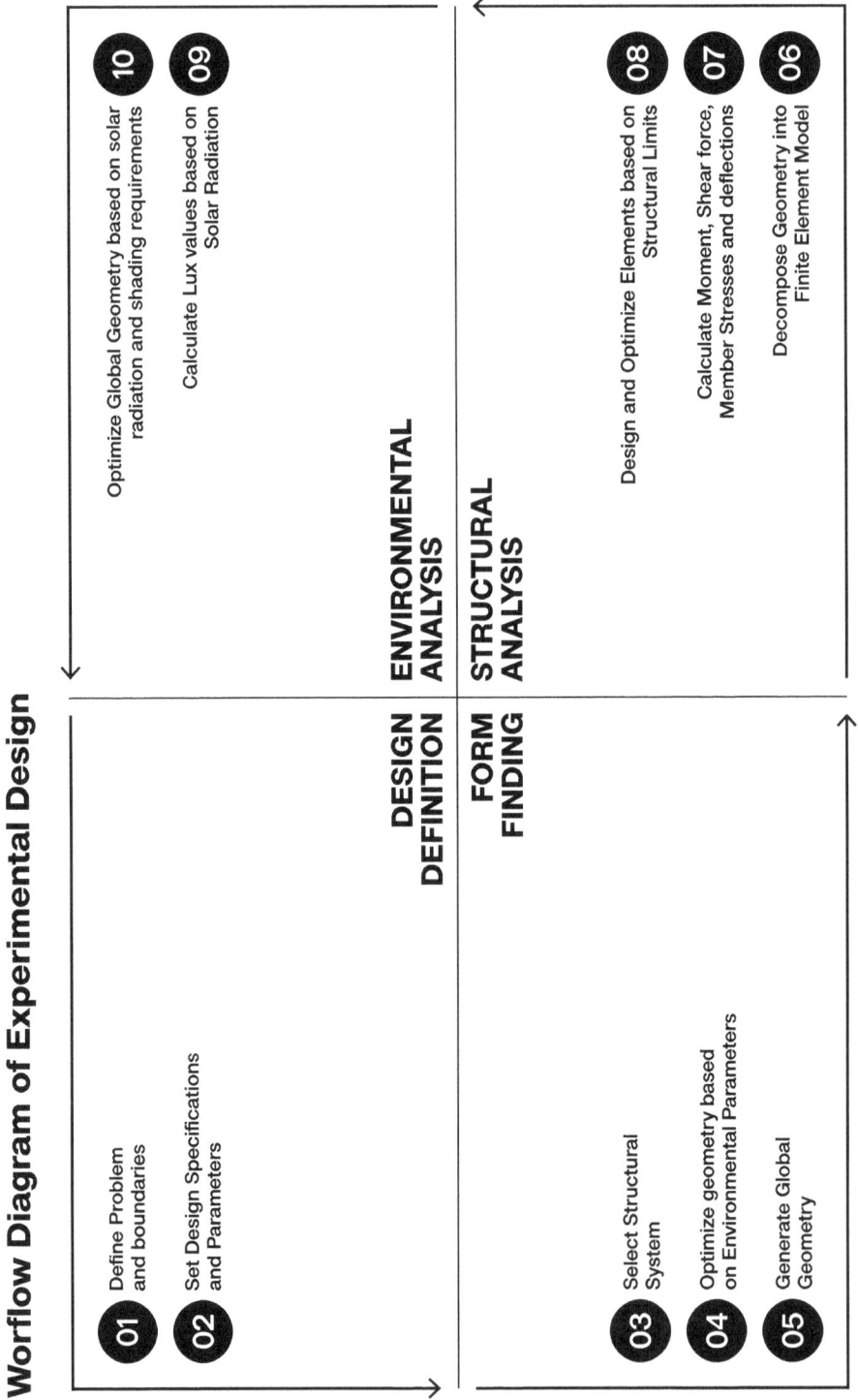

DESIGN DEFINITION

01 Define Problem and boundaries

02 Set Design Specifications and Parameters

ENVIRONMENTAL ANALYSIS

10 Optimize Global Geometry based on solar radiation and shading requirements

09 Calculate Lux values based on Solar Radiation

FORM FINDING

03 Select Structural System

04 Optimize geometry based on Environmental Parameters

05 Generate Global Geometry

STRUCTURAL ANALYSIS

08 Design and Optimize Elements based on Structural Limits

07 Calculate Moment, Shear force, Member Stresses and deflections

06 Decompose Geometry into Finite Element Model

Figure 4.56: Workflow diagram of experimental design.

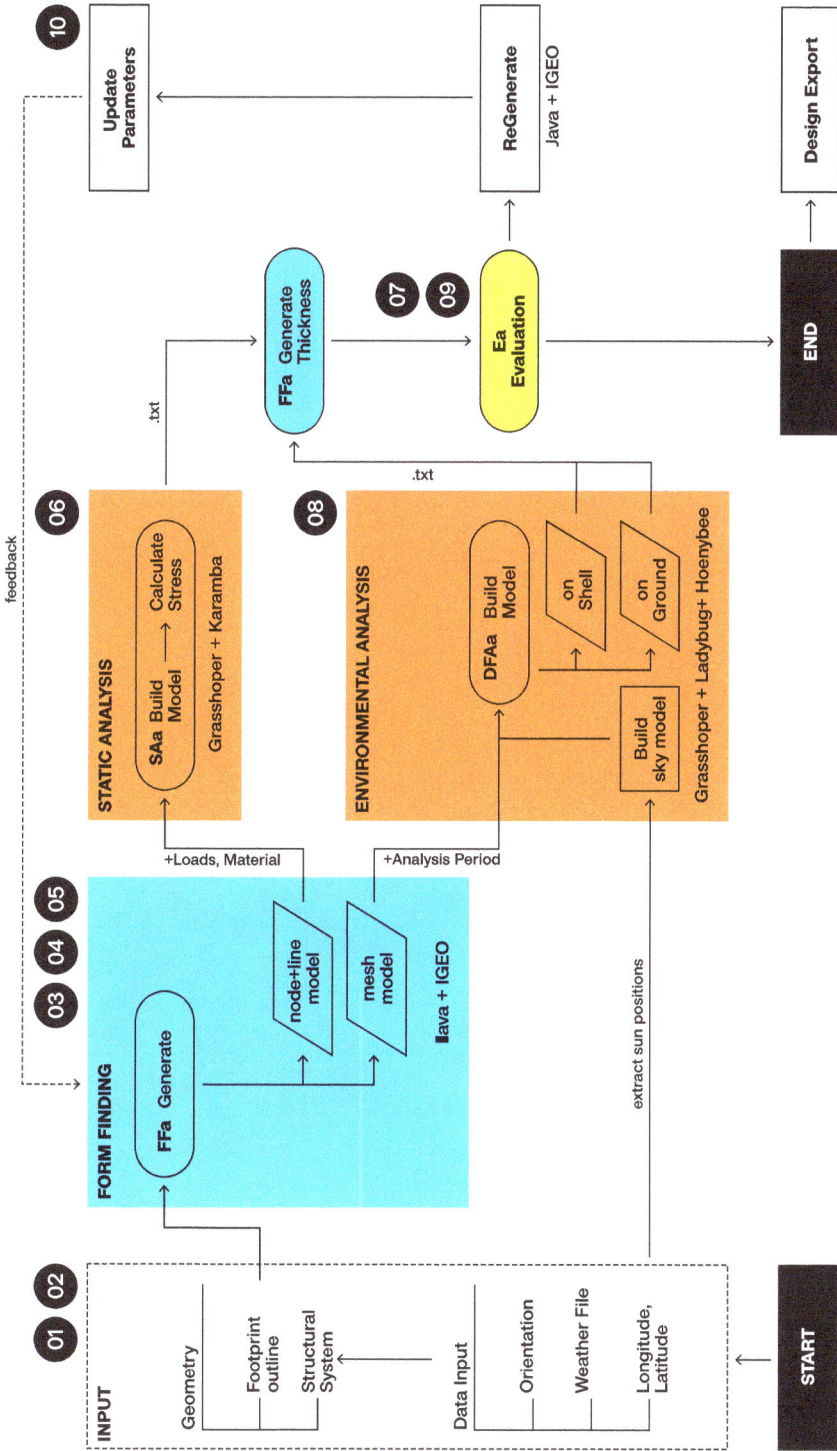

Figure 4.57: Flowchart diagram of experimental design.

Experimental Design Steps

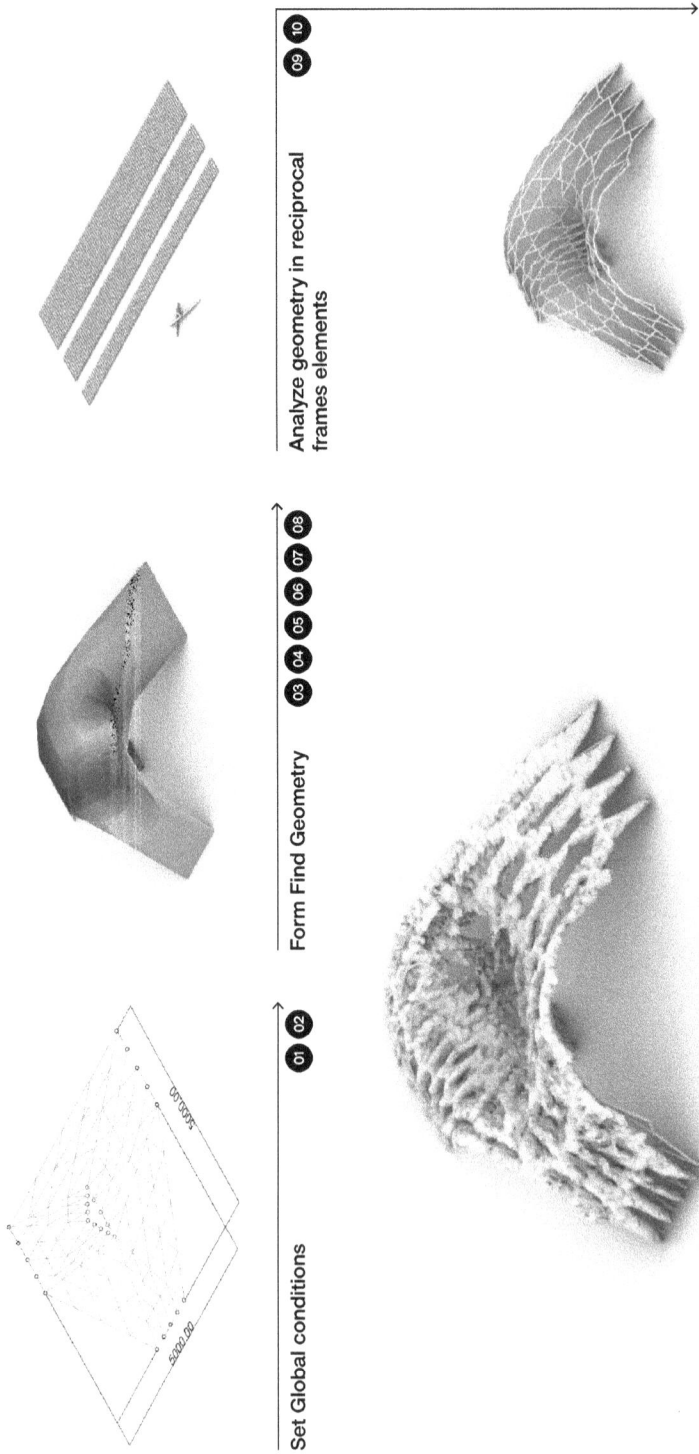

01 02 Set Global conditions

03 04 05 06 07 08 Form Find Geometry

09 10 Analyze geometry in reciprocal frames elements

Add fiber reinforced concrete on top of the reciprocal frame structure (possibility to reuse the form work after material sets)

Assemble a reciprocal frame structure to be used as form work (multiple configurations)

Figure 4.58: Illustration of the design steps of the case study.

Reciprocal Frames – Design Parameters

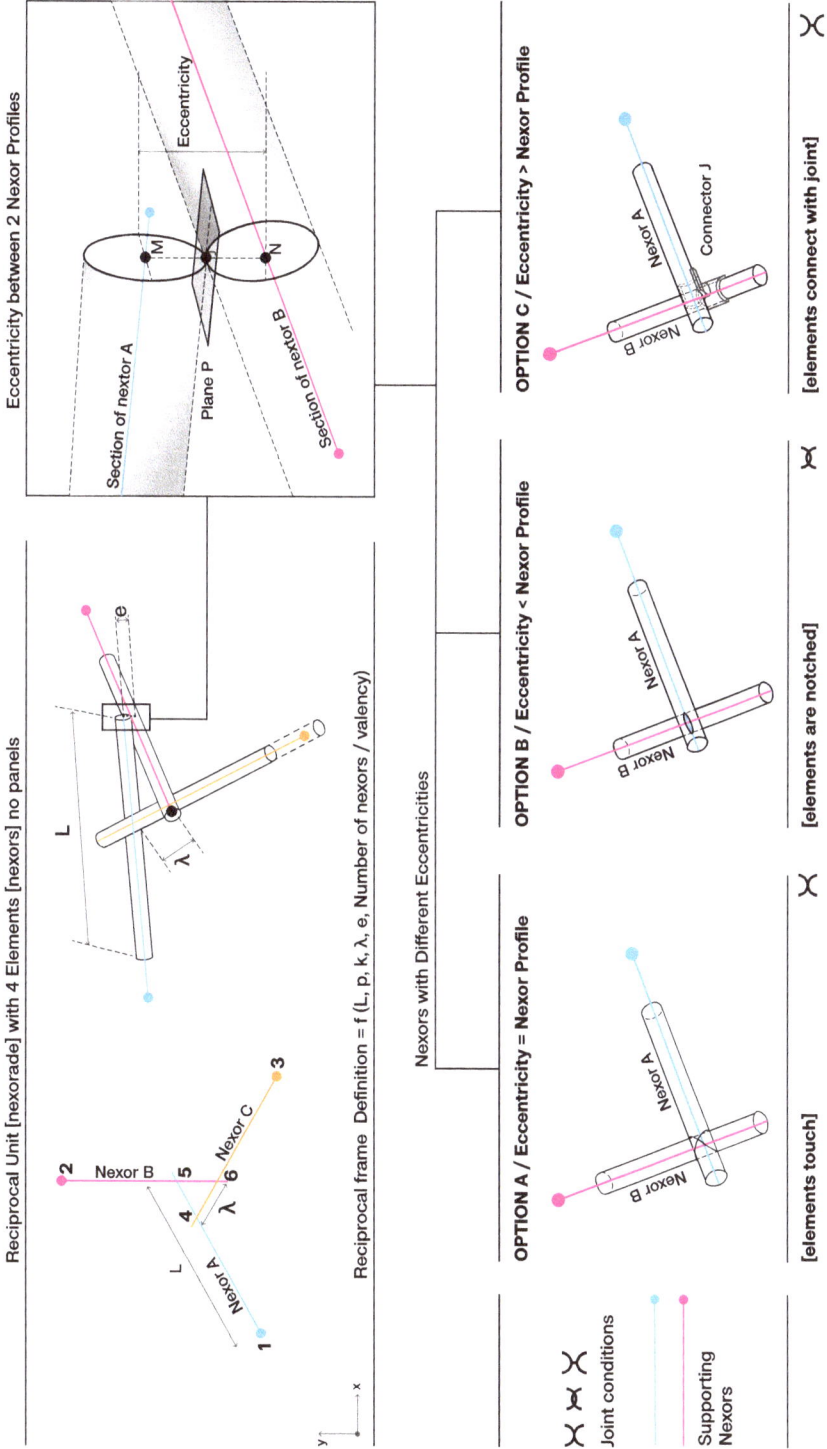

Reciprocal Unit [nexorade] with 4 Elements [nexors] no panels

Eccentricity between 2 Nexor Profiles

Reciprocal frame Definition = f (L, p, k, λ, e, Number of nexors / valency)

Nexors with Different Eccentricities

OPTION A / Eccentricity = Nexor Profile

[elements touch]

OPTION B / Eccentricity < Nexor Profile

[elements are notched]

OPTION C / Eccentricity > Nexor Profile

[elements connect with joint]

Joint conditions

Supporting Nexors

Figure 4.59: The basic design parameters of reciprocal frames with n>=3 nexors.

of sunlight (measured in lux) projected on and beneath the surface, as well as the maximum displacement of the shape (measured in millimeters).

4.3.4.2 Results and analysis

So far, the results of this work are on both a local and a global level. On the local level, different section profiles that range from standard to nonstandard were tested and it was observed that the structural performance of each reciprocal element is largely affected by its cross-sectional profile. Six different types of profiles were tested, including a circular pipe, a rectangular element and four types of planar elements with different proportions of width over height with one transformation (torsion along main axis and offset along the main axis). Figures 4.60–4.62 show a plot of the stresses on the reciprocal elements with different geometries in relation to varying lengths and thicknesses.

Only one type of element (twisted planar) fails to meet the requirements for allowable stress based on the American Building Code, while the rest of the profile types show an almost linear relationship between their length and thickness to the stress. When the length increases, so does the stress, but when the profile increases, the stresses on the element decrease. In terms of the unit, we can see in Figure 4.60 that when the number of elements increase (valency), the stress decreases.

For a unit (reciprocal frame) of $n = 3$ and a pipe profile, the stresses can be as high as 1,620 kN, while for $n = 6$ they drop as low as 703 kN (66%). At this stage, for the sake of simplicity and clarity, only one profile type and one reciprocal frame unit ($n = 3$) are tested and applied on global geometry. In terms of global geometry, eight different topological configurations were generated as shown in Figures 4.63 and 4.64, based on the support conditions and whether or not the geometry had an opening.

From these initial configurations, six were selected and the form-found as well as the environmentally influenced geometries were analyzed and compared. The following measures were observed: compression, stress, moments, displacement, cross-sectional utilization and solar radiation on and below the shell (see Figures 4.65–4.71). The utilization of the cross-section is the ratio of applied stress over the yield strength of the element; using a safety factor of 200%, the elements were sized in order to achieve a target utilization between −50% and +50%.

Additionally, the maximum displacement was measured and found to be within limits (0.5–5 cm). The structural analysis shows that geometries with a hole had a significant increase in moments (130–200% increase); it was observed that the amount of support conditions did not significantly affect the resulting principal stresses but rather their position and shape. The environmentally informed shells were subject to higher principal forces as their form was modified from the pure form-finding; however, they performed better in the environmental analysis (see Figures 4.72 and 4.73). By better, we mean that the shell where environmental behavior was applied achieved results that were intended by the designer (i.e., increase daylight availability on a specific part of the structure).

This case study demonstrates how the MAS framework presented can be applied to designing form-found shell structures that consider environmental parameters.

Local Geometry: Morphology and Structural Analysis of Reciprocal Frame units

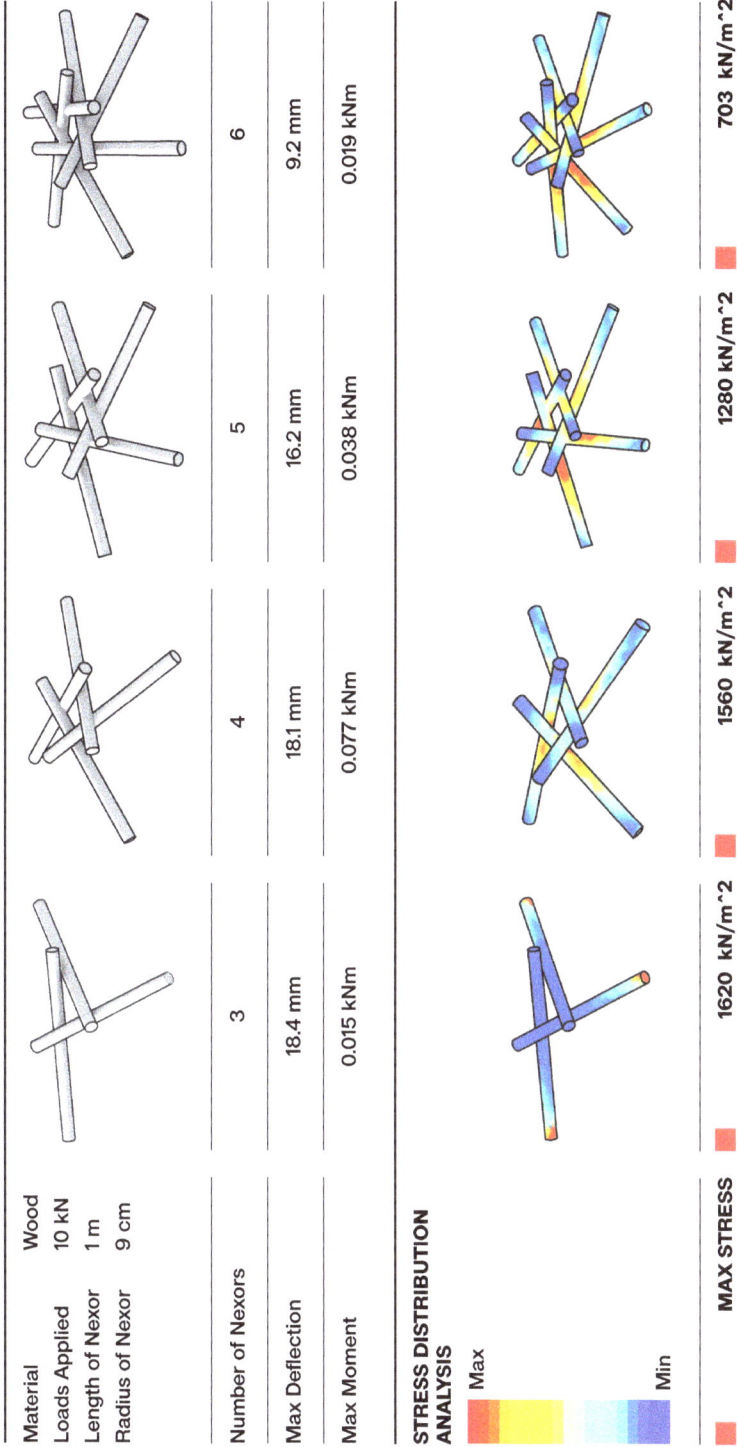

Material	Wood
Loads Applied	10 kN
Length of Nexor	1 m
Radius of Nexor	9 cm

Number of Nexors	3	4	5	6
Max Deflection	18.4 mm	18.1 mm	16.2 mm	9.2 mm
Max Moment	0.015 kNm	0.077 kNm	0.038 kNm	0.019 kNm

STRESS DISTRIBUTION ANALYSIS

Max

Min

MAX STRESS	1620 kN/m^2	1560 kN/m^2	1280 kN/m^2	703 kN/m^2

Figure 4.60: Nexorade with different valences and the corresponding stress distribution diagrams.

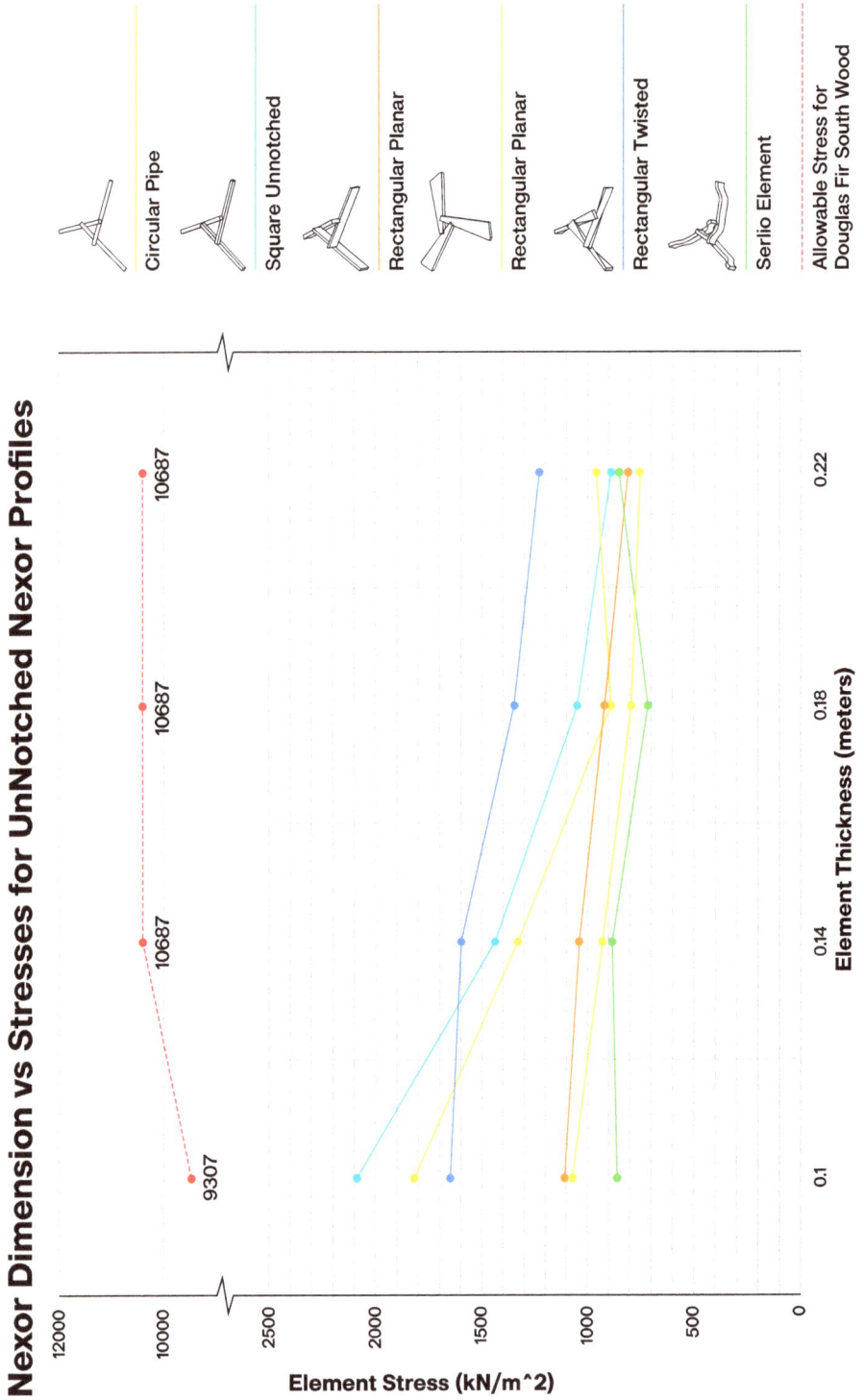

Figure 4.61: Plot of stress distribution with regard to nexor thickness for unnotched profiles.

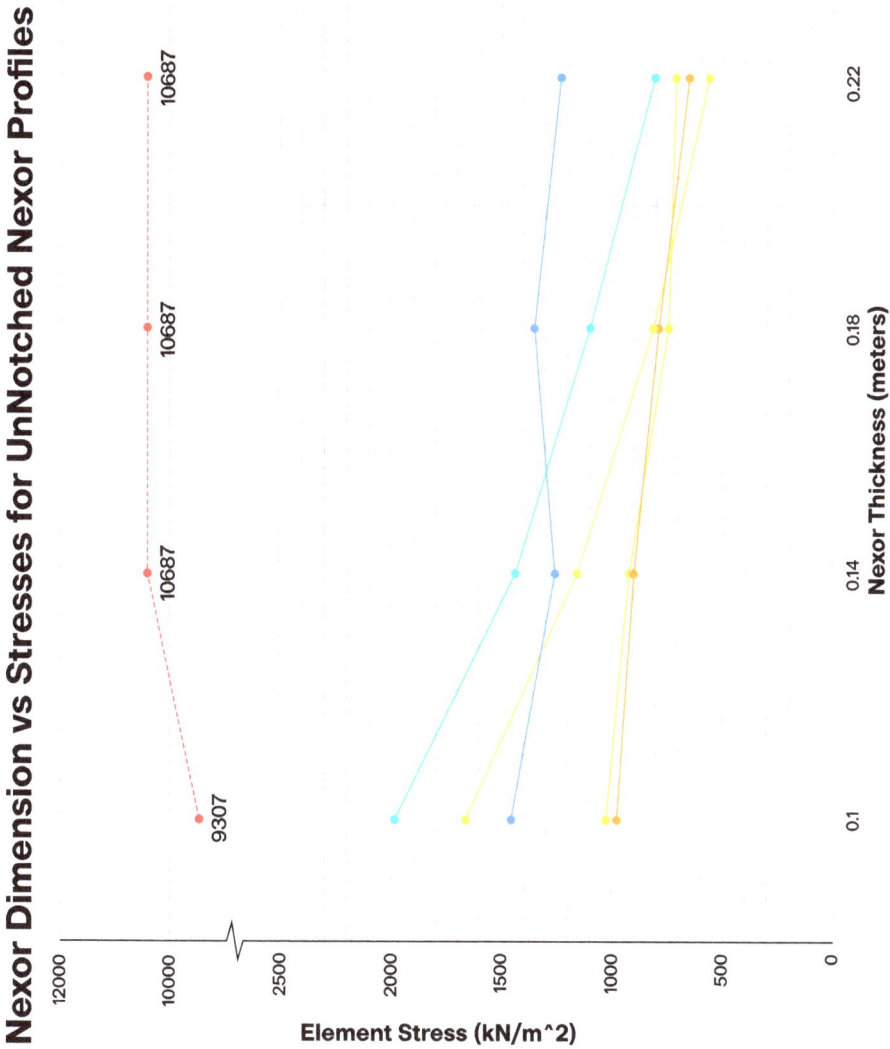

Figure 4.62: Plot of stress distribution with regard to nexor thickness for notched profiles.

Exploration of different topological configurations

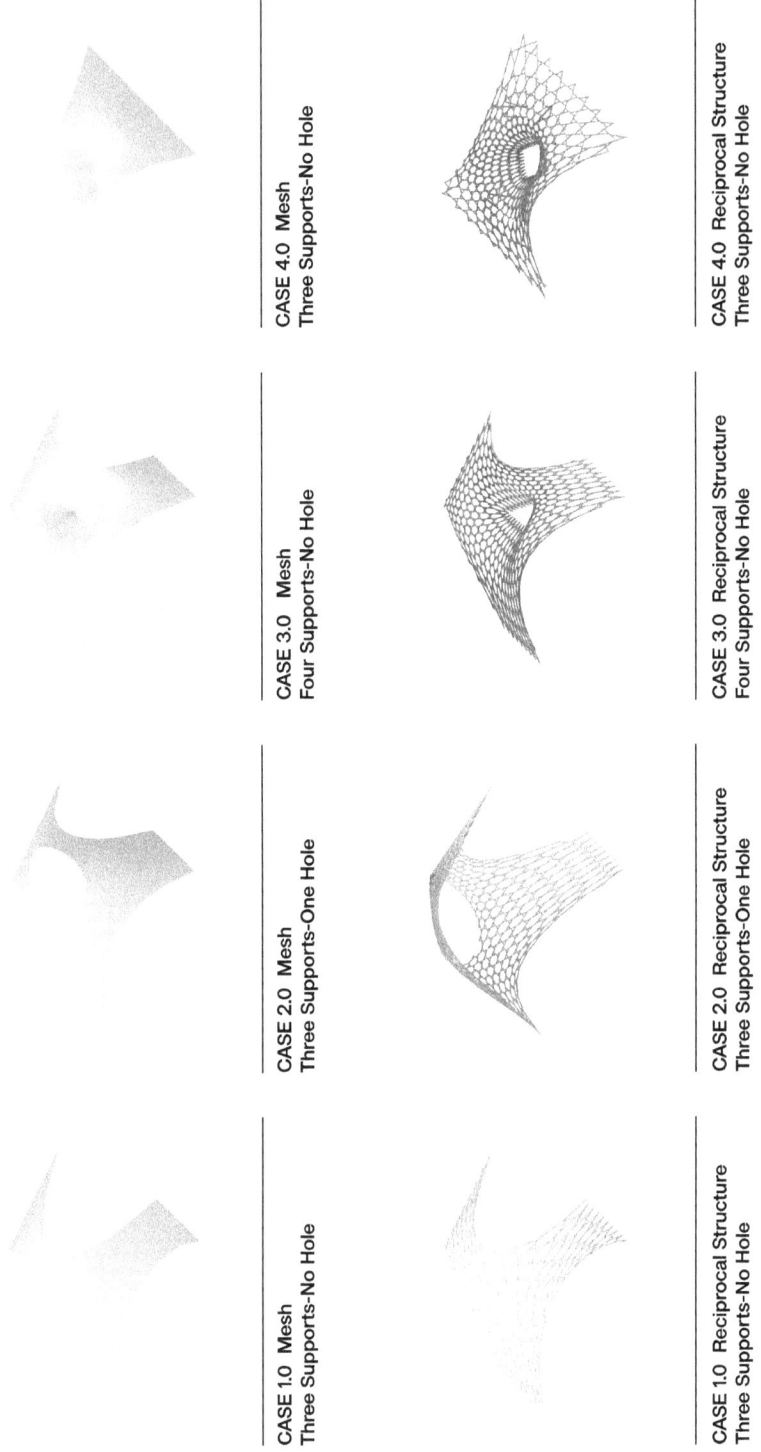

CASE 4.0 Mesh
Three Supports-No Hole

CASE 4.0 Reciprocal Structure
Three Supports-No Hole

CASE 3.0 Mesh
Four Supports-No Hole

CASE 3.0 Reciprocal Structure
Four Supports-No Hole

CASE 2.0 Mesh
Three Supports-One Hole

CASE 2.0 Reciprocal Structure
Three Supports-One Hole

CASE 1.0 Mesh
Three Supports-No Hole

CASE 1.0 Reciprocal Structure
Three Supports-No Hole

Figure 4.63: A subset of form-found shells and their analysis into reciprocal frames (set A).

Exploration of different topological configurations

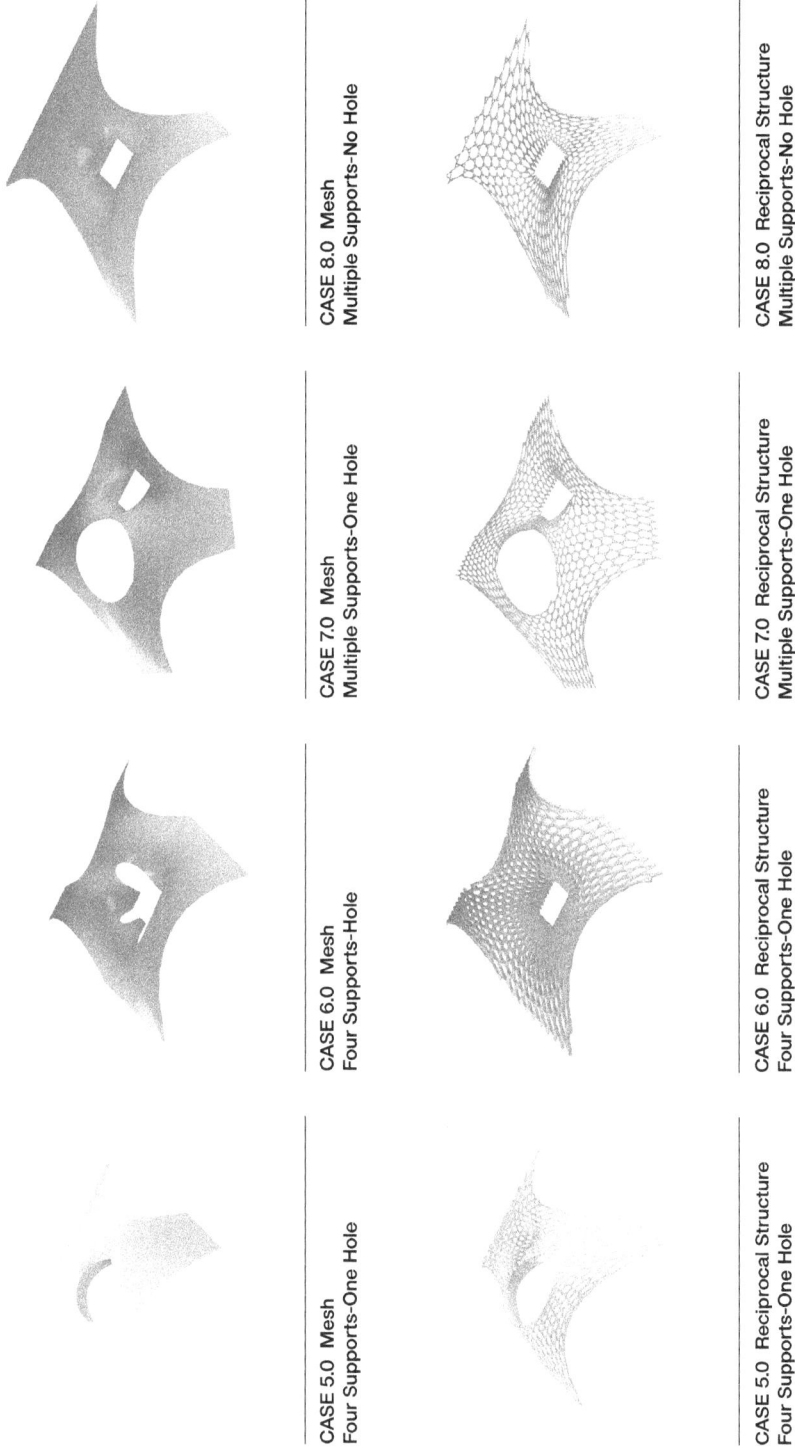

CASE 8.0 Mesh
Multiple Supports-No Hole

CASE 8.0 Reciprocal Structure
Multiple Supports-No Hole

CASE 7.0 Mesh
Multiple Supports-One Hole

CASE 7.0 Reciprocal Structure
Multiple Supports-One Hole

CASE 6.0 Mesh
Four Supports-Hole

CASE 6.0 Reciprocal Structure
Four Supports-One Hole

CASE 5.0 Mesh
Four Supports-One Hole

CASE 5.0 Reciprocal Structure
Four Supports-One Hole

Figure 4.64: A subset of form-found shells and their analysis into reciprocal frames (set B).

Form Found Geometries with Environmental Behavior

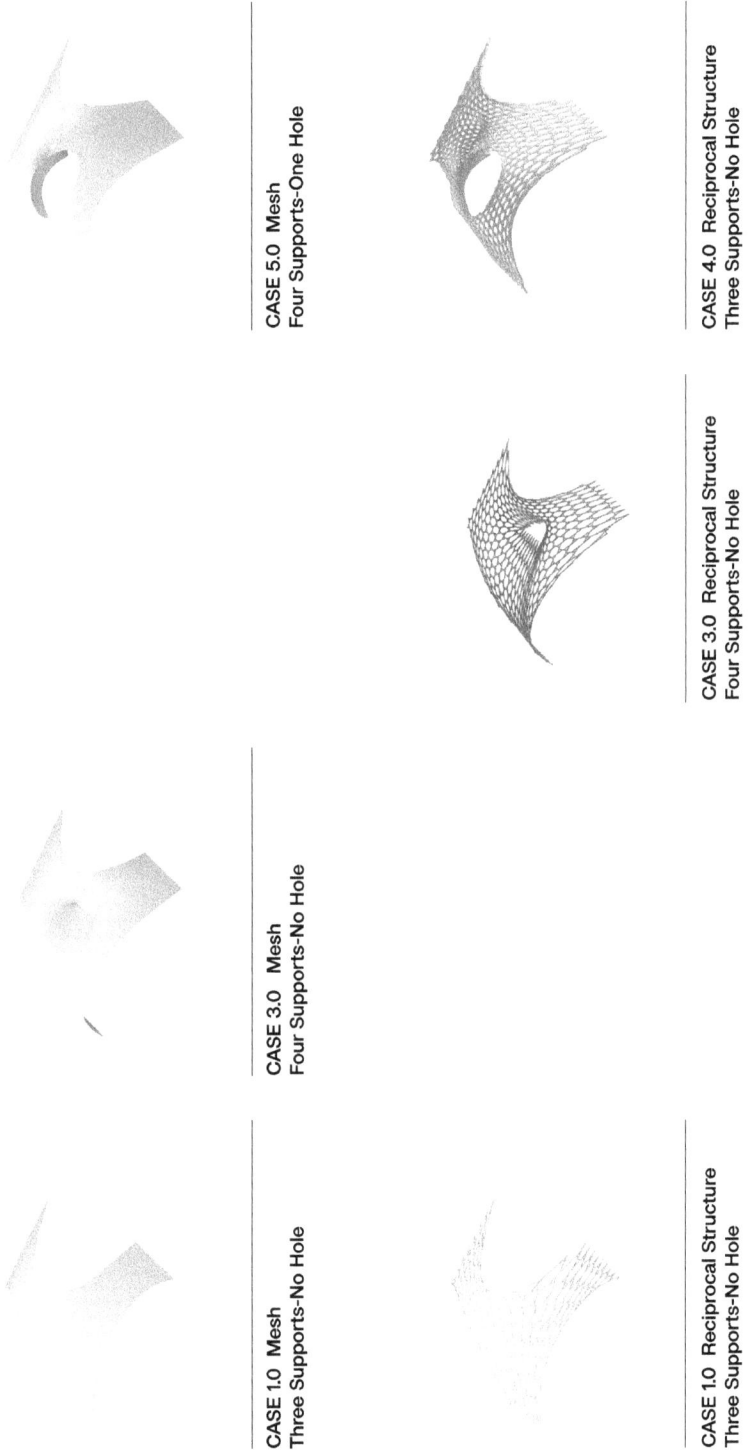

CASE 1.0 Mesh
Three Supports–No Hole

CASE 3.0 Mesh
Four Supports–No Hole

CASE 5.0 Mesh
Four Supports–One Hole

CASE 1.0 Reciprocal Structure
Three Supports–No Hole

CASE 3.0 Reciprocal Structure
Four Supports–No Hole

CASE 4.0 Reciprocal Structure
Three Supports–No Hole

Figure 4.65: Agent based form-found shells and their analysis into reciprocal frames.

The simulation results showed improvements in energy efficiency by form-finding shells whose global geometry increased the amount of available daylight. In terms of the design process, we provided the designer with a methodology to describe a structural system as an agent class and offered an approach to steer away from pure form-finding by integrating environmental parameters as behaviors. This approach was extended by implementing an agent class that enables form-finding and allows the creation of behaviors based on environmental parameters.

The experimental results are presented in the form of two scales and investigations: (1) on the local level from studying the critical design parameters of the reciprocal frame morphology (valency) and the cross-sectional and element lengths; and (2) on the global level, for the design of a shell structure. The geometric configurations of the design solutions vary, but the coupling of the generative with the analytical and optimizing processes in an agent-based logic ensured the satisfaction of the prescribed goals, even if the design setup changed.

The objective has remained: to demonstrate the value of the emergent, nonstandard and geometrically intricate structures as a viable post-fordist solution for form-finding. The MAS framework was able to generate a shell as a reciprocal frame configuration(s), which provided spans longer than the short self-similar structural elements. The experiments have demonstrated that architects and designers can benefit by implementing an integrated bottom-up approach to the design of building components that lead to optimal global configurations. By combining analytical methods with agent-based modeling, and by properly formulating and passing analytical data automatically to a generative process, the MAS system provides designers with a larger pool of complex, yet well-performing design solutions that could not be modeled manually.

To further emphasize the generative capacity of the system, a number of design experiments are presented, where the same method was applied on different topological configurations, and only the agents have an extra attribute related to additive manufacturing: they are moving on the generated reciprocal structure and adding/removing material based on the solar radiation under the structure and stress values on different points. The process is outlined in Figure 4.58, and the vision is to use reciprocal frames as substrate, where you can deposit the material. Once the deposited material settles, the reciprocal frame can be disassembled and reused.

The design intent with this exploration is to generate shells with nonuniform thickness that have variable porosities controlled by the amount of desired daylight availability. Variable porosities allow intricate lighting effects to happen underneath the structure. In Figures 4.74 and 4.75, a set of generated shell designs are presented and in Figures 4.76–4.79 one topological case is selected and the geometries are shown along the solar radiation analysis. In Figure 4.80, a prototype using additive manufacturing is presented (FDM 3d printing).

Although the designed workflow is not yet fully automated in terms of the generation and transfer of geometrical data for FEA, it lays the foundation for showcasing how behaviors can be used in order to drive design systems away from purely form-found geometries and how structural and environmental analysis can be used in the

Pure Form Finding / Structural Analysis

GLOBAL PARAMETERS

Max span = 50 m
Footprint Area = 2500 sqm
Element Section: Solid Pipe
Material: Wood
Number of reciprocal
elements in each unit: 3

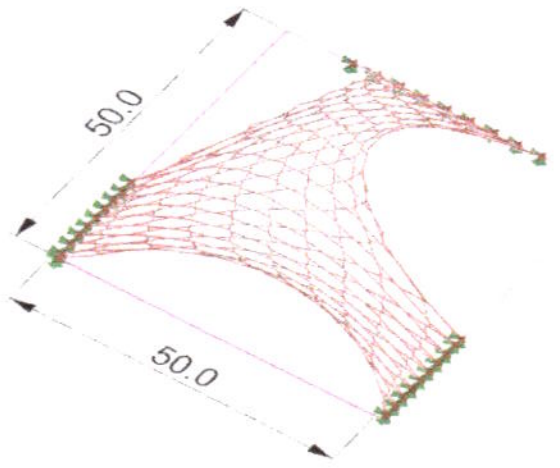

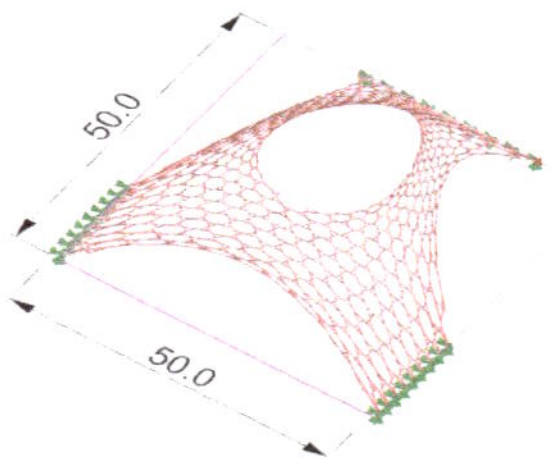

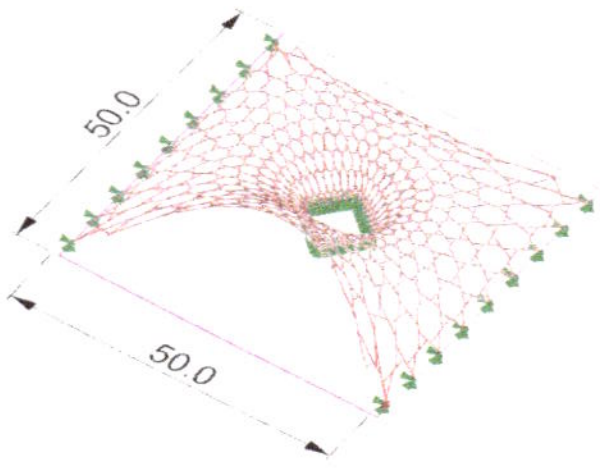

Topolgy / Support Condition	no Hole / 3 supports	1 Hole / 3 supports	1 Hole / 3 supports
Member Dimension	35 cm diameter	35 cm diameter	35 cm diameter
Dead Load	10 kN/sqm	10 kN/sqm	10 kN/sqm
Min & Max Principal Force (kN)	-176.62 to 20.51	-174.96 to 19.53	-166.88 to +41.08
Min & Max. Moment (kNm)	-4.60 to +10.36	-2.51 to +6.79	-5.42 to +12.78
Max Displacement (cm)	0.05	0.015	0.04
Cross Section Utilization	-44.5 to +30.0%	-36.2 to +21.8%	-77.4 to +53.2%

Figure 4.66: Structural analysis of purely form-found shells (set A).

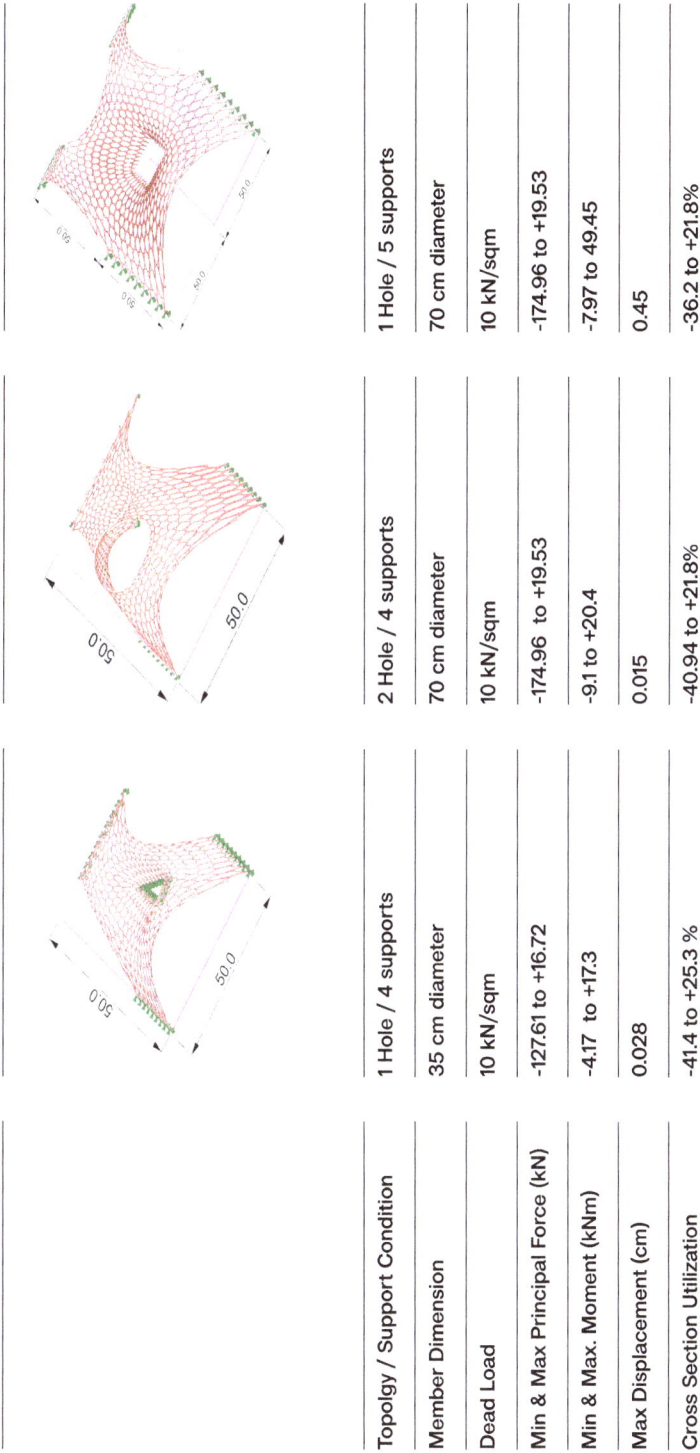

Topolgy / Support Condition	1 Hole / 4 supports	2 Hole / 4 supports	1 Hole / 5 supports
Member Dimension	35 cm diameter	70 cm diameter	70 cm diameter
Dead Load	10 kN/sqm	10 kN/sqm	10 kN/sqm
Min & Max Principal Force (kN)	-127.61 to +16.72	-174.96 to +19.53	-174.96 to +19.53
Min & Max. Moment (kNm)	-4.17 to +17.3	-9.1 to +20.4	-7.97 to 49.45
Max Displacement (cm)	0.028	0.015	0.45
Cross Section Utilization	-41.4 to +25.3 %	-40.94 to +21.8%	-36.2 to +21.8%

Figure 4.67: Structural analysis of purely form-found shells with different topological configurations (set B).

Form Finding with Environmental Behavior / Structural Analysis

GLOBAL PARAMETERS

Max span = 50 m
Footprint Area = 2500 sqm
Element Section: Solid Pipe
Material: Wood
Number of reciprocal
elements in each unit: 3

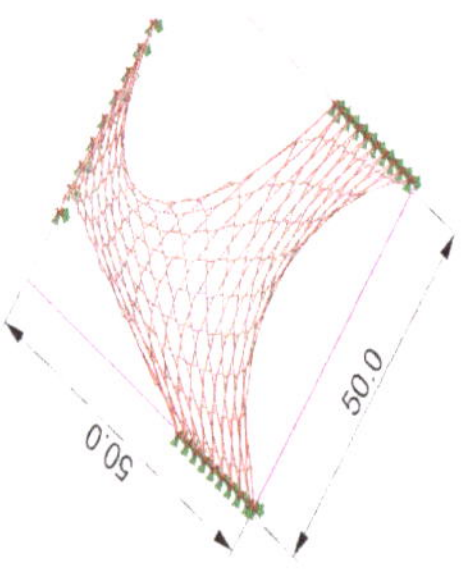

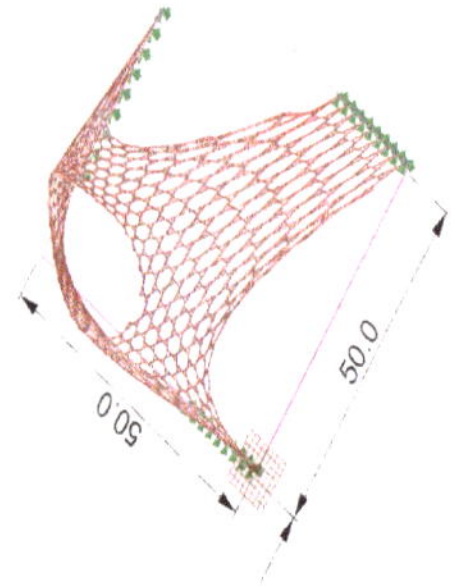

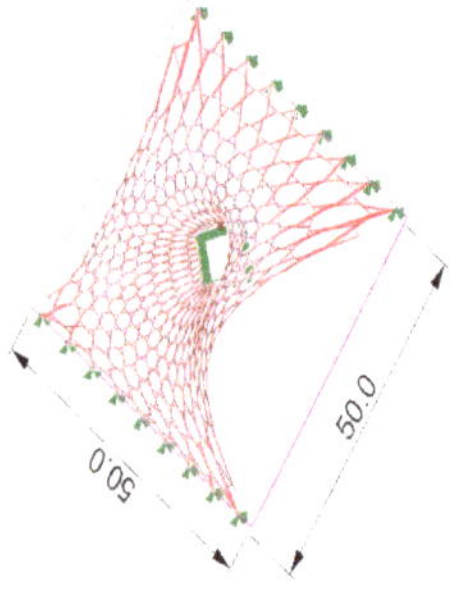

Topolgy / Support Condition	no Hole / 3 supports	1 Hole / 3 supports	1 Hole / 3 supports
Member Dimension	35 cm diameter	70 cm diameter	70 cm diameter
Dead Load	10 kN/sqm	10 kN/sqm	10 kN/sqm
Min & Max Principal Force (kN)	-176.62 to 20.51	-1904.23 to 1729.31	-1159.17 to 292.76
Min & Max. Moment (kNm)	-4.60 to 10.36	-8.51 to 12.79	-91.56 to 160.95
Max Displacement (cm)	0.05	0.015	1.02
Cross Section Utilization	-44.5 to -30.0%	-36.2 to -21.8%	-98.9 to -44.4%

Figure 4.68: Structural analysis of form-found shells with varying agent behaviors (set A).

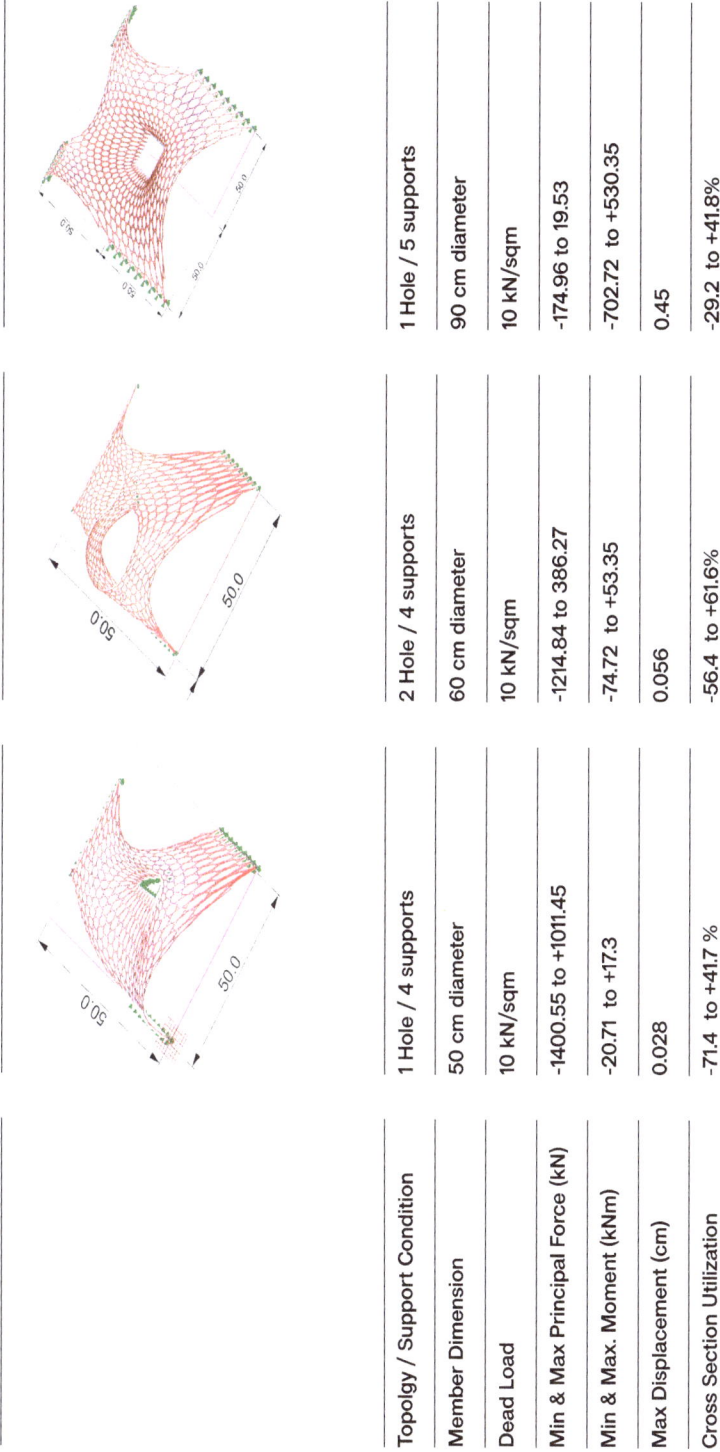

Topolgy / Support Condition	1 Hole / 4 supports	2 Hole / 4 supports	1 Hole / 5 supports
Member Dimension	50 cm diameter	60 cm diameter	90 cm diameter
Dead Load	10 kN/sqm	10 kN/sqm	10 kN/sqm
Min & Max Principal Force (kN)	-1400.55 to +1011.45	-1214.84 to 386.27	-174.96 to 19.53
Min & Max. Moment (kNm)	-20.71 to +17.3	-74.72 to +53.35	-702.72 to +530.35
Max Displacement (cm)	0.028	0.056	0.45
Cross Section Utilization	-71.4 to +41.7 %	-56.4 to +61.6%	-29.2 to +41.8%

Figure 4.69: Structural analysis of form-found shells with varying agent behaviors (set B).

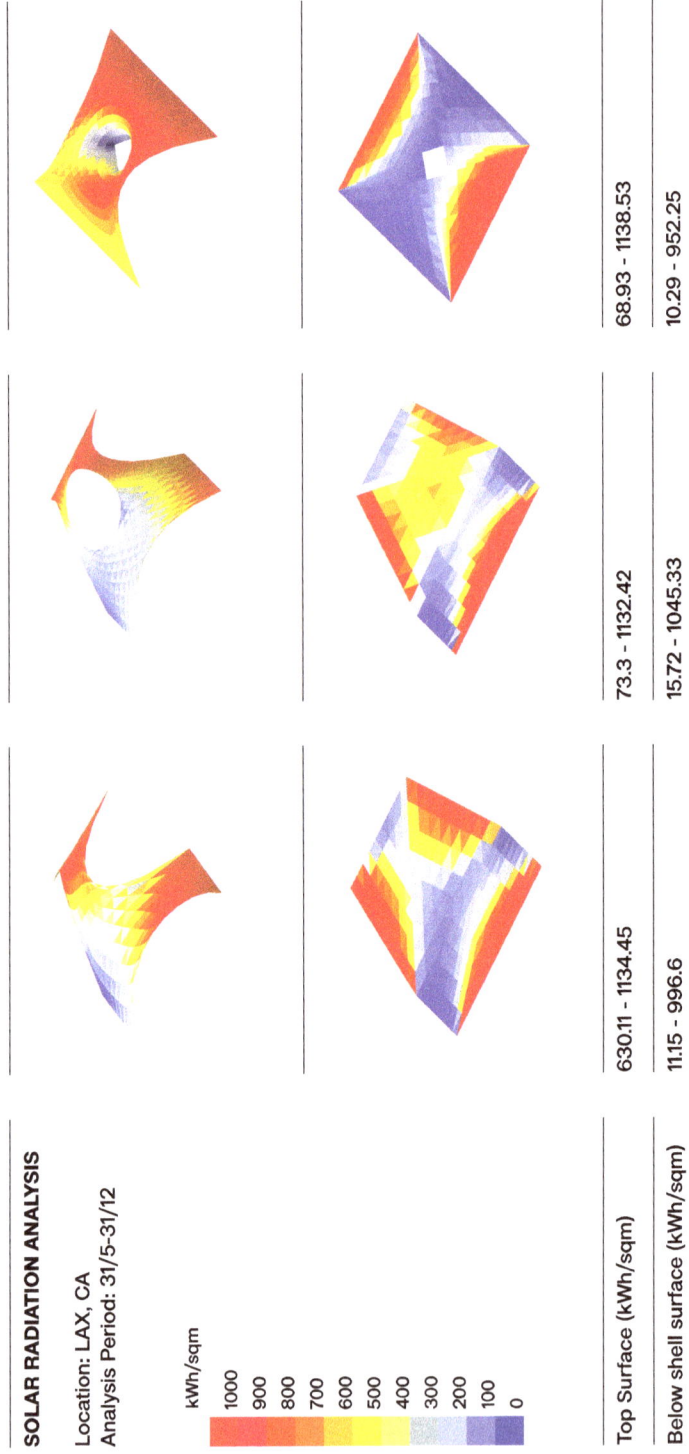

Pure Form Finding / Solar Radiation Analysis

SOLAR RADIATION ANALYSIS

Location: LAX, CA
Analysis Period: 31/5–31/12

kWh/sqm

1000
900
800
700
600
500
400
300
200
100
0

Top Surface (kWh/sqm)	630.11 - 1134.45	73.3 - 1132.42	68.93 - 1138.53
Below shell surface (kWh/sqm)	11.15 - 996.6	15.72 - 1045.33	10.29 - 952.25

Figure 4.70: Solar radiation analysis of purely form-found (set A).

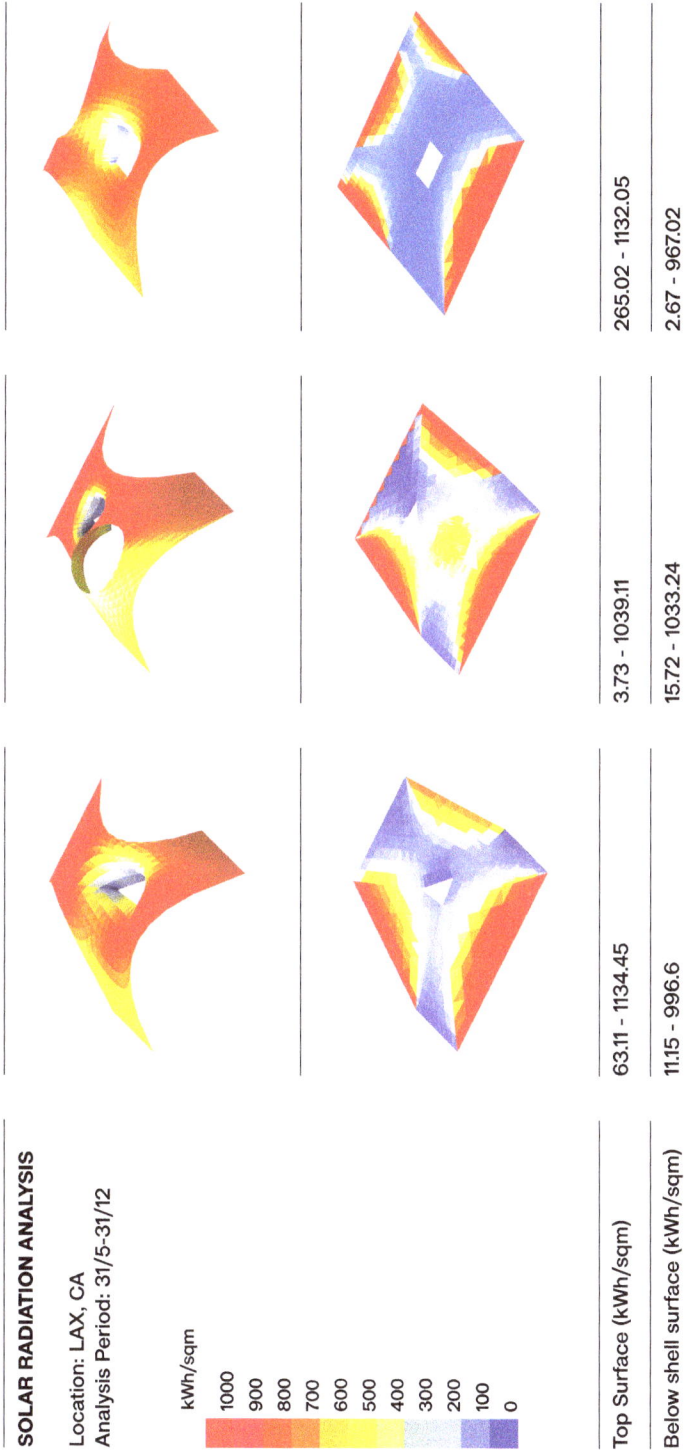

SOLAR RADIATION ANALYSIS

Location: LAX, CA
Analysis Period: 31/5-31/12

kWh/sqm

1000
900
800
700
600
500
400
300
200
100
0

Top Surface (kWh/sqm)

Below shell surface (kWh/sqm)

63.11 - 1134.45

11.15 - 996.6

3.73 - 1039.11

15.72 - 1033.24

265.02 - 1132.05

2.67 - 967.02

Figure 4.71: Solar radiation analysis of purely form-found shells (set B).

Form Finding with Environmental Behavior / Solar Radiation Analysis

SOLAR RADIATION ANALYSIS

Location: LAX, CA
Analysis Period: 31/5–31/12

kWh/sqm

1000
900
800
700
600
500
400
300
200
100
0

Top Surface (kWh/sqm)	34 - 1119.1	51.11 - 1127.33	51.11 - 1127.33
Below shell surface (kWh/sqm)	6.26 - 977.7	13.58 - 945.1	13.58 - 925.1

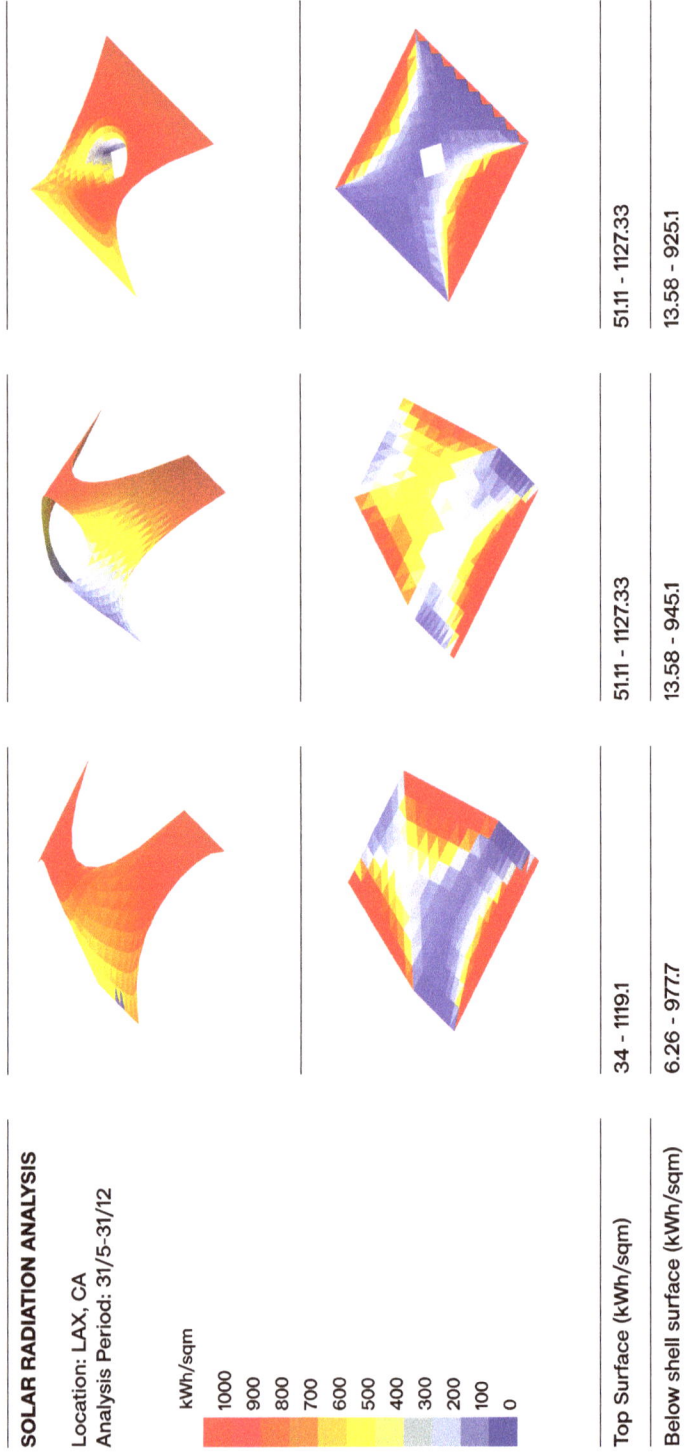

Figure 4.72: Solar radiation analysis of form-found shells with varying agent behaviors (set A).

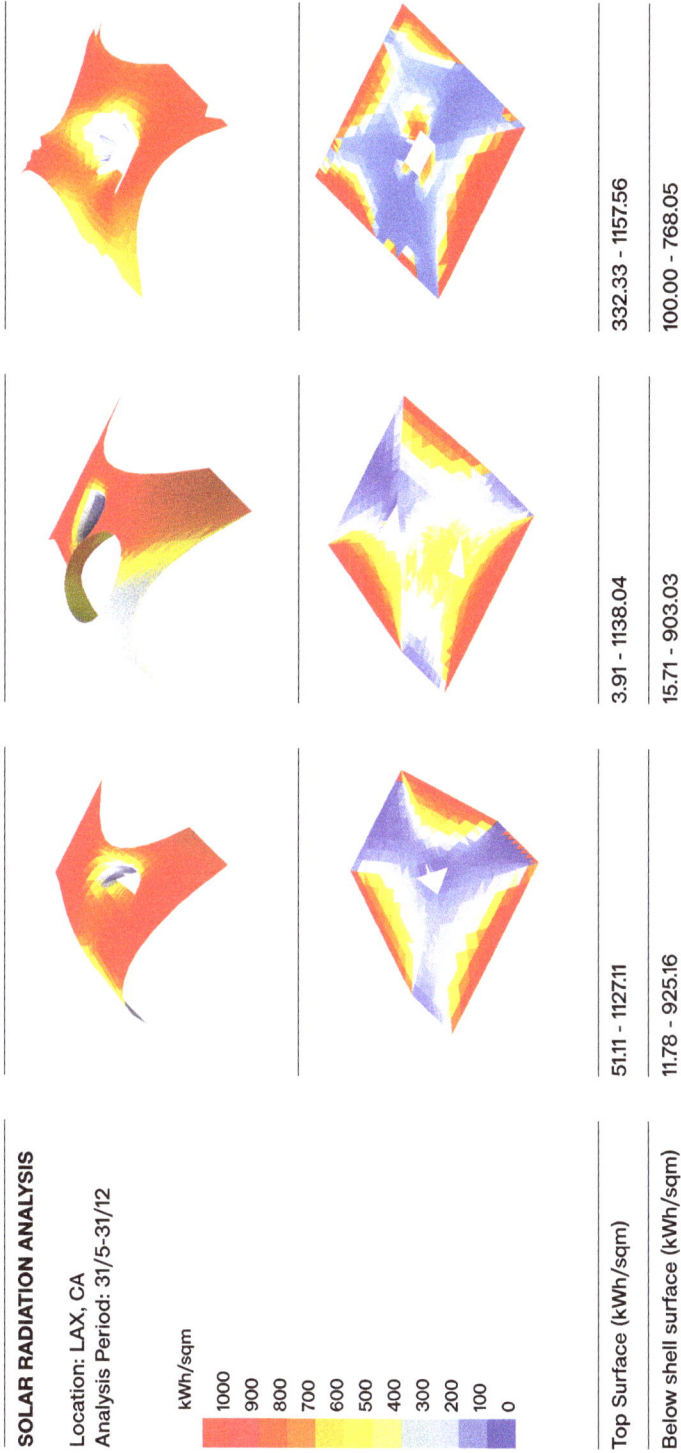

Figure 4.73: Solar radiation analysis of form-found shells with varying agent behaviors (set B).

Generation of geometries with variable porosity

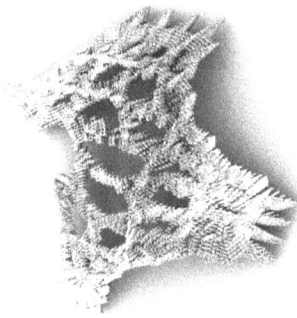

Case 1_A:
1ad0a1225e3cf5ae11f9a42464d8fd67

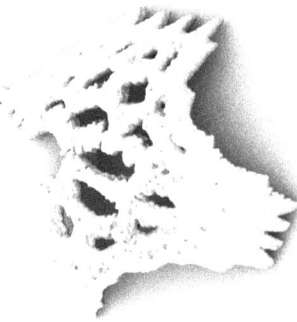

Case 1_B
1cd0a1225e3cf5ae11f9a42464d8fd69

Case 1_C:
1bd0a1225e3cf5ae11f9a42464d8fd68

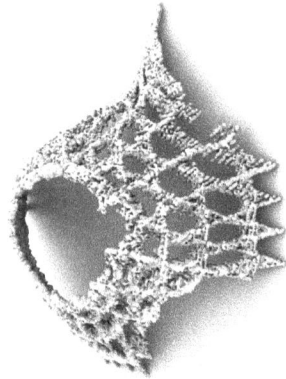

Case 2_A:
d53cdabfd877b1b299d90b592d24dc77

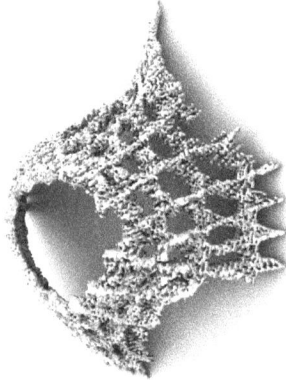

Case 2_B:
2b2cdabfd877b1b299d90b592d24db78

Case 2_C:
2c3cdabfd877b1b299d90b592d24dc79

Figure 4.74: Behavioral form-found geometries (set A) with agent-based thickness.

Case 3_C:
3cee02250cf5cc87edc3de8b802cd589

Case 4_C:
4cde03360cf5aa87edc3de8b802cd999

Case 3_B:
3bee03350cf3bb87edc3de8b802cd588

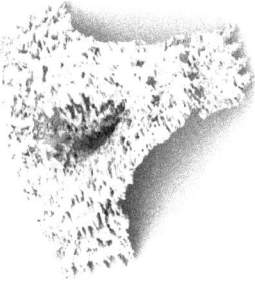

Case 4_B:
4bdee03360cf5aa87edc3de8b802cd998

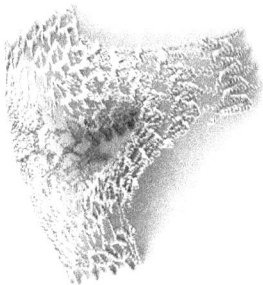

Case 3_A:
3bee03350cf5aa87edc3de8b802cd587

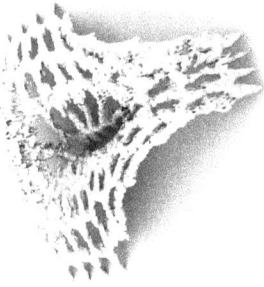

Case 4_A:
4ade03360cf5aa87edc3de8b802cd697

Figure 4.75: Behavioral form-found geometries (set B) with agent-based thickness.

Evaluation of generated geometries with variable porosity

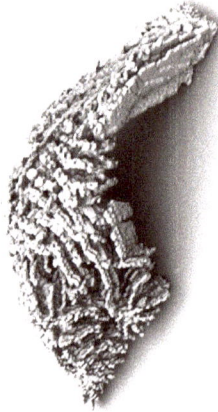

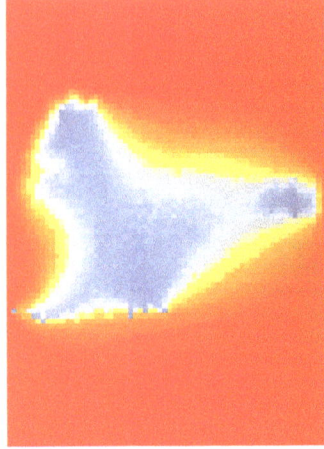

case 4_E
Radiation Analysis Score = 7.25/10
Material Volume: 547 m³

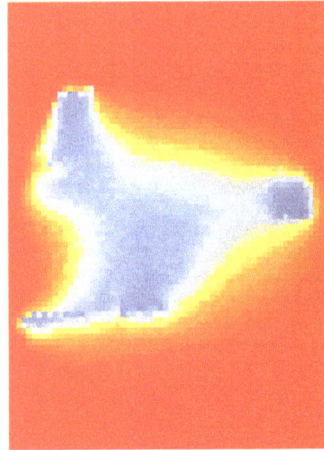

case 4_D
Radiation Analysis Score= 7.7/10
Material Volume: 674 m³

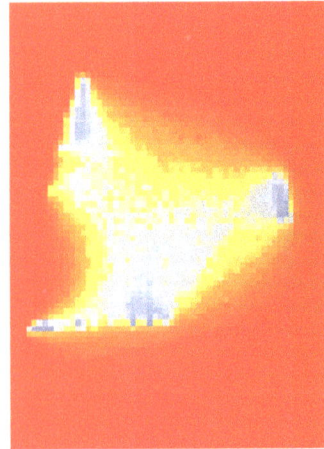

case 4_C
Radiation Analysis Score=6.7/10
Material Volume: 192 m³

Figure 4.76: Behavioral form-found geometries (set A) and the solar radiation analysis underneath.

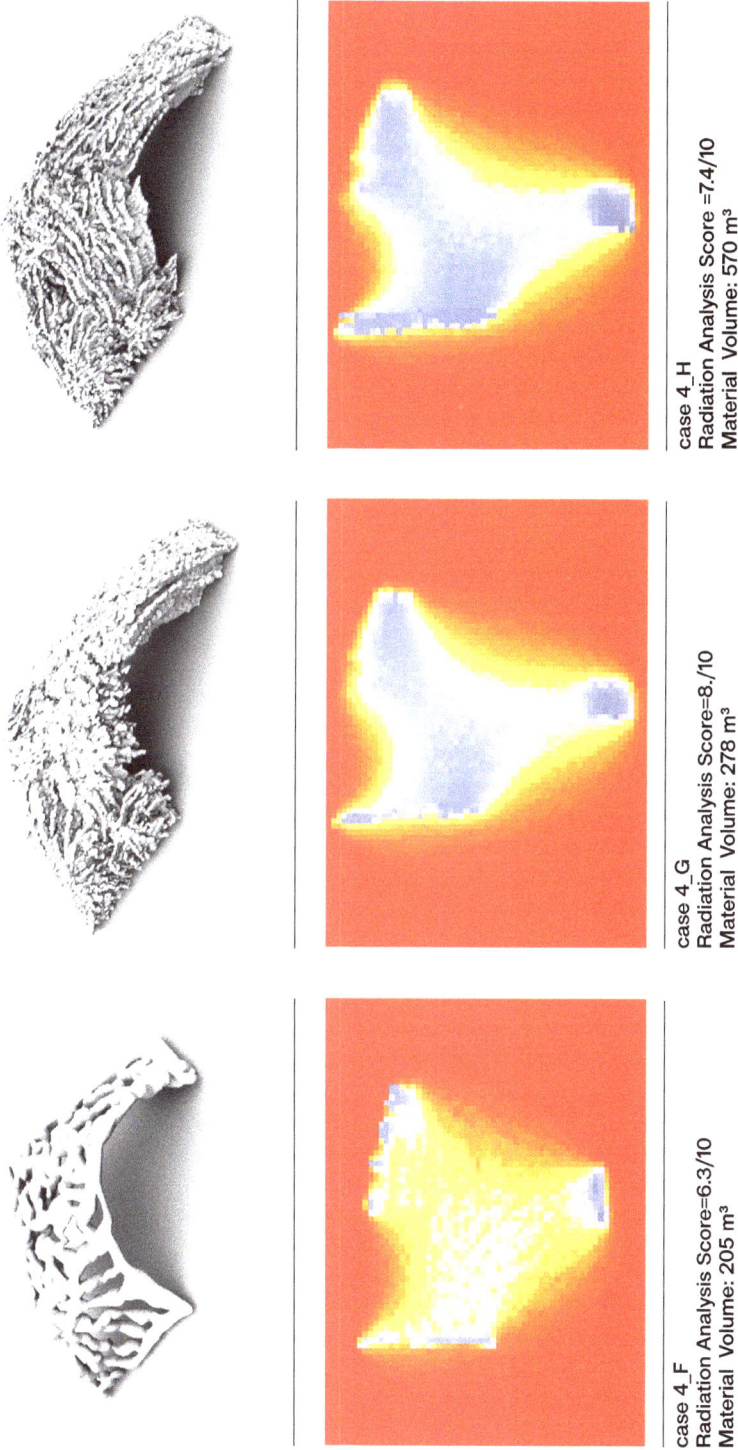

case 4_H
Radiation Analysis Score =7.4/10
Material Volume: 570 m^3

case 4_G
Radiation Analysis Score=8./10
Material Volume: 278 m^3

case 4_F
Radiation Analysis Score=6.3/10
Material Volume: 205 m^3

Figure 4.77: Behavioral form-found geometries (set B) and the solar radiation analysis underneath.

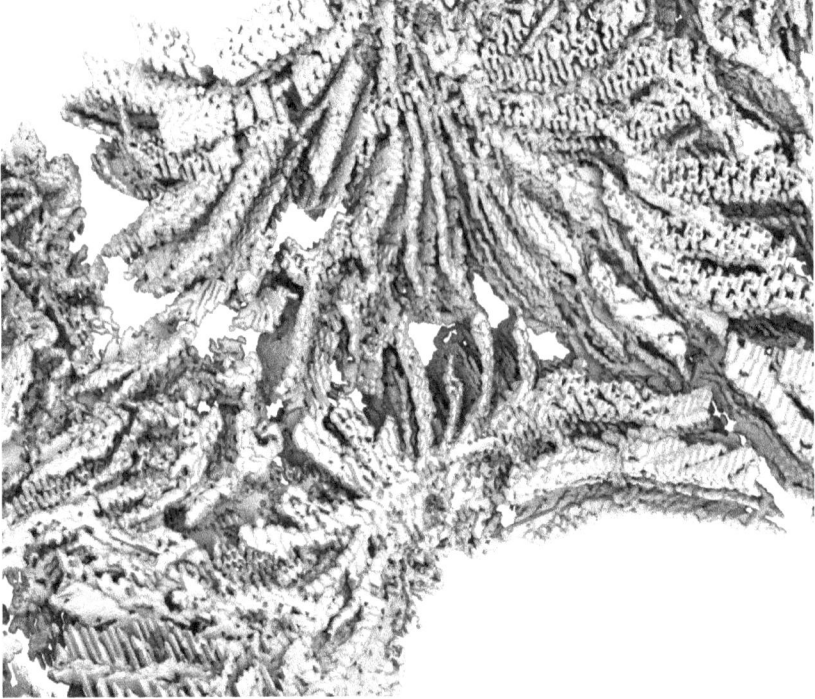

Close up of view of agent based thickness of shell

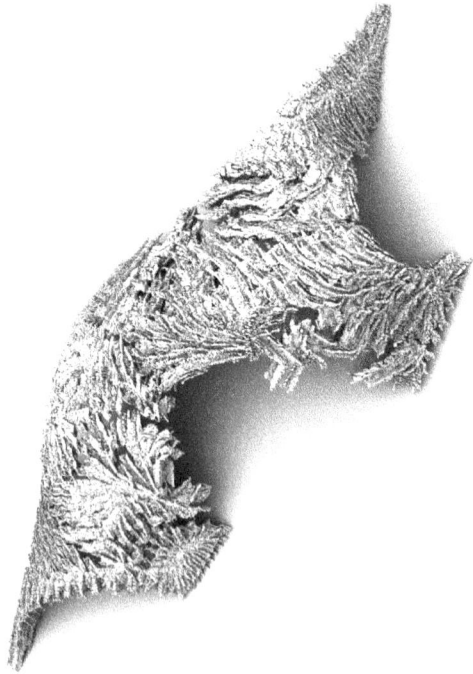

Perspective View

Figure 4.78: Behavioral form-found geometries (set B) and the solar radiation analysis underneath.

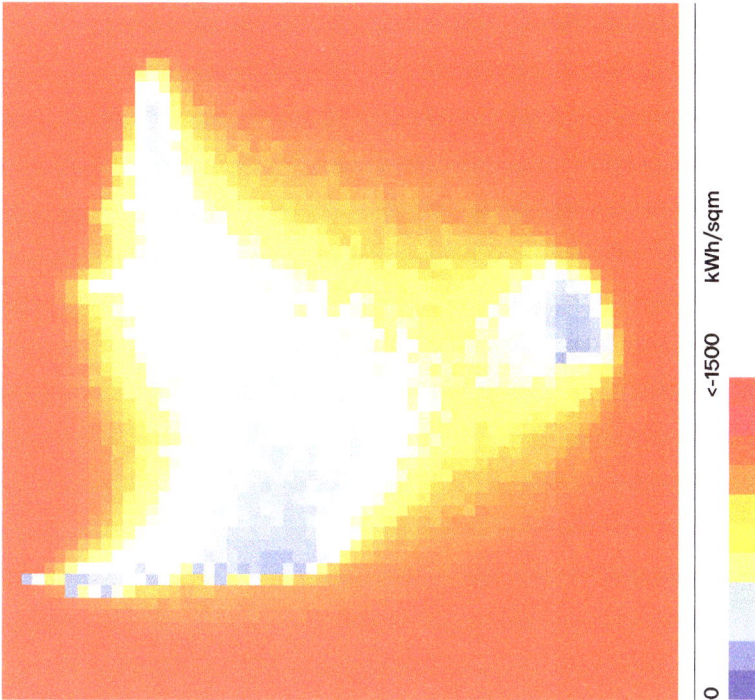

Figure 4.79: Selected (4b) geometry and the corresponding solar radiation analysis underneath.

Figure 4.80: Photo of a 3D printed prototype of a behaviorally form-found geometry with variable porosities (case 3B). Image by author from author's archive.

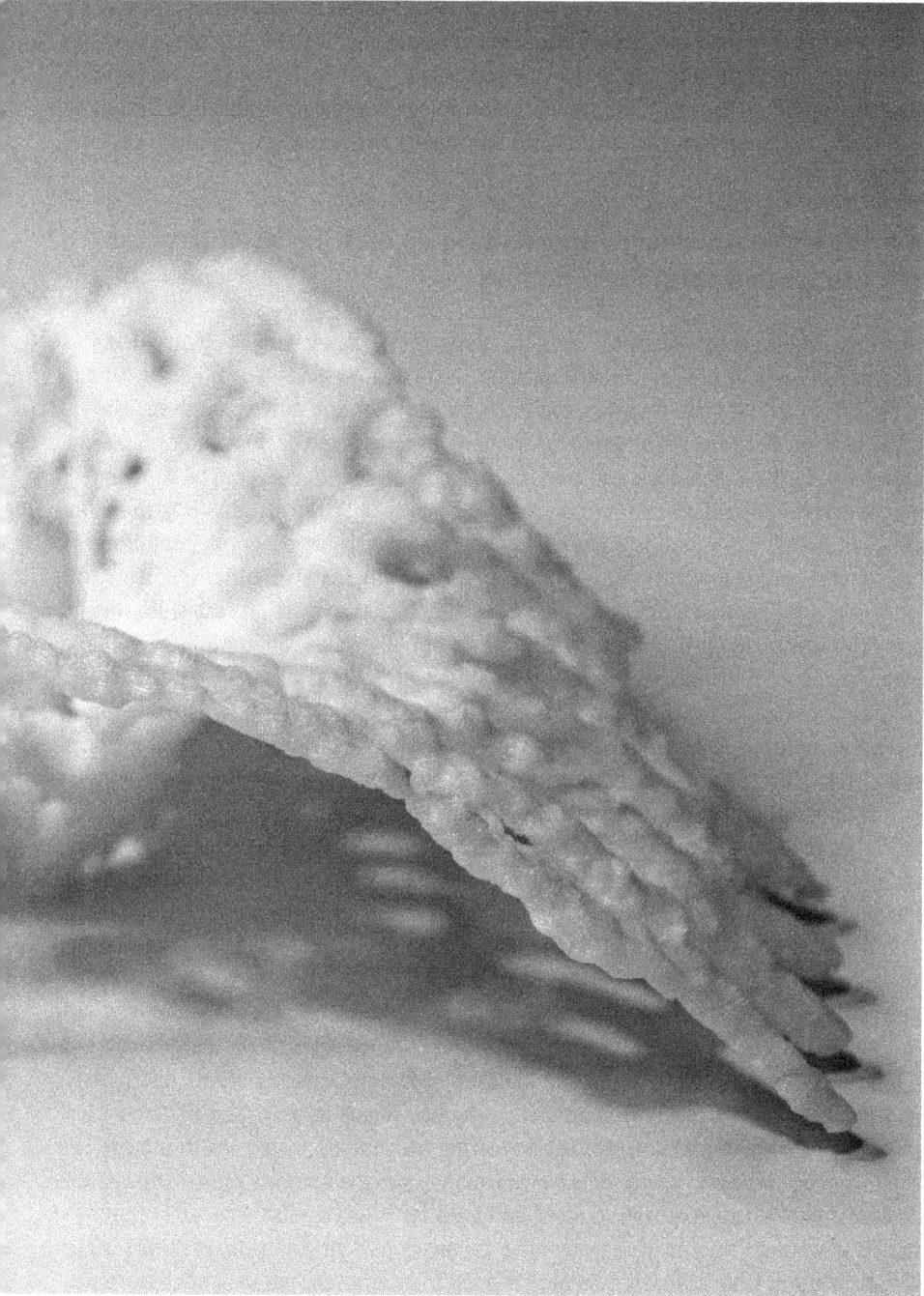

early design stage to drive the generation of design, rather than simply rationalize them after they have been generated. The next steps include the development of an autonomous structural agent, which will be able to analyze, visualize and evaluate structural analysis data from the generated geometries and communicate information to the generative agent in order to continually optimize the next generative iterations.

4.4 Experimental design 4: revisiting an existing shell structure using behavioral form-finding

This experiment tested the framework by applying it to an existing structure that was designed in the 1970s by Heinz Isler using form-finding methods. The objective was to extend the existing form-finding techniques by representing additional design intentions as agent behaviors. The aim was to allow designers to consider multiple design objectives, beyond structure and material in the early design stage. The topology of the structure, as well as behaviors relating to the positions of the Sun throughout the year, namely photophilic/photophobic behaviors, were investigated in addition to typical parameters like stiffness, support conditions and loads. Another objective was to evaluate the approach of Experiment 3 by applying it to an actual case study, evaluating the energy footprint of the structure (not only the solar radiation) and trying to compress form-finding and environmental optimization into one step.

Isler devised a model for physical form-finding that exploited the deformation of a flexible hanging cloth under gravity to generate a surface in pure tension under self-weight. The resulting form, when inverted, is in pure compression under equivalent loading. Although he had been experimenting with this method since the 1950s, Isler started actually building structures in the late 1960s [241]. The virtue of this technique is that it can generate an infinite number of potential surface forms, even with the same supports and loads, just by altering the type and orientation of the cloth. Another advantage of the technique is that instead of pushing the form toward a desired geometry, Isler was letting the form find itself based on the design criteria he had set.

Following that, precise measurements of the physical model – on many occasions, a large-scale model – were made and load tested to determine the stress distribution and buckling behavior. Since geometry is the key factor in assessing the structural behavior of a shell, this stage of the modeling process enabled Isler to modify the shell form according to its predicted performance. Once a working model showed the desirable performance, it was selected and used for other similar structures homogenously. Although this process proved to be rigorous and accurate, it is also time-consuming and labor-intensive, so only a small number of design iterations were possible within the limits of building design.

As was briefly described in Section 4.3, a few researchers have been inspired by Isler and have developed computational form-finding methods. When Isler presented his seminal work on his physical form-finding method at the First Congress of the International Association for Shell Structures (IASS) in 1959, he listed five key aspects of shell design: (1) the functional, (2) the shaping, (3) the architectural expression, (4) the statics and (5) others, such as acoustics and light [242]. The review of the literature indicates that the most emphasis has been placed on the first four aspects and in acoustics, but little research has been done on the impact of light. Therefore, a draw-back of Isler's and other contemporary form-finding methods is that they do not con-sider contextual and environmental parameters. Although computational tools offer designers the opportunity to simulate the behavior of fabrics digitally, they also give an opportunity to model additional behaviors that relate to the Sun, humidity or tem-perature, for instance.

Thus, the hypothesis here (similar to the previous experiment) is the following: by developing design behaviors based on environmental parameters in the early de-sign stage, designers can augment the purely form-found shapes in order to reduce annual energy consumption and improve the DFA of a structure without sacrificing its structural performance.

4.4.1 Case study: sports center by Heinz Isler

Isler used fabric and physical form-finding to design a number of structures [243]. One of the most widely used Isler's designs is the tennis and sports halls that he built in various locations in Switzerland (Figure 4.81). This case study revisits a tennis hall built in 1978 in Heimberg, a small town in Switzerland (Figure 4.82).

The thin shell structure has a span of 48 m, a length of 72.00 m and is supported in 10 points. It is made out of 100-mm-thick reinforced concrete, has a footprint of approximately 3,000 m^2 and is still in use. Isler developed different designs for one bay (48 × 18 m) to test different design parameters, such as the fabric density and ori-entation. He then evaluated the performance of the structure and selected one that was replicated four times to create the final structure. The structure served as an ar-chetype for three more shell structures that were designed and built in the next de-cade. The structures are all located in Switzerland and have exactly the same span, but their overall length and orientation vary. The material used in all the structures is untreated concrete cast on top of 50 mm insulated Styrofoam panels, while the open-ings are single panel curtain walls.

4.4.2 Design process

Although little information is publicly available about the detailed geometry of the Heimberg shell, by accessing information about the shell via an online database (www.structurae.com), one can see the basic design parameters, simulate the struc-ture and generate a 3D model using Rhinoceros 3D and the Kangaroo Particle Physics Solver. The first step after generating the 3D models is to simulate the existing struc-ture and analyze it structurally.

Heimberg sports center

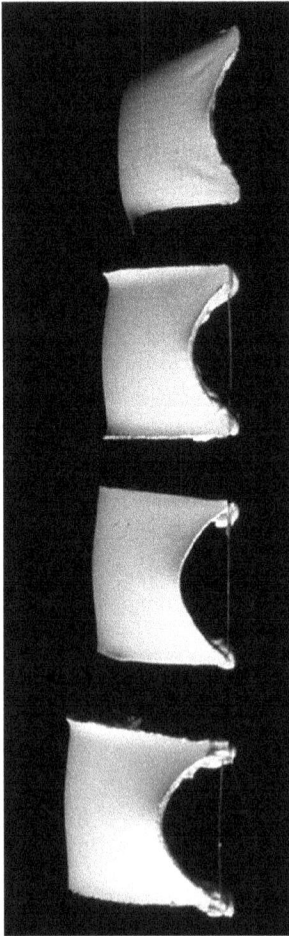

A. Physical (fabric) models (ca 1977)

B. Construction Phase (1978)

C. Interior View-Tennis Hall (2013)

D. Exterior View (2013)

Figure 4.81: Photos showing physical model, the construction phase and the current state of condition of the Heimberg sports center by H. Isler.

A. Heimberg, Berne, CH

B. Düdingen, Fribourg, CH

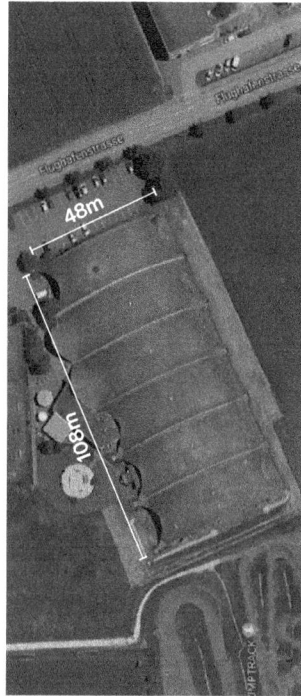

C. La Tène, Neuchâtel, CH

D. Grenchen, Solothurn, CH

LOCATION	PROGRAM	SPAN (m)	LENGTH (m)	TYPE	MATERIAL	COORDINATES
A. Heimberg, Berne, CH	Sports Center	48.00	72.00	Thin Shell	Reinforced concrete	46° 47' 33.77" N 7° 35' 43.33" E
B. Düdingen, Fribourg, CH	Sports Center	48.00	54.00	Thin Shell	Reinforced concrete	46° 51' 20.35" N 7° 11' 56.88" E
C. La Tène, Neuchâtel, CH	Sports Center	48.00	72.00	Thin Shell	Reinforced concrete	47° 01' 25.21" N 7° 01' 13.20" E
D. Grenchen, Solothurn, CH	Sports Center	48.00	108.00	Thin Shell	Reinforced concrete	47° 10' 57.20" N 7° 24' 32.25" E

Figure 4.82: Four different shell structures designed by H. Isler which were realized in different locations Switzerland during the 1970's.

The material was modeled and the structure was analyzed using Karamba, an FEA software geared toward interactive use in the visual scripting editor Grasshopper [210]. The same step is performed for all four structures that were designed based on the same model. Apart from modeling the structure parametrically using the Kangaroo Particle Physics Solver, the author also developed an agent-based modeling approach in order to explore more design alternatives. The established MAS framework was used to integrate the environmental parameters, in parallel to form-finding. This approach was based on the following assumptions: the generative agents (form-finding) are represented as particles that are interconnected to represent a mesh surface, and each connection among the agents is modeled as a linear elastic spring with variable stiffnesses. Figure 4.83 graphically illustrates this approach.

Forces are applied to each particle, but instead of only applying gravity and dead loads, as is typical in the existing form-finding methods, the "virtual forces" were also modeled, such as Sun's attraction or repulsion. The virtual forces were scaled with a weight factor to adjust its impact. The hypothesis is that agents can attain an equilibrium state, and by iterative calculations the form can be optimized not only for weight but also based on its environmental performance. The design process can be summarized as follows:

1. The typical parameters in structural design are defined, such as F, the outline of the provided footprint in the form of a polyline; P, as well as determining the number and type of support conditions S (n,t); the material stiffness (E); material properties (G); and loads (L).

2. Next, the maximum number of agents and the topology of their connections were determined. This was the discretization of the given input outline and set the initial geometric configuration of the mesh surface for the form-finding. In this case, four different topological variations were developed, namely orthogonal, triangular, hexagonal and rhomboidal. The rectangular variation was selected for the purpose of clarity.

3. The designer provided the location's longitude and latitude (LL), the orientation (O) of the structure, orientation (N, S, E, W) and the creation of a weather data (.epw file) file. Additionally, the designer provided a generic use of the space (i.e., school, office and gym).

4. Next, the designer developed an environmental behavior for the agent, that is, photophilic or photophobic behavior, depending on the design objective. In this case, the behaviors are very basic, such as getting attracted by selected Sun's positions to allow direct sunlight in the morning. What the behaviors do is to exert a force on a specific group of point.

5. External physical loading (self-weight, dead load) was applied on the surface, and external virtual loading (i.e., Sun's attraction) was used to derive the shape of the shell. The stiffness, weight and level of attraction of nodes were adjusted to find the equilibrium state. This is the main difference between the suggested approach and Isler's physical modeling or computational form-finding approaches. At each time, the position of the agents is updated based on the summation of gravity forces, as well as additional forces that act upon it. By introducing specific posi-

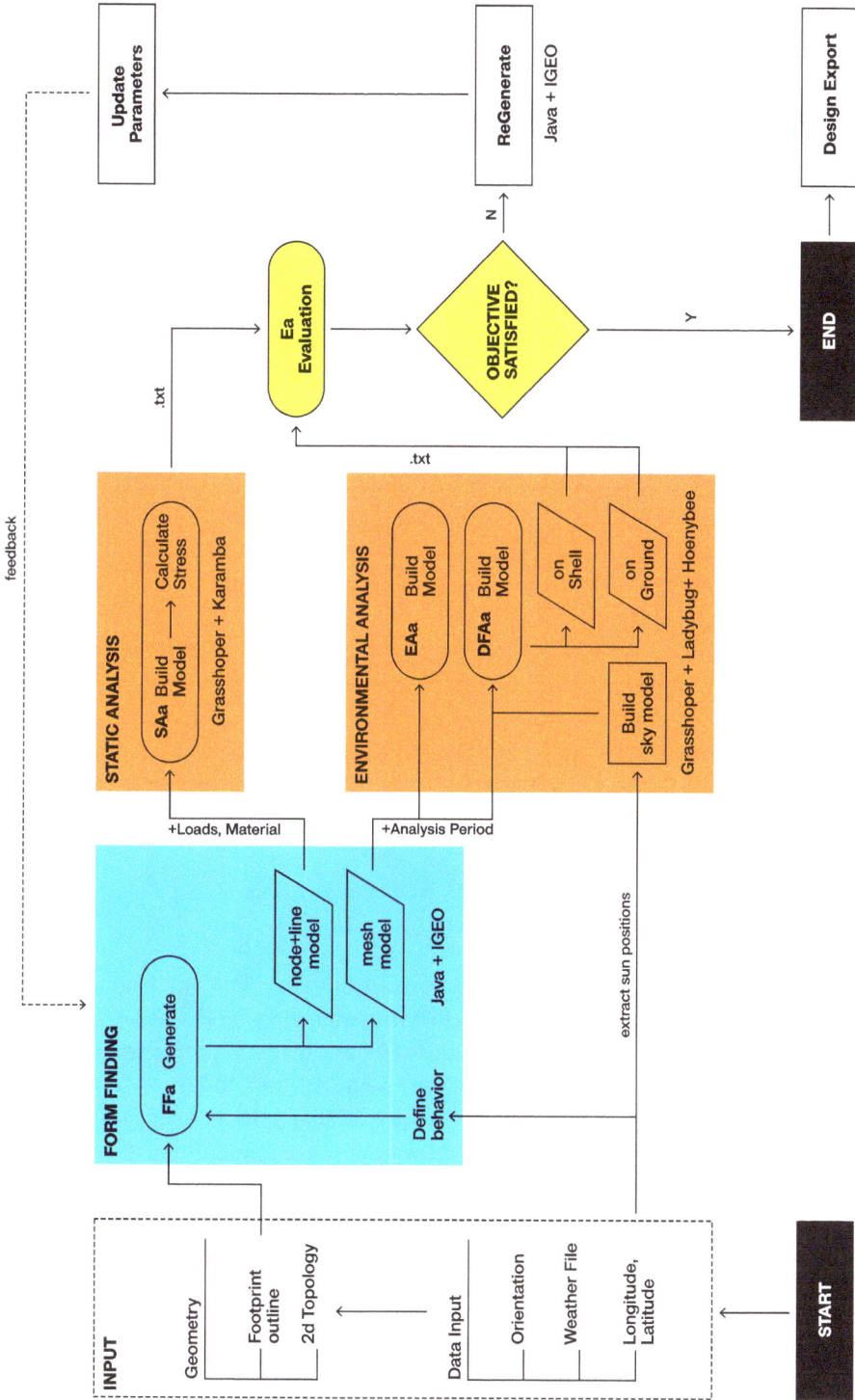

Figure 4.83: Flowchart diagram of the integrated workflow that combines form-finding with behaviors.

tions of the Sun as "virtual forces," this work directly links environmental parameters with form-finding, which the designer can adjust by providing "weights" for each force. In this step, she or he specifies the duration of the form-finding process and the "weight" of the photophilic or photophobic behavior (Figures 4.84 and 4.85).

6. Once a global equilibrium is reached and the velocity of each agent is close to 0, a NURBS geometry is generated based on the optimal force distribution. This can be directly exported to Rhinoceros 3D for further analysis.
7. The generated geometry is automatically passed to an analytical software for structural and environmental analysis using Grasshopper interface. The results are collected and used to inform the weight of the environmental behavior.
8. The process is repeated iteratively until the design objectives are met or the user stops when satisfied.

Three types of specialist agents were created for each of the analyses performed and communicated with the generative (form-finding) agent: the SAa, which accounts for displacement; an energy analysis agent, which accounts for the total thermal energy required annually for the structure; and a DFAa, which accounts for the amount of sunlight on and beneath the shell structure.

A coordinator agent ensures the passing of information between the agents and the different software platforms. The flowchart in Figure 4.83 graphically outlines the steps of the design process. In this case, the MAS framework is used to rapidly develop alternatives and enable the designer, via a GUI (Figure 4.86), to build a representation of how different behaviors and configurations affect both the structural and environmental performance of the shell design.

4.4.3 Results and analysis

I used the coordinates of the structure to extract the solar path and load a weather file for each location to analyze the structures environmentally. I provided data, such as the use of the space (program), schedules and basic material properties in order to build the energy analysis model and analyzed each structure. In Figure 4.87, the analytical results for each of Isler's structures are tabularized and compared. As one would expect, structurally they behave almost the same with only small differences between them due to their different sizes. However, their environmental performance varies quite significantly. The annual energy consumption, for instance in Case D, which is the longest structure and is also oriented along a southeast/northwest orientation axis, has the lowest average daylight factor.

Although simulations are not based on detailed 3D models, in order to validate the results, the author compared them against a survey from the Chair of Ecological System Design at ETH Zurich, which catalogs the embodied environmental impact of building stock in Switzerland since the 1920s [244]. This case study's annual energy simulation results fall within the survey's indicated range of average energy con-

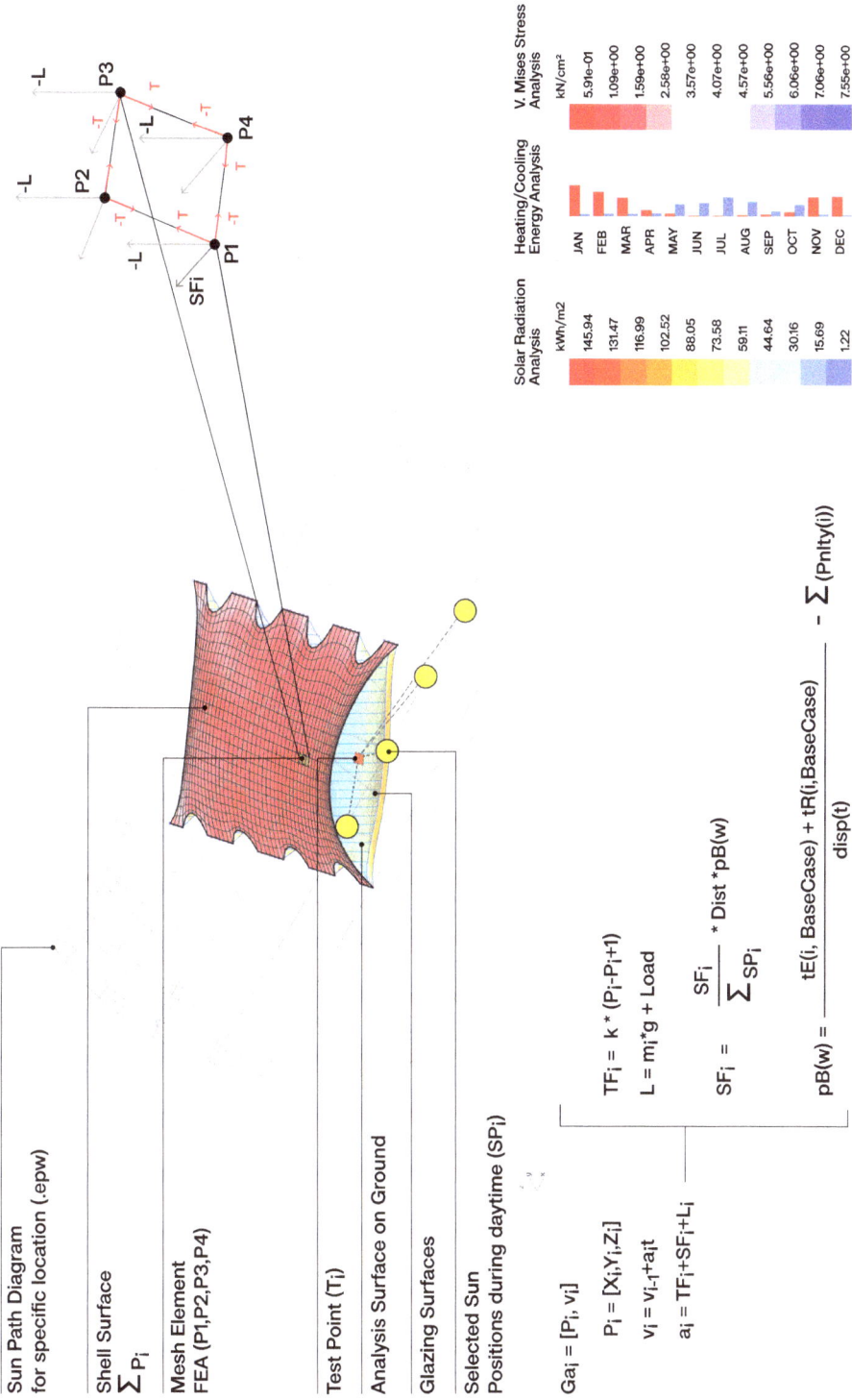

Solar Radiation Analysis	Heating/Cooling Energy Analysis	V. Mises Stress Analysis
kWh/m2		kN/cm²
145.94	JAN	5.9e-01
131.47	FEB	1.09e+00
116.99	MAR	1.59e+00
102.52	APR	2.58e+00
88.05	MAY	3.57e+00
73.58	JUN	4.07e+00
59.11	JUL	4.57e+00
44.64	AUG	5.56e+00
30.16	SEP	6.06e+00
15.69	OCT	7.06e+00
1.22	NOV	7.55e+00
	DEC	

Sun Path Diagram
for specific location (.epw)

Shell Surface
$\sum P_i$

Mesh Element
FEA (P1,P2,P3,P4)

Test Point (T$_i$)

Analysis Surface on Ground

Glazing Surfaces

Selected Sun
Positions during daytime (SP$_i$)

Ga$_i$ = [P$_i$; v$_i$]

$P_i = [X_i; Y_i; Z_i]$

$v_i = v_{i-1} + a_i t$

$a_i = TF_i + SF_i + L_i$

$TF_i = k * (P_i - P_i + 1)$

$L = m_i * g + Load$

$SF_i = \dfrac{SF_i}{\sum SP_i} * Dist * pB(w)$

$pB(w) = \dfrac{tE(i, BaseCase) + tR(i, BaseCase)}{disp(t)} - \sum (Pnlty(i))$

Figure 4.84: Experimental design setup that describes the environment, parameters and heuristic function.

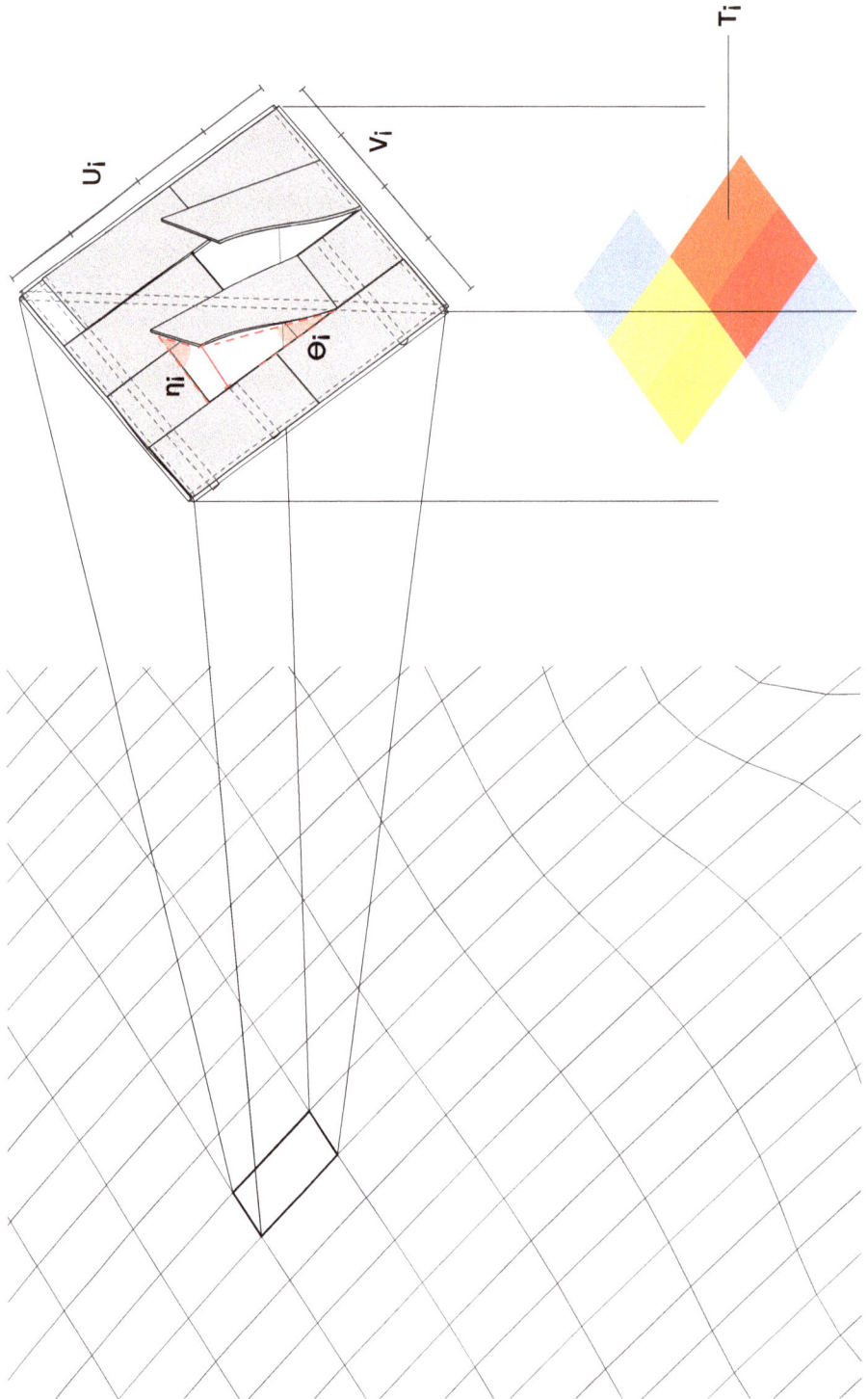

Figure 4.85: Diagram showing the logic for creating openings on the shell based on daylight availability.

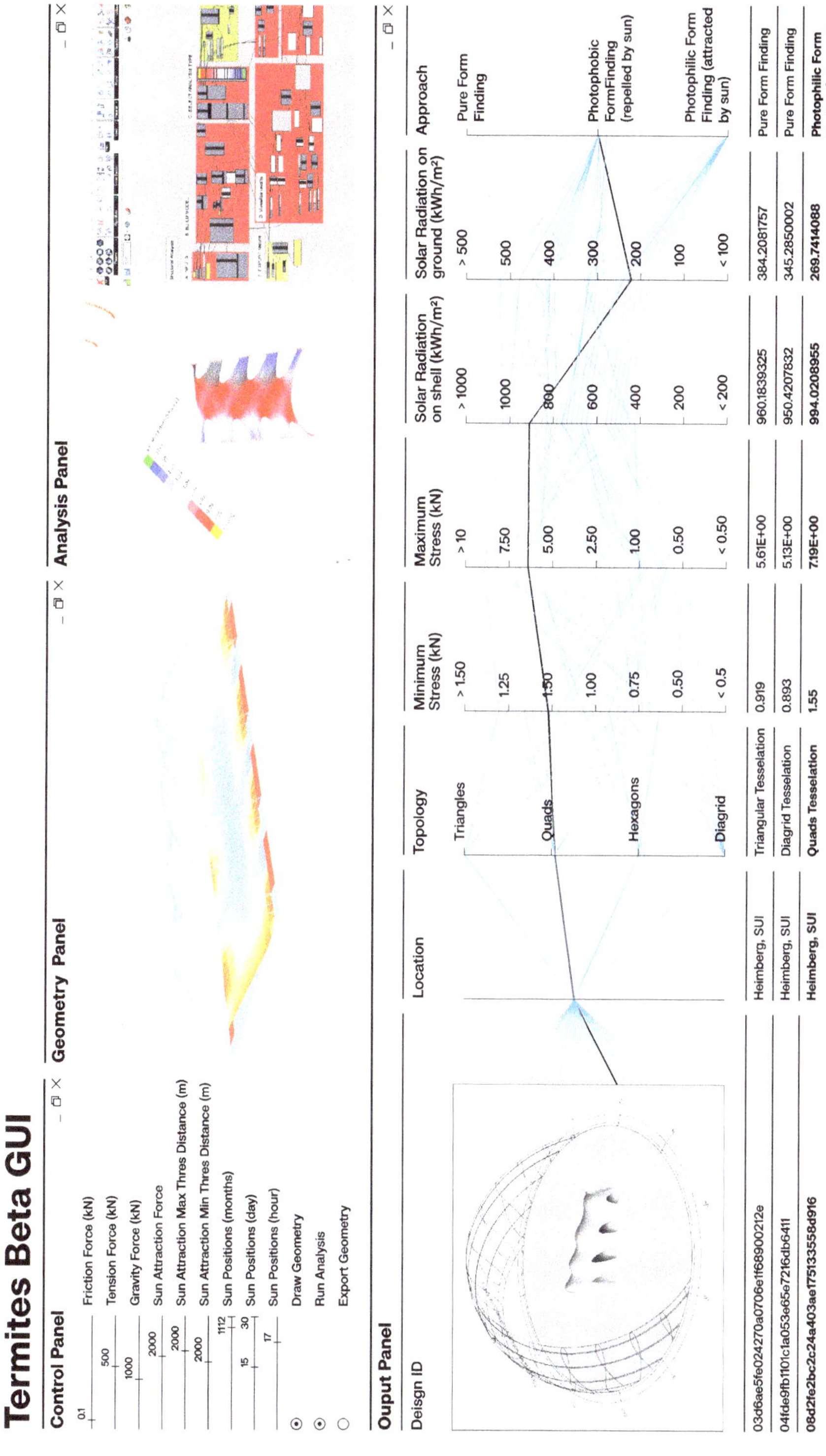

Figure 4.86: Graphical user interface (GUI) of the alpha version of the MAS design tool.

Evaluation of H. Isler's Shells

Information	Case A	Case B	Case C	Case D
Location (Switzerland)	Heimberg, Berne	Düdingen, Fribourg	La Têne, Neuchâtel	Grenchen, Solothurn
Span (m)	48.00	48.00	48.00	48.00
Length (m)	72.00	54.00	72.00	108.00
Footprint Area (sqm)	3,079.00	2,266.00	3,777.00	4,670.00
Shell area (sqm)	3,253.00	2,398.00	4,005.00	4,941.00
Structural Analysis				
Max displacement [cm]	**1.38**	**1.44**	**1.33**	**2.00**
Min Stress [kN/cm2]	1.27	1.25	0.28	1.20
Max Stress [kN/cm2]	8.37	8.40	10.00	8.70
Solar Radiation Analysis				
Annual solar Radiation on shell (kwh/m2)	1,131,400.00	1,403,500.00	1,403,500.00	1,699,900.00
Annual solar Radiation below shell (kwh/m2)	418,107.55	298,344.45	726,504.19	401,665.20
Annual solar Radiation on shell /sqm	367.45	585.30	350.44	344.00
Annual solar Radiation below shell /sqm	**135.80**	**131.70**	**192.35**	**86.00**
Enegry Analysis				
Annual Total Thermal Energy (kW/h)	786,400.15	439,080.40	767,175.45	780,620.70
Annual Cooling Energy Needed (kW/h)	843.45	1005.40	28,815.25	832.45
Annual Heating Energy Nedeed (kW/h)	777,960.70	429,030.00	738,360.20	772,300.30
Annual Daylight Factor(%)	95.90	94.85	95.50	96.70
Total Thermal Energy per area (kW/h/sqm)	**255.40**	**193.75**	**203.10**	**167.15**

Figure 4.87: Analytical results of the four different case studies.

sumption of buildings, according to their age. These results were then used as a base-line, and Case A was selected. To apply my approach, the results of the other three cases were used as design targets and the MAS system was used to generate new shell shapes. In order to augment the purely form-found shapes, a photophilic or photopho-bic behavior was assigned to specific points that corresponded to the Sun's positions of a given location that were selected by the designer. To clarify this, depending on the location, the designer may assign a photophilic behavior to the positions of the Sun during the morning hours, which increases the amount of daylight, and assign a photophobic behavior to the positions of the Sun during the afternoon hours, increas-ing the amount of daylight but also significantly increasing heat gain, with a poten-tially negative impact on the structure's total energy consumption. In order to ensure that the behavior does not lead to undesired results, the light attraction force is scaled according to the number of the attractor points, the distance of the point to the struc-ture and a weight $pb(w)$, as seen in the following equation:

$$F = \frac{\text{Sun attaction force}}{N(pt)} * D * pb(w).$$

The weight w is related to a heuristic function in which the analytical results are used to calcu-late its value. The heuristic function uses the total energy consumption (tE), the total radiation (tR) underneath the structure and the maximum displacement in order to calculate the value of the weight. The heuristic function is used to adjust the value of the force according to the desired result. The closer to the desired analytical results a solution is, as compared to a base case (i.e., the purely form-found case), the higher the weight of the photophilic behavior becomes. A pen-alty is added to reduce the weight if a point is moved to an undesired location:

$$pb(w) = \frac{tE(i, \text{base case}) + tR(i, \text{base case})}{\text{disp}(t)} - \sum (\text{Pnlty}(i)).$$

I ran the system using this heuristic. The designer can interactively change the value of the behavior based on the assessment of the analytical results and the geometry. Once she or he sets a value for the behavior, the system runs for a specified number of iterations in order to generate design alternatives. Figures 4.88 and 4.89 present a set of form-found shells where an environmental behavior has been applied. Fig-ure 4.90 displays a parallel line plot to show a subset of the behaviorally form-found shapes. The blue lines indicate the values of the four existing case studies, while the red lines show the design alternatives of case study A after applying a photophilic be-havior. The system was able to generate alternatives that have as much maximum displacement as the base case, yet they decrease the average annual consumption by 12%, increase the DFA by 9% and increase the average solar radiation underneath the structure by 102%.

Form finding with behaviors

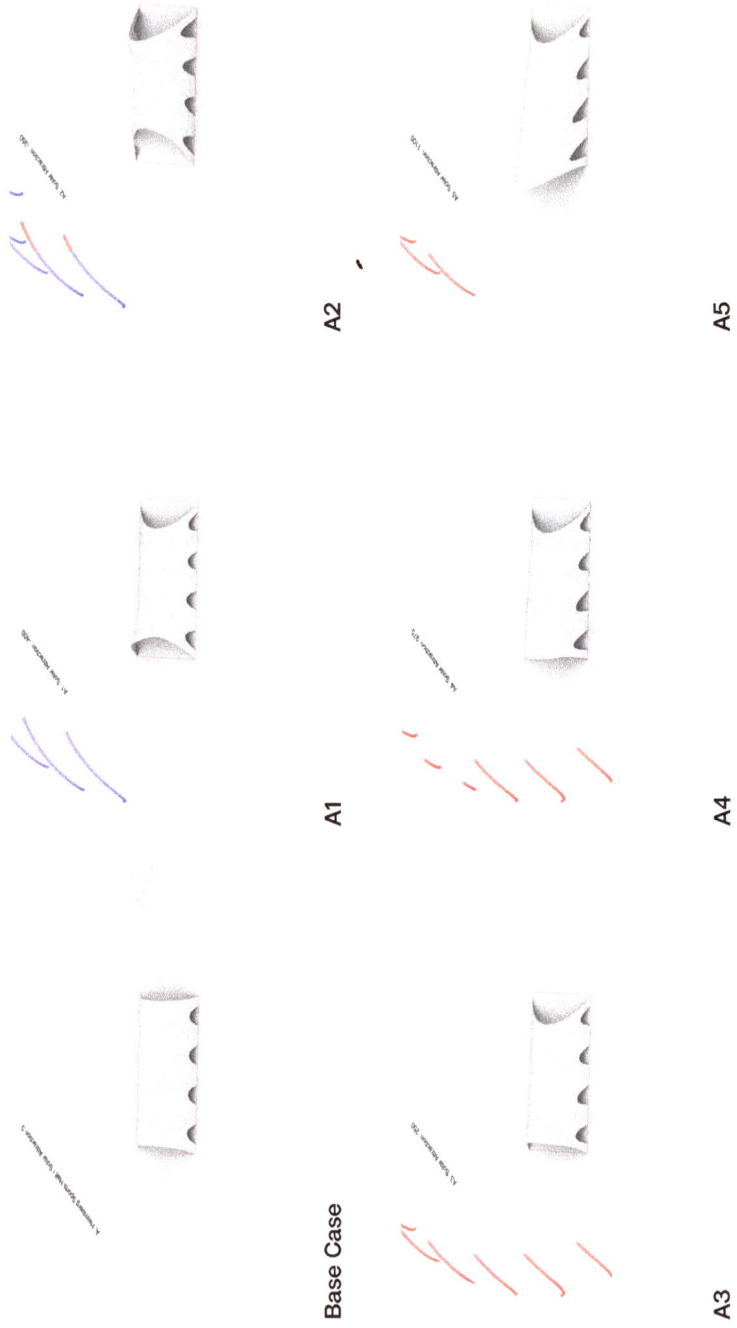

Base Case

A1

A2

A3

A4

A5

Figure 4.88: Base case geometry (Heimberg Sports center) and generated design alternatives (Set A0–A5).

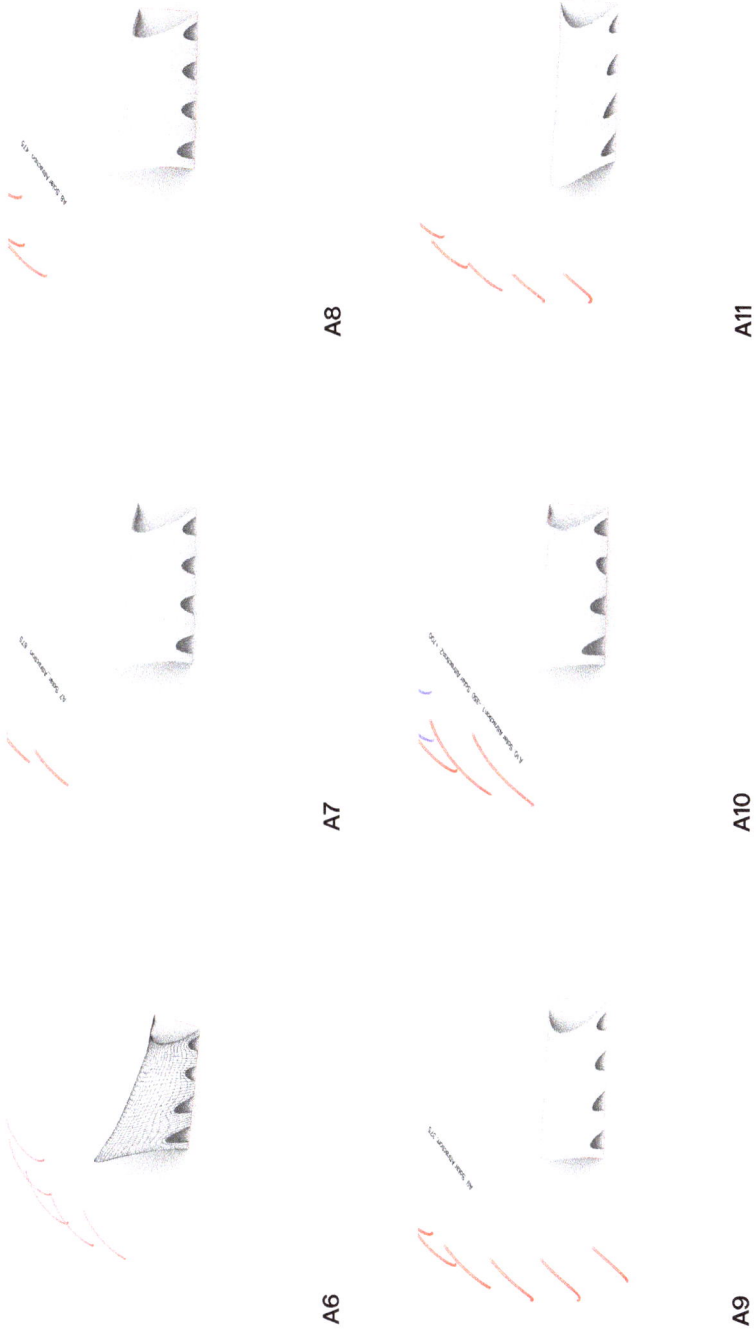

Figure 4.89: Generated design alternatives with different behaviors applied to it (Set A6–A11).

Termites Beta – Output Panel

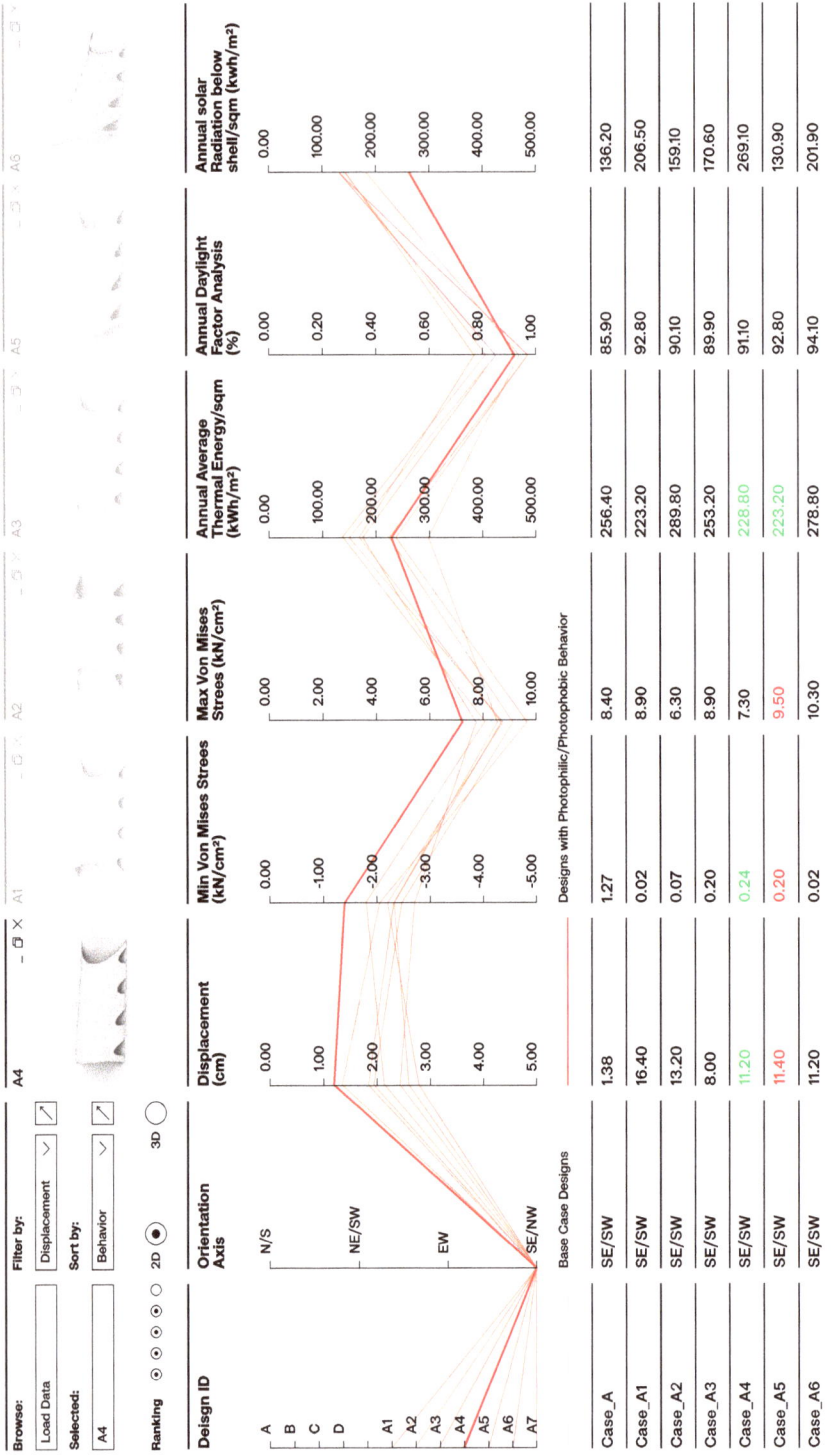

Browse: Load Data

Filter by: Displacement

Selected: A4

Sort by: Behavior

Ranking ○ ⊙ ○ ○ **2D** ⊙ **3D** ○

_ ⊡ × A4   _ ⊡ × A1   A2   A3   A5   A6

Design ID / Orientation Axis	Displacement (cm)	Min Von Mises Stress (kN/cm²)	Max Von Mises Stress (kN/cm²)	Annual Average Thermal Energy/sqm (kWh/m²)	Annual Daylight Factor Analysis (%)	Annual solar Radiation below shell/sqm (kwh/m²)
N/S	0.00	0.00	0.00	0.00	0.00	0.00
	1.00	-1.00	2.00	100.00	0.20	100.00
NE/SW	2.00	-2.00	4.00	200.00	0.40	200.00
	3.00	-3.00	6.00	300.00	0.60	300.00
EW	4.00	-4.00	8.00	400.00	0.80	400.00
SE/NW	5.00	-5.00	10.00	500.00	1.00	500.00

Orientation Axis: A, B, C, D / A1, A2, A3, A4, A5, A6, A7

Base Case Designs ———
Designs with Photophilic/Photophobic Behavior ———

		Displacement	Min Von Mises Stress	Max Von Mises Stress	Annual Average Thermal Energy/sqm	Annual Daylight Factor Analysis	Annual solar Radiation below shell/sqm
Case_A	SE/SW	1.38	1.27	8.40	256.40	85.90	136.20
Case_A1	SE/SW	16.40	0.02	8.90	223.20	92.80	206.50
Case_A2	SE/SW	13.20	0.07	6.30	289.80	90.10	159.10
Case_A3	SE/SW	8.00	0.20	8.90	253.20	89.90	170.60
Case_A4	SE/SW	11.20	0.24	7.30	228.80	91.10	269.10
Case_A5	SE/SW	11.40	0.20	9.50	223.20	92.80	130.90
Case_A6	SE/SW	11.20	0.02	10.30	278.80	94.10	201.90

Figure 4.90: A subset of design alternatives presented to the designer via the GUI with a parallel line plot.

Part V: **Evaluating alternatives**

5 Overall results and analysis

5.1 Summary of results

The multi-agent system (MAS) framework presented in this book has a tremendous potential to develop expert design systems in which the designer defines the components and low-level relationships, and the system generates and evaluates design alternatives based on the designer's interactive feedback. The open-ended character of the framework allows the designer to integrate design intentions, constraints and objectives in an abstract way by defining different behaviors. The results are discussed in two parts:

1. The general development of the methodology based upon MASs
2. The necessity of developing an understanding of complexity and new kinds of abstractions and holistic approaches for managing architectural complexity

5.1.1 Multi-agent systems framework for architectural design

This work presents a novel design methodology for architectural design using agent-based modeling and simulation. The methodology is titled "Behavioral Form Finding" and has been manifested with the introduction of a framework that allows the user to model building components as agents based on a decomposed architectural design problem (i.e., facade design, shell design). Through the integration of data sources and the combined use of analytical tools and construction constraints, the designer controls the behavior of the agents that, in turn, generate the design alternatives. The framework was applied on a set of design experiments implementing several agent classes that dealt with the generation of geometry and the translation of the geometric models into structural, environmental and construction models, as well as through the data passing between them. These classes can be combined into teams of agents in order to facilitate design exploration and to help designers explore larger solution spaces that could potentially lead to solutions impossible to attain via conventional design and building methods. A number of heuristic methods were investigated for informing the agents and predicting better behaviors, based on a set of objectives. It is important to note that a MAS approach differs from conventional design approaches in that the designer does not draw the geometry directly, but rather develops and adjusts agent behaviors that lead to the generation of geometries.

Behavioral form-finding can be considered a viable extension of existing agent-based modeling techniques in that it offers a framework for combining agent behaviors with analytical results and is also a way to model design objectives as forces that act upon the agents. Figure 5.1 illustrates the implementation of the suggested framework in the form of a toolkit. The Termite toolkit provides a way to combine top-down and bottom-up design approaches and provides a library of abstract and physical agents, as well as a number of behaviors. It also provides methods for forming

https://doi.org/10.1515/9783110797435-005

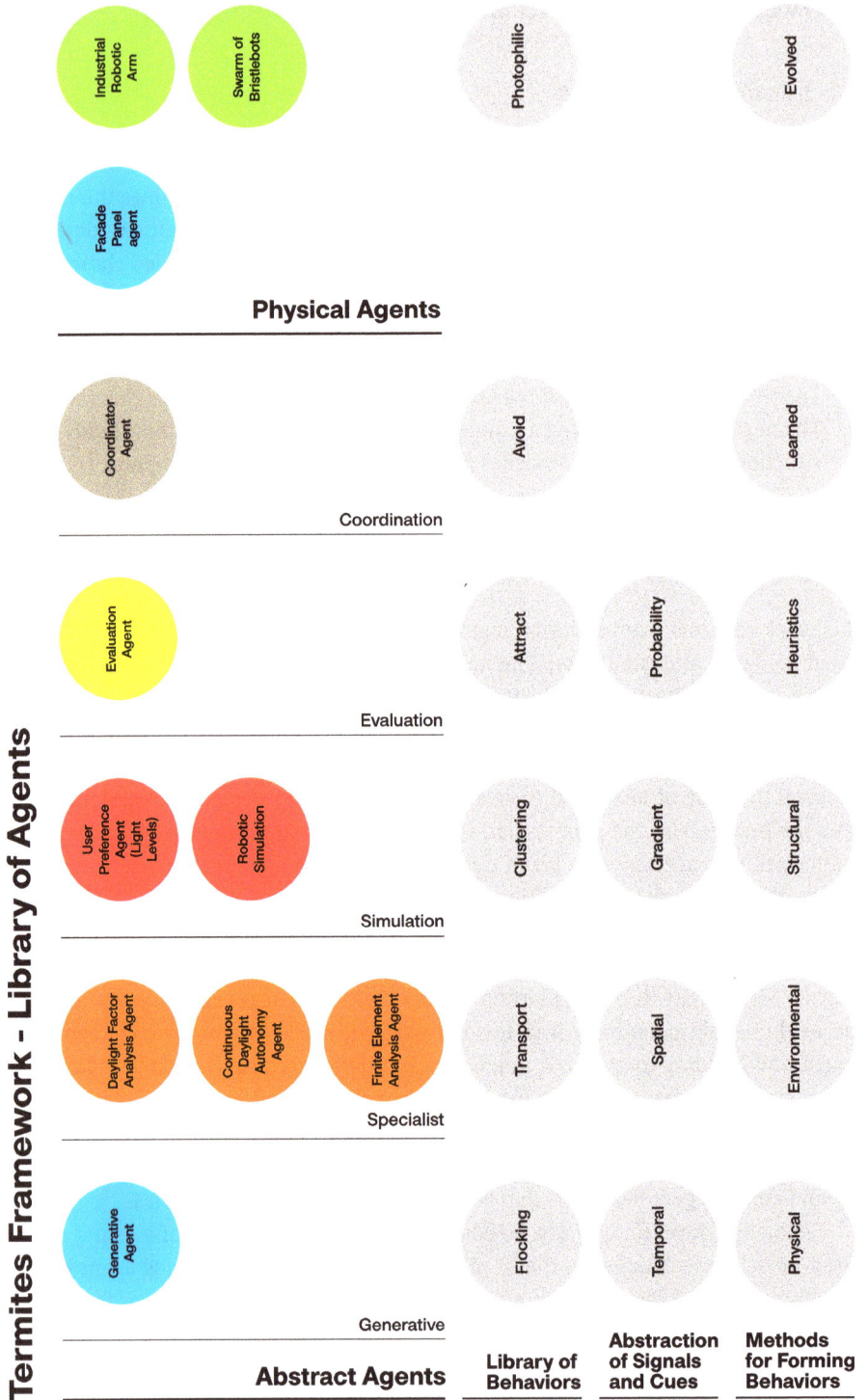

Figure 5.1: MAS framework library (Termites) with the developed agent classes, behaviors and methods.

agent behavior using computational methods that vary from geometric rules to heuristic methods and agent team formation. Lastly, the MAS framework provides a set of abstractions for mapping signals and cues that can be collected via sensors.

The MASs, whether physical or digital, that have been investigated here are scalable because additional agents (i.e., robots) do not increase interaction complexity. Instead of directly dictating commands to each robot individually, this framework allows the designer to specify high-level rules and spatially varying design parameters. In the proposed MAS framework, the designer does not need to design the final structure explicitly but is more of a conductor: by creating a robotic construction setup, from which geometric rules can be deducted, the designer runs the system and controls the behaviors of the players (robots), adding or removing robots as needed. High-level constraints can be site-specific, such as to avoid a specific area (the window opening in Experiment 2) or move toward areas with more solar radiation or even include preferences, such as regions where material density is unwanted.

Experiment 1 demonstrated how simple behaviors can be deduced from a very simple robotic system to develop simulations based on it. This was a pilot study to demonstrate how kinematic and dynamic constraints can be coupled with the physical limitation of a specific robotic system. By introducing these constraints, the design tool prevents the generation of physically nonfeasible design and design alternatives that would be impossible to fabricate. Due to the lack of tool-specific design tools, generative design examples to date have been complex to construct. The typical process has included generating a design digitally and exporting it to a design platform that would use a top-down approach to break it down into pieces that can be fabricated and assembled into the final structure. This required a significant logistic effort and coordination that increases the complexity of construction.

By coupling the design tool with a robotic system from the early design stage, as shown in Experiment 2, the design process is simplified while allowing the generation of complex and site-specific design alternatives. In Experiments 3 and 4, the author showed how the MAS framework and the concept of behaviors can be used in parallel with typical parametric modeling methods from the early design stage in order to augment purely form-finding techniques by incorporating environmental parameters. By tuning the type and magnitude of behaviors based on analytical data, we can generate design alternatives that satisfy multiple objectives, not only structural ones.

The advent of technologies, such as digital fabrication and additive manufacturing as well as Building Information Modeling (BIM), has had a great impact on the architectural practice either directly or indirectly and has clearly indicated the necessity for developing new kinds of abstractions in order to be able to handle design complexity. We are currently experiencing another radical shift in the domain of design with the rapid growth of adversarial neural networks and large language artificial intelligence (AI) models. Such models hold the promise of completely transforming the field of generative design.

As we move toward the Fourth Industrial Revolution, the interaction of disrupting new technologies with established workflows will become even more emphasized, and the underlying principles of complexity theory outlined in Section 2 become ever more relevant if architects and engineers wish to continue playing a deciding role in formulating their disciplines in the age of information. Emerging technologies such as AI and robotics should be viewed from a holistic perspective. For example, if we wish to talk about robots in construction sites we need to consider them as a group of dynamically interacting robots and not just as single operating industrial robotic arms. Current robotic simulations and design practices require the designer to provide geometry and a related toolpath for a robotic arm to construct a structure within a controlled environment (cage). It is obvious that when we scale this up to the level of a construction site, this motion planning easily becomes an intractable design task if we have multiple robots operating across different locations on a site. With each additional robotic arm, the overall complexity increases exponentially due to robot-to-robot and robot-to-structure interactions. Therefore, new approaches should be considered for tackling such problems, approaches that as we suggest in Experiment 1 rely on low-level interactions.

5.1.2 Toward new kinds of abstractions: a building as a finely tuned orchestra

In an attempt to summarize how the principles of holism and complexity (that were reviewed in Section 2.2) and concepts, such as abstraction and modularity, can be used to develop tools that enable designers to produce cohesive architectural proposals, let us draw a metaphor between designing and constructing a building and preparing and delivering a symphonic concert. Consider a symphonic orchestra with our "abstraction hats" on (see Figure 5.2).

An orchestra is an instance of hierarchy and abstraction at the same time. It is analogous to many things we see every day, including information technology systems, games, factories, cars, buildings and cities. They all have a common thread: they are collections and layers of rules, specifications, archetypes and abstractions. In an orchestra, the conductor leads the system. She or he is not playing an instrument but directs the events and acts as the quality controller. The conductor is the highest authority in a hierarchy of abstractions. If we look at a musical score, which is similar to architectural drawings, we see embedded instructions. The notes are written, but you cannot hear them; they have no sound. They encapsulate the rules of the music – the sequence, pitch, volume and timing – but nothing happens until we get to the executive layer: the individual musicians and their instruments.

The score contains the high-level sequential instructions necessary to execute in specific ways the lower level instructions required to get the intended result. The musicians do that job and perform a role analogous to that of architectural designers. The musician (designers) will execute parts (automation) within the score (high-level commands) on a particular instrument (operating system/design tool). The musician translates the encoded information into sounds (a tool) necessary for the audience to appreciate the outcome (business function), all managed by the top layer (orchestration) and intended by the composer (IT service owner) to please the audience (clients).

Structural Hierarchy

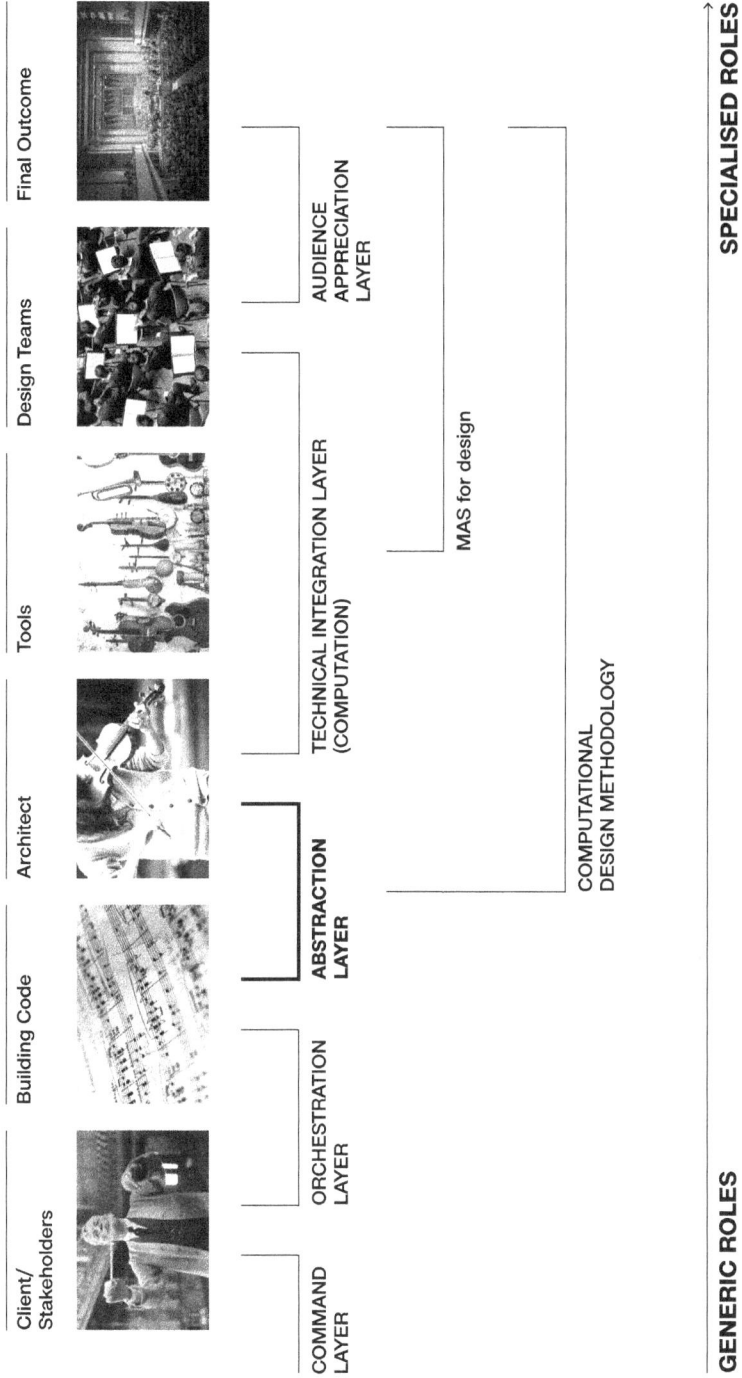

Figure 5.2: An illustration showing how structural hierarchies and different layers of abstraction can come together to produce a complex outcome in the case of an orchestra (top) and a building (bottom) (i.e., a concert and a building). Photos by Marius Masalar, Austrian National Library, Manuel Nageli, Clem Onojeghuo and Samuel Sianipar on Unsplash.com.

These abstractions enable useful hierarchies. Abstractions use interfaces between layers and are built on trusted behaviors. The system works because each layer is aware of the rules agreed upon by the adjacent layers. The common language between them is music. When something more specific is added, a transformation takes place within each layer. If the conductor had to play all the instruments, the outcome would be impossible. This allows us to see specialization and abstraction working to form a useful hierarchy, from a general to a specific case. There are technical elements, encoding elements and transformational elements. Without these structures, there would be no orchestra. Abstract layers are linked together via a common language, taking care of any translation internally.

While we are all aware of this example on a surface level, it is worth examining it in the light of our new awareness. This idea is as natural to us as breathing, but the sophistication of the abstractions would not have happened without recognizing the incremental disciplines involved. Humans achieved this musical result after years of evolved thinking. It is a great example of how much has been realized, thanks to the structural rules imposed. It has yielded far greater returns than if we simply sat the entire group of musicians in a room and shouted, "Play!"

Instead, it is as if we sat the musicians down and had them agree to a set of operating principles to reach a common goal. Then we filled the agreed structure with the rules of musical notation (software) that we had enabled and tested a priori, and then executed the sequence within them. The result is robust, recognizable and repeatable, and can deliver far more impact than random sounds or even one musician trying to play each instrument individually. If we drew an analogy between the delivery of an opera and a building project, we would observe that a lot of complexities and issues arise in the latter because the "musicians" of the construction industry (i.e., architects, engineers and construction crew) – despite agreeing to a set of operating principles (project brief, contracts, etc.) – are sitting in different rooms; each one is using their own notation and playing by themselves. The digitization and virtualization of the AEC holds the promise of bringing everyone in the same room but so far small steps have been taken toward developing "standard notation" which will render this possible. Additionally, despite the discourse around the potential of generative AI to change the industry, there is little discussion on how such technologies can be leveraged to better tackle long-standing problems of interoperability, lack of open standards, novel ways of better structuring unstructured construction data, among other issues.

5.2 Epilogue

This book attempts to provide some insight into methodologies and techniques that are considered necessary for navigating the complex landscape of the Information Age and are really emphasized in the architectural and design computing communities. The main contribution of this work is a computational design methodology for the early design stage in the form of a MAS framework. In this proposed methodology, the agents – modular programming blocks – are used to represent building elements and enable the seamless combination of generative design with environmental and structural analysis, as well as construction constraints. The concept of "behaviors" is introduced as a way to formally express intentions and constraints in an abstract way. The proposed methodology differs qualitatively from conventional parametric design approaches; instead of drawing explicit geometry, the architect develops agents and designs behaviors that generate the geometry. The role of the designer then becomes (a) to decompose design problems appropriately so that solutions can be generated by a group of agents and (b) to evaluate the behaviors developed based on a set of design targets.

The framework is manifested with Termite, an agent-based design and simulation toolkit, which serves as an apparatus for the developing and prototyping of behavior and studying the relationship of low-level agent behaviors with emerging high-level system behaviors. As developing and testing agent behaviors is an open-ended task that can be tedious to control, it was investigated how we can use analytical methods and statistics to evaluate both user-defined agent behaviors and the emerging system behavior. Due to both visual and numerical feedback, the designer can intuitively understand the impact of the agents' behaviors on the different types of performance metrics that she or he has defined. Thanks to the integration of domain-specific data (i.e., weather file, material characteristics and geometric constraints) and the communication with external analysis software (such as Finite Element Analysis and Energy Analysis software), the system can quickly evaluate designs and present the designer with the designs that meet the defined targets.

The framework is applied to three experimental designs that deal with a variety of problems. To be able to draw conclusions, the case studies focused on the subproblems of building designs that have traditionally required the close collaboration of architects and engineers to achieve cohesive solutions. Facade design as well as free-form shell design are typical problems of that class, and the experiments have demonstrated how low-level agent behaviors can lead to emergent design outputs, globally. The research also shows how, by developing and controlling behaviors that relate to building elements or construction processes, better performing design alternatives can be achieved both quantitatively and qualitatively (Figure 5.4). Due to its unique ability to accommodate change, the proposed agent-based framework can be effectively employed in the early design stage, thus enabling architects to explore design alternatives by considering geometrical, environmental, and structural and construction aspects.

Another important contribution of this work is bridging the gap between digitally developed behaviors and the actual behavior of physical robots, by establishing an experimental setup that relies on both physical experimentation and computational analysis. The first experiment demonstrated how swarm-like behaviors can be physically implemented via the manipulation of geometric robot features and lead to results similar to the ones observed in digital simulations. Specifically, digital tools were developed for gathering data from physical experiments with low-level robots and used it to design and test how the shape of the agents can lead to the formation of emergent structures. I consider this the very first step toward the development of a systemic approach

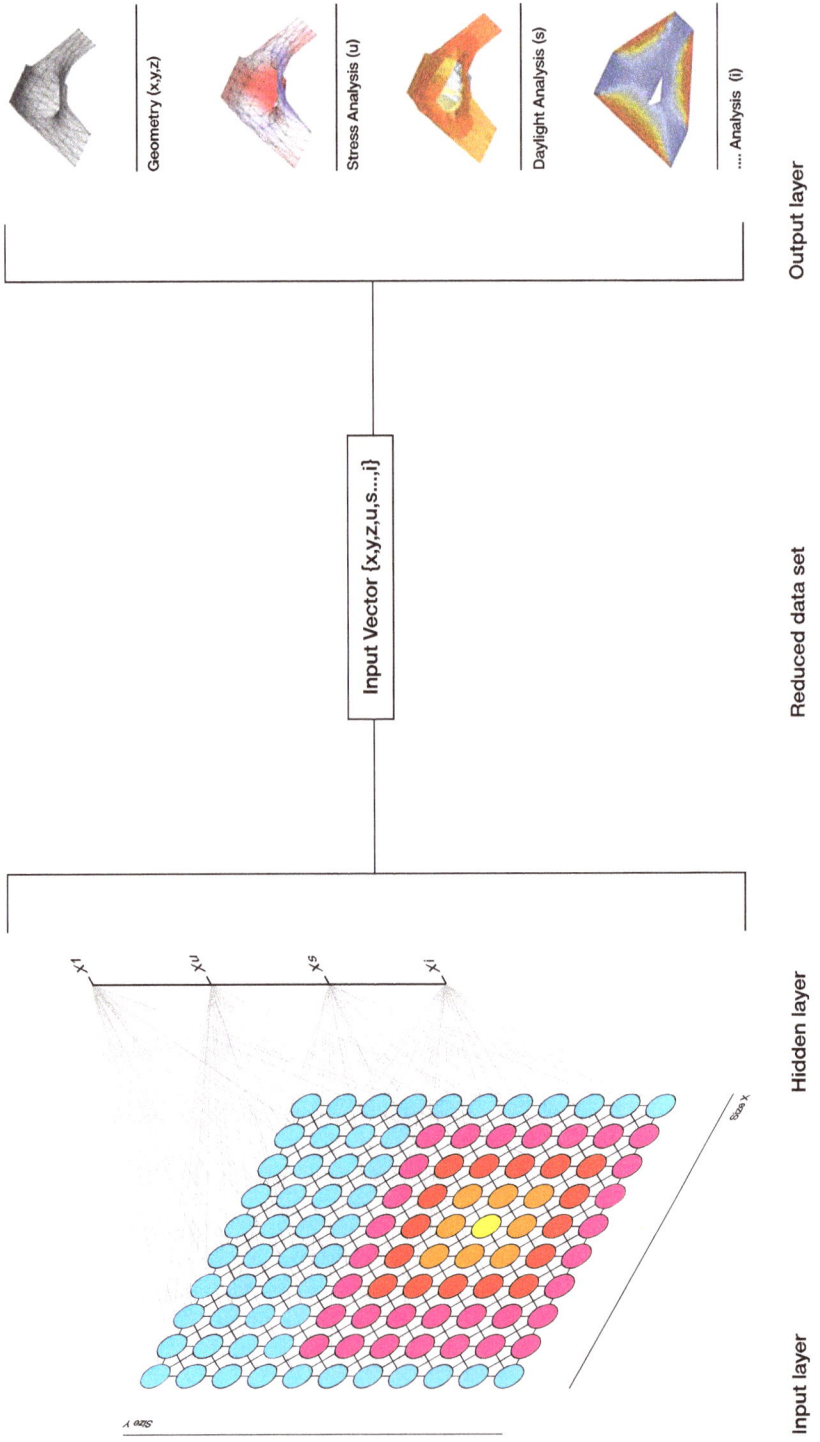

Figure 5.3: A diagram showing how SOM can be integrated into the MAS framework.

in building design for investigating novel building systems and autonomous robotic construction processes. This step has been motivated by the increasing integration of sensors in robotic construction, which has opened up the possibility of expanding MAS approaches from the digital realm to actual physical implementations.

5.3 Future research

In the short term, the MAS toolkit may be improved by several additions. The algorithmic framework could incorporate more agent classes and further facilitate the formation of teams of agents by retrieving data from existing projects. A generic building component class, in which the designer could directly import a building element from BIM software instead of designing it, would greatly improve the integration of the toolkit with BIM platforms. Additionally, to capture a building's future performance holistically, a specialist agent class focusing on human comfort could be developed. Another improvement would be a more rigorous integration of multiple user preferences through interactive visualization of design alternatives, aided by augmented reality.

Last but not least, although designers can inform the decision-making by coupling the assessment of geometric features with the visualization of analytical results, correlating all the high-dimensional datasets generated is challenging for human cognition [245]. Implementing dimensionality reduction methods is considered crucial for aiding designers by reducing high-dimensional data, while maintaining nonlinear associations between the design parameters. Therefore, extending the framework to include nonlinear approaches for feature extraction, such as self-organizing maps (SOMs) [246], is an essential step for integrating more types of design analysis. A diagram of how SOM can be integrated into the proposed MAS framework is shown in Figure 5.3.

In terms of the Termite toolkit, in the long term, it will benefit from further user testing and application to real-life cases; this will improve the reliability of the framework in producing meaningful solutions, and also test its scalability. Another major advance would be to create a web interface to visualize both results and geometry, as well as a cloud-based database for storing and easily querying all the solutions generated. The way the framework has been implemented allows for this extension, which is an important step toward developing AI tools. For example, architects could develop their own library of designs, with all the analytical data structured in a way that could be easily accessed from machine learning algorithms, so that when they are working on a new project the system could easily access previous projects that are similar and use existing data to generate design alternatives.

Lastly, based on the preliminary study with bristlebots and given the open-ended character of the MAS framework, a long-term goal is to model, simulate and control a more sophisticated cyber-physical robotic construction system and apply it to the construction of 1:1 scale structure. Current research in novel building methods using robotic arms and additive manufacturing has opened up the possibilities of developing new building methods, while also making strides toward building code compliance [247]. It is my firm belief that viable autonomous robotic construction in the real

Behavior as a confluence of Form-Function and Context

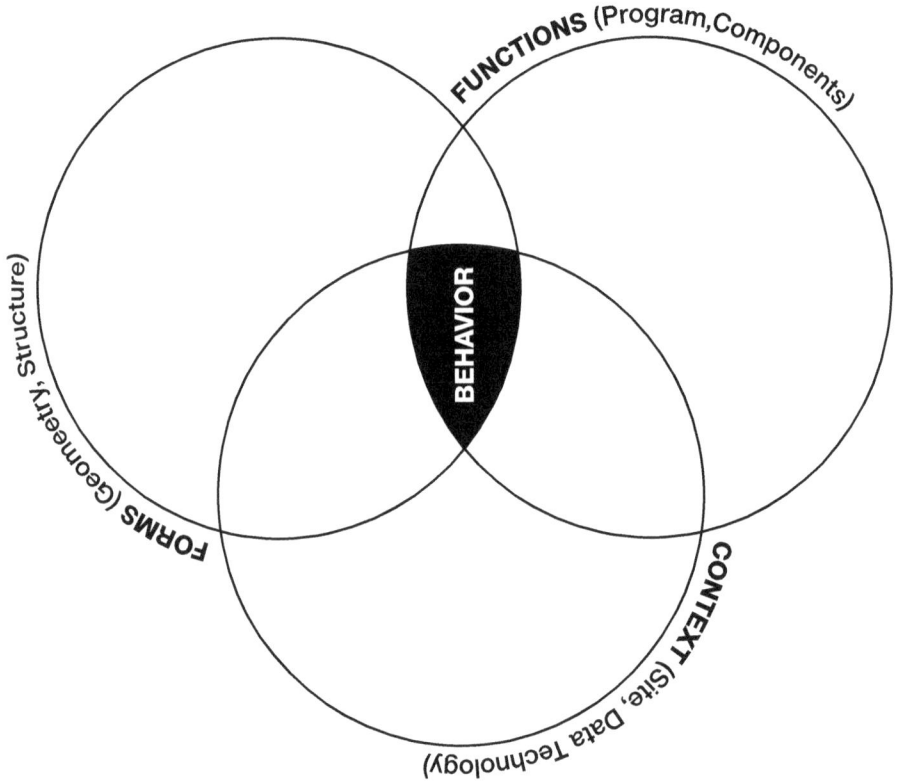

Figure 5.4: A behavior is shown as the confluence of form, context and function.

world will be achieved through the combined research of abstract agents, physical robots and vernacular building systems. My vision, supported by a few peer roboticists and design researchers, is to develop robotic platforms that operate autonomously, can navigate 3D space while dealing with external disturbances, like uneven surfaces, dust and the dynamically changing environment of construction sites, and can manipulate generic building blocks in its environment in order to build aggregate structures.

A lot of discussion in the last years has revolved around the disruptive character of technologies and how it will take out jobs and potentially render designers useless. Although it is true the recent advancements in generative AI indicate that it will radically change the nature or work for many (similar to how digital design tools eliminated the need for draftsmen), my perspective is more optimistic: Despite all the technological advancements, designers are still relevant or at least as relevant as positions of design technologists and BIM managers are today.

In fact, in this future, designers can become more powerful as they will have more time for actual design thinking and will be able to focus more of their energy on creative instead of laborious tasks. In this future, we will not be competing with objects or AI systems, but we will use them to extend our reach and capacity as we have always done. But to get there, we need not forget that the focus is not technology but the problem we are trying to solve which is providing environments for habitation. While doing this we need not forget that tools are useless without skilled craftsmen. More importantly, we need to remind ourselves that new tools also require new methodologies and techniques, and in order to propel forward and redefine our discipline we need to first solve basic problems that have existed since the introduction of digital design tools. Tools are not meant to make our lives easier, not necessarily. They are meant to extend our capacities so that we can push ourselves to new limits. Tools are used to lift rocks and, nowadays, to compute millions of bites in a split second, but people are the ones who build cities, museums and cathedrals.

Acknowledgements: A work of this nature owes a large debt of gratitude to the many people who have directly or indirectly contributed to it. During the period that this research was taking shape, I had the privilege to work closely with talented architects, engineers, computer scientists and design researchers from both academia and the industry.

Firstly, I would like to express my gratitude to my advisor Professor David Jason Gerber, with whom I embarked on this journey to investigate this topic. I owe a big thanks to Professors E. Chatzi and L. Hovestadt who were the first to support me in pursuing graduate studies.

I also want to thank the members of my Doctoral Committee, Professors Lucio Soibelman, Burcin Becerik Gerber, Marcos Novak and Yan Jin, for their guidance and support in what has become a meandering path to the completion of this thesis.

I would like to thank the academics at USC who I came to know either by being their student or by interacting though my teaching or other academic activities namely: A. McDowell, B. Khoshnevis, E. Johnson, M. Tambe, H. Koffman, P. Lynett, F. De Barros, M Triffunac, W. Wood.

I am thankful to my colleagues and co-authors Arsalan Heydarian, Leandro Soriano Marcolino and Alan Wang. Our collaboration taught me a lot and gave shape to my research. I also want to thank my fellow graduate students from USC who we shared similar stresses namely Biayna Bogosian, Nikos Kalligeris, Eyup Koc, Ashrant Aryal, Meida Chen, Joao Carneiro, Ali Kazemian, and Pouyan Hosseini. I would like to thank S. Sugihara, J. Sanchez, J. Haymaker, A. Bohrani, N. Kalantar, A. Pistofidou, N. Napp, K Petersen, P. Jenning, S.Jokic, J.Dawson, Adam P. Mahardy, V. Toulkeridou, S. Koumlis, for discussing and exchanging their thoughts with me. A big thanks goes to my close friends and colleagues, including C. Miltiadis, R. Morgan, C. Jakob, O. Carasco, C.Schinas, and S.Panagiotopoulos for dedicating time to reading and commenting during the literature review process.

I would like to thank the graduate and undergraduate students for their contributions in the experiment case studies namely: Ye Tian, Yuze Liu, Jan Jingbo, Justin Yang, Alejandro Medina, Nick Morof, Rheseok Kim, Kayla Ching, Vivian Wang, Brian Herrera, Kevin Daley, Punit Das, and Francine Ngo. I owe a big thank you my close friends Anastasios Dedes and Viktoras Gogas for their invaluable help with the graphics and Despoina Pavlaki's help with the text. Their input was critical in making the text more easily legible and the existing graphical content visually more coherent.

I am both literally and metaphorically indebted to all the organizations, institutes and companies that accepted my research proposals and believed in my vision. Therefore, I would like to thank: the National Science Foundation in the USA, which funded my research (grant no.1231001), the Onassis Foundation in Greece and especially Ioanna Kailani, the IKEA foundation in Switzerland, the Myronis Fellowship and the Gerondelis Foundation, the Institute of Advanced Architecture of Catalonia (IAAC) and especially A. Markopolou and A. Dubor as well as Autodesk, Inc and especially G. Katz, E. Bradner, R. Rundell, A. Moore for supporting financially my research efforts.

I owe the foremost debt of gratitude to my parents Spiros Pantazis and Eleni Margaritidou and my siblings, Magdalena and Iason, who have always shone a bright light on my path and supported me on my academic endeavors. Last but not least, I am more than grateful to Christina for her patience and love throughout this journey.

Part VI: **Bibliography**

6 List of relevant publications

Findings of the work presented in this book have been published in peer-reviewed conferences and journals as part of my Ph.D. studies. There are in total 21 peer-reviewed international publications, 4 of which are journal papers and the remaining 17 are long conference papers. Lastly, I have released Termite, the alpha version of a multi-agent systems toolkit for architectural design which can be found on Github (https://github.com/TopoVague). An initial implementation of the MAS in Grasshopper has been published on food4rhino.com under the name of "TermitesABMStoolkit" (https://github.com/TopoVague/termitesABMS_toolkit).

Peer-reviewed journal papers

[1] **Pantazis**, E. and Gerber, D. (2018): "A framework for generating and evaluating façade designs using a multi-agent systems approach", *International Journal of Architectural Computing*, Vol 16, pp. 248–270.
[2] Gerber, D., **Pantazis**, E., and Wang, A. (2017): "A multi-agent approach for performance-based architecture: Design exploring geometry, user, and environmental agencies in façades", *Automation in Construction*, Vol 76, pp. 45–58.
[3] **Pantazis**, E. and Gerber, D. (2019): "Beyond Geometric Complexity: A Critical Review of Complexity Theory in Architecture", *Architecture Science Review Journal*, Vol 60, 7 (accepted minor reviews).
[4] **Pantazis**, E. and Gerber, D. "Behavioral Form Finding: Interactive Shell Design Using Multi-Agent Systems", *ArchiDoc* (accepted minor reviews).

Peer-reviewed conference papers

[1] **Pantazis, E.** and Gerber, D. (2017) "Emergent order through swarm fluctuations – a framework for exploring self-organizing structures using swarm robotics, in ShoCK: Sharing computational knowledge!" Proceedings of the 35th eCAADe Conference, A. Fioravanti, et al., Editors, Sapienza University of Rome, Italy, pp. 75–84.
[2] Heydarian, A., **Pantazis, E.**, Gerber, D., and Becerik-Gerber, B. (2016). "Defining Lighting Settings to Accommodate End-User Preferences While Reducing Energy Consumption in Buildings", *Proc. Construction Research Congress*, pp. 1122–1132.
[3] **Pantazis, E.** and Gerber, D. (2016): "Design Exploring Complexity in Architectural Shells – Interactive Form Finding of Reciprocal Frames Through a Multi-Agent System", *in 34th eCAADe: Complexity and Simplicity*, A. Herneoja, et al., Editors, Oulu, Finland.
[4] Marcolino, L.S., Xu, H., Gerber, D., Kolev, B., Price, S., **Pantazis, E.**, and Tambe, M. (2016): "Multi-Agent Team Formation for Design Problems", in *Coordination, Organizations, Institutions, and Normes in Agent Systems XI*, V. Dignum, et al., Springer.
[5] **Pantazis, E.**, Gerber, D., and Wang, A. (2016) "A Multi-Agent System for Design: Geometric Complexity in Support of Building Performance", *in Symposium for Modelling and Simulation in Architecture and Urban Design (SIMAUD)*, Ramtin, A., et al., Editors, pp. 137–143, London, UK.
[6] Gerber, D. and **Pantazis**, E. (2016) "A Multi-Agent System for Design: A Design Methodology for Design Exploration, Analysis and Simulated Robotic Fabrication", *in ACADIA16: Posthuman Frontiers*, Velikov, K., et al., Editors, pp. 12–21, University of Michigan, Ann Arbor, MI.

https://doi.org/10.1515/9783110797435-006

[7]   Gerber, D., **Pantazis**, **E.**, and Marcolino L.S. (2015) "Design Agency", *in Computer Aided Architectural Design Futures. The Next City – New Technologies and the Future of the Build Environment*, Celani, G., et al., Editors, Springer, pp. 213–235, Sao Paulo, Brazil.

[8]   Heydarian, A., Carneiro, J.P., **Pantazis**, **E.**, Gerber, D., and Becerik-Gerber, B. (2015). "Default Conditions: A Reason for Design to Integrate Human Factors", *Sustainable Human-Building Ecosystems*, pp. 54–62.

[9]   Heydarian, A., **Pantazis**, **E.**, Carneiro, J.P., Gerber, D., and Becerik-Gerber, B. (2015): "Towards Understanding End-user Lighting Preferences in Office Spaces by Using Immersive Virtual Environments", *The International Workshop on Computing in Civil Engineering*, pp. 475–482.

[10]   Gerber, D., **Pantazis**, **E.**, Marcolino, L., and Heydarian, A. (2015): "A Multi-Agent Framework for Simulation of Cyber Physical Social Feedback for Architecture", *The Symposium on Simulation for Architecture and Urban Design (SimAUD)*, pp. 205–212.

[11]   Heydarian, A., Carneiro, J.P., **Pantazis**, **E.**, Gerber, D., and Becerik-Gerber, B. (2015): "Default Conditions: A Reason for Design to Integrate Human Factors" in *Sustainable Human-Building Ecosystems*, pp. 54–62.

[12]   Heydarian, A., **Pantazis**, **E.**, Carneiro, J.P., Gerber, D., and Becerik-Gerber, B. (2015): "Towards Understanding End-User Lighting Preferences in Office Spaces by Using Immersive Virtual Environments", in Computing in Civil Engineering. ASCE, pp. 475–482.

[13]   Heydarian, A., **Pantazis**, **E.**, Gerber D., and Becerik-Gerber B. (2016): "Defining Lighting Settings to Accommodate End-User Preferences While Reducing Energy Consumption in Buildings", in Construction Research Congress 2016. pp. 1122–1132.

[14]   Marcolino, L., Price, S., **Pantazis**, **E.**, and Tambe, M. (2015): "Multi-Agent Team Formation for Design Problems", in Coordination, Organizations, Institutions, and Normes in Agent Systems XI: COIN 2015 International Workshops, COIN@ AAMAS. Springer, 354, Istanbul, Turkey.

[15]   Marcolino, L.S., Gerber, D., Kolev, B., Price, S., **Pantazis**, **E.**, Tian, Y., and Tambe, M. (2015): "Agents Vote for the Environment: Designing Energy-Efficient Architecture", in AAAI Workshop on Computational Sustainability.

[16]   Marcolino, L.S., Xu, H., Gerber, D., Kolev, B., Price, S., **Pantazis**, **E.**, and Tambe M. (2015): "Agent Teams for Design Problems", *in COIN@AAMAS*, pp. 189–195, Istanbul, Turkey.

[17]   **Pantazis**, **E.** and Gerber, D. (2014): "Material Swarm Articulations – New View Reciprocal Frame Canopy", *in 32nd eCaade: Fusion*, E. M. Thompson, et al., Editors, pp. 463–473, Newcastle, England.

References

[1]   Wolfram, S., *A new kind of science*. Vol. 5. 2002: Wolfram media Champaign.

[2]   Mitchell, W.J., *Constructing complexity*, in *Computer aided architectural design futures 2005*. 2005, Springer. pp. 41–50.

[3]   Malkawi, A., *Performance simulation: research and tools*, in *Performative architecture: beyond instrumentality*, B. Kolarevic and A. Malkawi, Editors. 2005, Spon Press: New York. pp. 85–96.

[4]   Rahman, M., *Complexity in building design*, in *Re-inventing Construction*, I. Ruby and A. Ruby, Editors. 2010, Ruby Press: Berlin. pp. 440.

[5]   Kalay, Y.E., *Architecture's new media: principles, theories, and methods of computer-aided design*. 2004: MIT Press.

[6]   Jencks, C., *Architecture 2000 and Beyond: Success in the art of prediction*. 2000: Wiley.

[7]   Block, P., Thrust Network Analysis: Exploring Three-Dimensional Equilibrium, Cambridge, MA, 2009, Doctor of Philosophy 156

[8]   Woodbury, R., et al., Typed feature structures and design space exploration. AI EDAM, 1999. 13(4), pp. 287–302.

[9]   Gero, J.S. and R. Sosa, Complexity measures as a basis for mass customization of novel designs. Environment and Planning B: Planning and Design, 2008. 35(1), pp. 3–15.

[10]   Scheurer, F., Materialising Complexity. Architectural Design, 2010. 80(4), pp. 86–93.

[11]   Oosterhuis, K., *Towards a New Kind of Building: A Designers Guide for Non-Standard Architecture*. 2011: The Netherlands: NAi Uitgevers/Publishers Stichting.

[12]   Oxman, R., Performance-based design: Current practices and research issues. International Journal of Architectural Computing, 2008. 6(1), pp. 1–17.

[13]   Carpo, M., *The digital turn in architecture 1992-2012*. 2013: John Wiley & Sons.

[14]   Oesterreich, T.D. and F. Teuteberg, Understanding the implications of digitisation and automation in the context of Industry 4.0: A triangulation approach and elements of a research agenda for the construction industry. Computers in Industry, 2016. 83, pp. 121–139.

[15]   Oxman, R., Theory and design in the first digital age. Design Studies, 2006. 27(3), pp. 229–265.

[16]   Adriaenssens, S., et al., *Shell Structures for Architecture: Form Finding and Optimization*, ed. S. Adriaenssens, et al. Vol. I. 2014, London and New York: Taylor & Francis – Routledge.

[17]   Menges, A. *Computational morphogenesis*, in *Proceedings for 3rd International ASCAAD Conference*. 2007. Alexandria, Egypt.

[18]   Oxman, N. *Digital craft: Fabrication-based design in the age of digital production*. in *UbiComp 2007*. 2007. Innsbruck, Austria.

[19]   Fricker, P., et al., *Organised Complexity*. Predicting the Future. 2007, Frankfurt am Main, Germany: 25th eCAADe Conference Proceedings.

[20]   Simon, H., *The Architecture of Complexity*, in *Facets of Systems Science*. 1991, Springer: New York, USA. pp. 457–476.

[21]   Halfawy, M. and T. Froese, Building Integrated Architecture/Engineering/Construction Systems Using Smart Objects: Methodology and Implementation 1. Journal of Computing in Civil Engineering, 2005. 19(2), pp. 172–181.

[22]   Economist, T., Rise of the robots; New roles for technology, London, UK, 2014, 1 (13),

[23]   Bonabeau, E., M. Dorigo, and G. Theraulaz, *Swarm intelligence: from natural to artificial systems*. Santa Fe Institute Studies in the Sciences of Complexity. 1999, New York, USA: Oxford University Press.

[24]   Werfel, J., K. Petersen, and R. Nagpal, Designing Collective Behavior in a Termite-Inspired Robot Construction Team. Science, 2014. 343, pp. 754–758.

[25]   Fischer, T. *Wiener's refiguring of a cybernetic design theory*. in *Norbert Wiener in the 21st Century (21CW), 2014 IEEE Conference on*. 2014. IEEE.

[26]   Archer, A., Multimodal semiotic resources in an Engineering curriculum. Academic Literacy and the Languages of Change, 2006, p. 130.

[27]   Newell, A. and H.A. Simon, *Human problem solving*. Vol. 104. 1972: Prentice-Hall Englewood Cliffs, NJ.

https://doi.org/10.1515/9783110797435-007

[28]   Asimow, M., *Introduction to design*. 1962: Prentice-Hall.

[29]   Moneo, R., On Typology. Oppositions summer, 1978, 13.

[30]   Von Bülow, P., An intelligent genetic design tool (IGDT) applied to the exploration of architectural trussed structural systems, 2007, Ph.D.

[31]   Simon, H.A., The structure of ill-structured problems. Artificial Intelligence 1973, Models of Discovery (4), pp. 181–201.

[32]   Cryer, J.N., Design Team Agreements 3.43. The Architect's Handbook of Professional Practice, 1994. 2.

[33]   Plesk, P. and T. Wilson, Complexity, leadership, and management in healthcare organizations. BMJ, 2001. 323(7315), pp. 746–749.

[34]   Senatore, G. and D. Piker, Interactive real-time physics: an intuitive approach to form-finding and structural analysis for design and education. Computer-Aided Design, 2015. 61, pp. 32–41.

[35]   Roudsari, M., M. Pak, and A. Smith, Ladybug: A Parametric Environmental Plugin for Grasshopper to Help Designers Create an Environmentally-Conscious Design. 2014.

[36]   Chen, C.-H. *A prototype using Multi-Agent Based Simulation in Spatial Analysis and Planning*. in *14th Annual Conference of the Association of Computer Aided Architectural Design (CAADRIA)*. 2009. Douliu, Taiwan.

[37]   Eastman, C.M., et al., *BIM handbook: A guide to building information modeling for owners, managers, designers, engineers and contractors*. 2011: John Wiley & Sons.

[38]   Keough, I. and D. Benjamin. *Multi-objective optimization in architectural design*. in *Proceedings of the 2010 Spring Simulation Multiconference*. 2010. Society for Computer Simulation International.

[39]   Rutten, D. Navigating multi-dimensional landscapes in foggy weather as an analogy for generic problem solving, Innsbruck, Austria, 2014, 14.

[40]   Gerber, D.J. and S.-H.E. Lin, Designing in complexity: Simulation, integration, and multidisciplinary design optimization for architecture. Simulation, 2013, pp. 1–24.

[41]   Gero, J.S., Computational models of creative designing based on situated cognition, Loughborough, UK, 2002, pp. 3–10.

[42]   Gero, J.S., Computational models of innovative and creative design processes. Technological Forecasting and Social Change, 2000. 64(2–3), pp. 183–196.

[43]   Gero, J.S., Creativity, emergence and evolution in design. Knowledge-Based Systems, 1996. 9(7), pp. 435–448.

[44]   Kaelbling, L.P., M.L. Littman, and A.R. Cassandra, Planning and acting in partially observable stochastic domains. Artificial Intelligence, 1998. 101(1–2), pp. 99–134.

[45]   Simon, H.A., *The sciences of the artificial*. 1996: MIT Press.

[46]   Mitchell, J., J. Wong, and J. Plume. *Design collaboration using IFC*. in *Computer-Aided Architectural Design Futures (CAADFutures 2007)*. 2007. Dordrecht: Springer Netherlands.

[47]   Kotnik, T., Digital architectural design as exploration of computable functions. International Journal of Architectural Computing, 2010. 8(1), pp. 1–16.

[48]   Kalay, Y.E., Performance-based design. Automation in Construction, 1999. 8(4), pp. 395–409.

[49]   Duro-Royo, J., L. Mogas-Soldevila, and N. Oxman, Flow-based fabrication: An integrated computational workflow for design and digital additive manufacturing of multifunctional heterogeneously structured objects. Computer-Aided Design, 2015. 69, pp. 143–154.

[50]   Rippmann, M., L. Lachauer, and P. Block, Interactive vault design. International Journal of Space Structures, 2012. 27(4), pp. 219–230.

[51]   Gerber, D.J. and S.-H.E. Lin. *Designing-in performance through parameterization, automation, and evolutionary algorithms: 'H.D.S. BEAGLE 1.0'*. in *CAADRIA 2012: Beyond Codes and Pixels*. 2012. Chennai, India.

[52]   Caldas, L., Generation of energy-efficient architecture solutions applying GENE_ARCH: An evolution-based generative design system. Advanced Engineering Informatics, 2008. 22(1), pp. 59–70.

[53]   Marincic, N., Towards communication in CAAD: Spectral Characterisation and Modelling with Conjugate Symbolic Domains, Zurich, Switzerland, 2016, Doctor of Sciences 231.

[54]   Caldas, L.G. and L.K. Norford, A design optimization tool based on a genetic algorithm. Automation in Construction, 2002. 11(2), pp. 173–184.

[55]   Bukhari, F., J.H. Frazer, and R. Drogemuller. *Evolutionary algorithms for sustainable building design*. in *The 2nd International Conference on Sustainable Architecture and Urban Development*. 2010. Amman, Jordan.

[56]   Dimcic, M.,Structural optimization of grid shells based on genetic algorithms, Stuttgart, Germany, 2012, Ph.D.

[57]   Gerber, D.J., et al. *Design optioneering: Multi-disciplinary design optimization through parameterization, domain integration and automation of a genetic algorithm*. in *SimAUD 2012*. 2012. Orlando, FL, USA.

[58]   Malkawi, A.M., et al., Decision support and design evolution: integrating genetic algorithms, CFD and visualization. Automation in Construction, 2005. 14(1), pp. 33–44.

[59]   Pugnale, A. and M. Sassone, Morphogenesis and structural optimization of shell structures with the aid of a genetic algorithm. Journal-International Association for Shell and Spatial Structures, 2007. 155, p. 161.

[60]   Kilian, A., Design Exploration through bidirectional modeling of Constraints, Cambridge, MA, 2006, Doctor of Philosophy in Architecture, Design and Computation.

[61]   Alexander, C., *Notes on the synthesis of form*. 1964: Harvard University Press. 224.

[62]   Stiny, G., Introduction to shape and shape grammars. Environment and Planning B: Planning and Design, 1980. 7(3), pp. 343–351.

[63]   Prusinkiewicz, P. and A. Lindenmayer, *The algorithmic beauty of plants*. 2012: Springer Science & Business Media.

[64]   Müller, P., et al. *Procedural modeling of buildings*. in *ACM Transactions on Graphics (Tog)*. 2006. ACM.

[65]   Popov, N., Generative Sub-division Morphogenesis with Cellular Automata and Agent-Based Modelling, University of Ljubljana, Faculty of Architecture (Slovenia), 2011, 166–174.

[66]   Frazer, J.H., An Evolutionary Architecture. Themes, 1995.

[67]   Minsky, M., Steps toward artificial intelligence. Proceedings of the IRE, 1961. 49(1), pp. 8–30.

[68]   Negroponte, N., *The architecture machine: towards a more human environment*. 1970: MIT Press Cambridge, MA.

[69]   Hebron, P. *Rethinking Design Tools in the Age of Machine Learning*. Artists and machine intelligence, 2017.

[70]   Marcolino, L.S., X.J. Jiang, and M. Tambe, Multi-agent Team Formation: Diversity Beats Strength?, Beijing, China, 2013.

[71]   Kolarevic, B., Simplexity (and Complicity) in Architecture. 2016.

[72]   Bundy, A., Computational thinking is pervasive. Journal of Scientific and Practical Computing, 2007. 1(2), pp. 67–69.

[73]   Kauffman, S.A., *The origins of order: Self organization and selection in evolution*. 1993: Oxford university press.

[74]   Perna, A. and G. Theraulaz, When social behaviour is moulded in clay: on growth and form of social insect nests. Journal of Experimental Biology, 2017. 220(1), pp. 83–91.

[75]   Weiss, A., Google N-gram viewer. The Complete Guide to Using Google in Libraries: Instruction, Administration, and Staff Productivity, 2015. 1, p. 183.

[76]   Venturi, R., *Complexity and contradiction in architecture*. Vol. 1. 1977, New York, USA: The Museum of Modern Art.

[77]   Gell-Mann, M. *Complexity and complex adaptive systems*. in *Proceedings of the Santa Fe Institute Studies in the Sciences of Complexity*. 1992. California Institute of Technology, Pasadena: Addison- Wesley Publishing Co.

[78]   Holland, J.H., *Adaptation in natural and artificial systems: An introductory analysis with applications to biology, control, and artificial intelligence*. Vol. 1. 1992, Cambridge, MA: MIT Press. 211.

[79]   Turing, A.M., On computable numbers, with an application to the Entscheidungsproblem. J. of Math, 1936. 58(345–363), p. 5.

[80]   Shannon, C.E., A mathematical theory of communication. Mobile Computing and Communication Review (SIGMOBILE), 1948. 5(1), pp. 3–55.

[81]   Bertalanffy, L.V., An Outline of General System Theory. The British Journal for the Philosophy of Science, 1950. 1(2), pp. 134–165.

[82] Von Neumann, J., The general and logical theory of automata. Cerebral mechanisms in behavior, 1951, pp. 1–41.

[83] Bennett, C.H., *Logical depth and physical complexity*. 1995: Springer.

[84] Feldman, D.P. and J. Crutchfield, A survey of complexity measures. Santa Fe Institute, USA, 1998. 11.

[85] Crutchfield, J.P., The calculi of emergence: computation, dynamics and induction. Physica D: Nonlinear Phenomena, 1994. 75(1), pp. 11–54.

[86] Lloyd, S., Measures of complexity: a nonexhaustive list. IEEE Control Systems Magazine, 2001. 21(4), pp. 7–8.

[87] Gell-Mann, M., *What is complexity?*, in *The Quark and the Jaguar*. 1995, John Wiley & Sons Ltd: MA, USA. pp. 13–24.

[88] Kauffman, S.A., Antichaos and adaptation. Scientific American, 1991. 265(2), pp. 78–84.

[89] Goldenfeld, N. and L.P. Kadanoff, Simple Lessons from Complexity. Science, 1999. 284(5411), pp. 87–89.

[90] Yates, F.E., Complexity and the limits to knowledge. American Journal of Physiology-Regulatory, Integrative and Comparative Physiology, 1978. 235(5), pp. R201–R204.

[91] Holland, J.H., *Adaptation in natural and artificial systems: An introductory analysis with applications to biology, control, and artificial intelligence*. 1992, Ann Arbor: A Bradford Book. 211.

[92] Klir, G., *Architecture of systems problem solving*. 1985, New York: Plenum Press.

[93] Jacobs, J., *The death and life of great American cities*. 1961: Vintage.

[94] Salingaros, N., Complexity and Urban Coherence. Journal of Urban Design, 2000. 5, pp. 291–316.

[95] Dent, E.B., Complexity science: A worldview shift. Emergence, 1999. 1(4), pp. 5–19.

[96] Wiener, N., *Cybernetics or Control and Communication in the Animal and the Machine*. 2nd ed. Vol. 25. 1961, Cambridge, Massachusetts: MIT Press.

[97] Fisher, R.A., Statistical methods and scientific inference. 1956.

[98] Von Bertalanffy, L., The meaning of general system theory. General System Theory: Foundations, Development, Applications, 1973, 1, pp. 30–53.

[99] Rapoport, A., *General system theory: essential concepts & applications*. Vol. 10. 1986: CRC Press.

[100] Gerard, R., Concepts and principles of biology. Initial Working Paper. Behavioral Science, 1958. 3(1), pp. 95–102.

[101] Heylighen, F. and C. Joslyn, Cybernetics and second order cybernetics. Encyclopedia of Physical Science & Technology, 2001. 4, pp. 155–170.

[102] Maturana, H.R. and F.J. Varela, *The tree of knowledge: The biological roots of human understanding*. 1987, Boston, MA: New Science Library/Shambhala Publications.

[103] Weaver, W., Science and complexity. American Scientist, 1948. 36(536), pp. 449–456.

[104] Schuh, G. and W. Eversheim, Release-Engineering – An Approach to Control Rising System-Complexity. CIRP Annals-Manufacturing Technology, 2004. 53(1), pp. 167–170.

[105] Suh, N.P., *Complexity: Theory and application*. 2005, New York: Oxford University Press.

[106] Suh, N.P., Complexity in Engineering. CIRP Annals – Manufacturing Technology, 2005. 54(2), pp. 46–63.

[107] Stefan Wrona, A.G. *Complexity in Architecture – How CAAD can be involved to Deal with it*. in *AVOCAAD – Added Value of Computer Aided Architectural Design*. 2001. ogeschool voor Wetenschap en Kunst – Departement Architectuur Sint-Lucas, Campus Brussel.

[108] Kolmogorov, A.N., Three approaches to the definition of the concept "quantity of information". Problems of Information Translation, 1965. 1(1), pp. 3–11.

[109] Chaitin, G.J., *Information, randomness & incompleteness: papers on algorithmic information theory*. Series in Computer Science. Vol. 8. 1990, New York, USA: World Scientific.

[110] Lloyd, S. and H. Pagels, Complexity as thermodynamic depth. Annals of Physics, 1988. 188(1), pp. 186–213.

[111] Cover, T.M. and J.A. Thomas, *Elements of information theory*. 2012: John Wiley & Sons.

[112] Traub, J.F., G.W. Wasilkowski, and H. Woźniakowski, *Information, uncertainty, complexity*. 1983: Addison-Wesley Publishing Company, Advanced Book Program/World Science Division.

[113] Horgan, J., From complexity to perplexity: Can science achieve a unified theory of complex systems? Scientific American, 1995. 284, pp. 104–109.

[114] Barton, G.E., Berrywick, R.C. and Ristad, E.S., *Computational Complexity and Natural Language*. 1987, MIT Press, Cambridge MA.

[115] Cobham, A. *The intrinsic computational difficulty of functions*. in *Logic, Methodology and Philosophy of Science: Studies in Logic and the Foundations of Mathematics*. 1965. North-Holland Publishing.

[116] Mitchell, W.J., *The logic of architecture: Design, computation, and cognition*. 1990, Cambridge, MA: MIT Press.

[117] Suh, N.P., *The principles of design*. Vol. 990. 1990: Oxford University Press New York.

[118] Baccarini, D., The concept of project complexity – a review. International Journal of Project Management, 1996. 14(4), pp. 201–204.

[119] Morris, P.W. and G. Hough, The anatomy of major projects, 1987.

[120] Levin, L.A., Different Measures of Complexity of finite objects (Axiomatic Description). Doklady Akademii Nauk SSSR, 1976. 227(4), pp. 804–807.

[121] Gell-Mann, M. and S. Lloyd, Information measures, effective complexity, and total information. Complexity, 1996. 2(1), pp. 44–52.

[122] Rittel, H. and M.M. Webber, 2.3 planning problems are wicked. Polity, 1973. 4, pp. 155–69.

[123] Glanville, R., Designing complexity. Performance Improvement Quarterly, 2007. 20(2), pp. 75–96.

[124] Glanville, R., An intelligent architecture. Convergence: The International Journal of Research into New Media Technologies, 2001. 7(2), pp. 12–24.

[125] Terzidis, K., *Algorithmic architecture*. 2006, Oxford; Burlington, MA: Architectural Press.

[126] Simon, H.A., *The structure of ill-structured problems*, in *Models of discovery*. 1977, Springer. pp. 304–325.

[127] Theraulaz, G. and E. Bonabeau, Coordination in distributed building. Science, 1995. 269(5224), p. 686.

[128] Cross, N. and N. Roozenburg, Modelling the design process in engineering and in architecture. Journal of Engineering Design, 1992. 3(4), pp. 325–337.

[129] Scheurer, F., Getting complexity organised Using self-organisation in architectural construction. Automation in Construction, 2007(16), pp. 78–85.

[130] Soibelman, L., et al., Management and analysis of unstructured construction data types. Advanced Engineering Informatics, 2008. 22(1), pp. 15–27.

[131] Bennett, J., *International construction project management: general theory and practice*. 1991, New Hampshire, USA: Butterworth-Heinemann.

[132] Beyer, J.M. and H.M. Trice, A reexamination of the relations between size and various components of organizational complexity. Administrative Science Quarterly, 1979, pp. 48–64.

[133] Gidado, K., Numerical index of complexity in building construction to its effect on production time,1993.

[134] Teicholz, P. *Vision of future practice*. in *Berkeley-Stanford Workshop on Defining a Research Agenda for AEC Process/Product Development in*. 2000.

[135] Caldas, C.H. and L. Soibelman, Automating hierarchical document classification for construction management information systems. Automation in Construction, 2003. 12(4), pp. 395–406.

[136] Shen, W., et al., Systems integration and collaboration in architecture, engineering, construction, and facilities management: A review. Advanced Engineering Informatics, 2010. 24(2), pp. 196–207.

[137] Rahaman, H. and B.-K. Tan, Interpreting digital heritage: A conceptual model with end-users' perspective. International Journal of Architectural Computing, 2011. 9(1), pp. 99–114.

[138] Jazizadeh, F., et al. *Human-building interaction for energy conservation in office buildings*. in *Proc. of the Construction Research Congress*. 2012.

[139] Clements-Croome, D., *Intelligent buildings: design, management and operation*. 2004: Thomas Telford.

[140] Ferreira, P., Tracing Complexity Theory, Pittsburgh, PA, 2001.

[141] Petersen, K.H., Collective Construction by Termite-Inspired Robots. 2014.

[142] Macal, C.M. and M.J. North. *Agent-based modeling and simulation*. in *Winter simulation conference*. 2009. Winter Simulation Conference.

[143] Hansell, M.H., Animal architecture and building behaviour. Animal architecture and building behaviour, 1984.

[144] Kilian, A., Steering of form. Shell Structures for Architecture: Form Finding and Optimization, 2014, p. 131.

[145] Beni, G. and J. Wang, *Swarm intelligence in cellular robotic systems*, in *Robots and Biological Systems: Towards a New Bionics?* 1993, Springer. pp. 703–712.

[146] Weiss, G., Multiagent Systems: A Modern Approach to Distributed Artificial Intelligence. 1999, p. 619.

[147] Kasabov, N. and R. Kozma, Introduction: Hybrid intelligent adaptive systems. International Journal of Intelligent Systems, 1998. 13(6), pp. 453–454.

[148] Tambe, M., Implementing agent teams in dynamic multiagent environments. Applied Artificial Intelligence, 1998. 12(2–3), pp. 189–210.

[149] Gero, J.S. and F.M.T. Brazier, Intelligent agents in design. Artificial Intelligence for Engineering Design, Analysis and Manufacturing (AIEDAM), 2004. 18, p. 113.

[150] Russell, S.J., et al., *Artificial intelligence: a modern approach*. Vol. 2. 2003, Upper Saddle River: Prentice Hall.

[151] Weinstein, P., et al. *Agents swarming in semantic spaces to corroborate hypotheses*. in *Proceedings of the Third International Joint Conference on Autonomous Agents and Multiagent Systems-Volume 3*. 2004. IEEE Computer Society.

[152] Reynolds, C.W., Flocks, herds and schools: A distributed behavioral model. ACM SIGGRAPH Computer Graphics, 1987. 21(4), pp. 25–34.

[153] Yezioro, A., D. Bing, and F. Leite, An applied artificial intelligence approach towards assessing building performance simulation tools. Energy and Buildings, 2008. 40(4), pp. 612–620.

[154] Woodridge, M. and N.R. Jennings, Intelligent Agents: Theory and Practice. Knowledge Engineering Review, 1995. 10, pp. 115–152.

[155] Tambe, M., *Teamwork in real-world, dynamic environments*. 1996: University of Southern California, Information Sciences Institute.

[156] Jin, Y. and W. Li, Design concept generation: a hierarchical coevolutionary approach. Journal of Mechanical Design, 2007. 129(10), pp. 1012–1022.

[157] Holland, J.H., *Hidden order: How adaptation builds complexity*. 1995: Basic Books.

[158] Sycara, K., Multiagent systems. AI Magazine, 1998. 19(2), pp. 79–92.

[159] Beetz, J., J. Van Leeuwen, and B. De Vries, *Towards a Multi Agent System for the Support of Collaborative Design*. 2004, Developments in Design & Decision Support Systems in Architecture and Urban Planning: Eindhoven University of Technology.

[160] Anumba, C.J., et al., *Intelligent Agent Applications in Construction Engineering, Creative Systems in Structural and Construction Engineering*. 2001: CRC Press.

[161] Leach, N., Swarm Urbanism. Architectural Design, 2009. 79, pp. 56–63.

[162] Jennings, N.R., J.M. Corera, and I. Laresgoiti. *Developing Industrial Multi-Agent Systems*. in *ICMAS*. 1995.

[163] Werfel, J., et al., Distributed Construction by Mobile Robots with Enhanced Building Blocks, New England, 2006.

[164] Schwinn, T., A. Menges, and D.O. Krieg, Behavioral Strategies: Synthesizing design computation and robotic fabrication of lightweight timber plate structures, Los Angeles, 2014.

[165] Herr, C.M., Generative Architectural Design and Complexity Theory, Milan, Italy, 2002, 16 1–13

[166] Simeone, D., et al. *Modelling and Simulating Use Processes in Buildings*. in *eCAADe 2013: Computation and Performance*. 2013. Delft, The Netherlands: Faculty of Architecture, Delft University of Technology; eCAADe (Education and research in Computer Aided Architectural Design in Europe).

[167] Sugihara, S. *Comparison between Top-Down and Bottom-Up Algorithms in Computational Design Practice*. in *International Symposium on Algorithmic Design for Architecture and Urban Design, ALGODE* 2011. Tokyo.

[168] Sugihara, S. *iGeo: Algorithm Development Environment for Computational Design Coders with Integration of NURBS Geometry Modeling and Agent Based Modeling.* in *ACADIA 14: Design Agency.* 2014. Los Angeles: eVolo.

[169] Maher, M.L. and M. Kim. *Supporting Design Using Self-Organizing Design Knowledge.* in *CAADRIA 2004 – Proceedings of the 9th International Conference on Computer Aided Architectural Design Research in Asia.* 2004. Seoul, Korea.

[170] Aranda, B. and C. Lasch, Flocking, New York, 2006, 27.

[171] Ednie-Brown, P. and A. Andrasek, Continuum: A Self-Engineering Creature-Culture. Architectural Design, 2006. 76(5), pp. 18–25.

[172] Carranza, P.M. and P. Coates. *Swarm modelling. the use of Swarm Intelligence to generate architectural form.* in *3rd Generative Art Conference. Generative Art.* 2000. Milan, Italy: AleaDesign Publisher.

[173] Ireland, T. *Emergent space diagrams: The application of swarm intelligence to the problem of automatic plan generation.* in *CAADFutures: Joining Languages, Cultures and Visions.* 2009. Montreal, Canada.

[174] Tsiliakos, M., Swarm Materiality: A multi-agent approach to stress driven material organization, 2012, 1, 301–309.

[175] Snooks, R. *Encoding Behavioral Matter.* in *International Symposium on Algorithmic Design for Architecture and Urban Design (ALGODE).* 2011. Tokyo, Japan.

[176] Leach, N., D. Turnbull, and C. Williams, *Digital tectonics.* 2004, University of Michigan: Wiley.

[177] Achten, H. and J. Jessurun. *An Agent Framework for Recognition of Graphic Units in Drawings.* in *20th eCAADe Conference Proceedings: Connecting the Real and the Virtual – design e-ducation.* 2002. Warsaw, Poland.

[178] Parascho, S., et al., Design Tools for Integrative Planning, Delft, The Netherlands, 2013, 2 237–246.

[179] Groenewolt, A., Schwinn, T., Nguyen, L. et al. An interactive agent-based framework for materialization-informed architectural design. Swarm Intell, 2018. 12, pp. 155–186. https://doi.org/10.1007/s11721-017-0151-8

[180] Soibelman, L. and F. Pena-Mora, Distributed multi-reasoning mechanism to support conceptual structural design. Journal of Structural Engineering, 2000. 126(6), pp. 733–742.

[181] Dijkstra, J., H.J. Timmermans, and A. Jessurun, *A multi-agent cellular automata system for visualising simulated pedestrian activity*, in *Theory and Practical Issues on Cellular Automata.* 2001, Springer. pp. 29–36.

[182] Meissner, U., U. Rüppel, and M. Theiss. *Network-Based Fire Engineering Supported by Agents.* in *Proceedings of the Xth International Conference on Computing in Civil and Building Engineering (ICCCBE-2004).* 2004. Weimar, Germany.

[183] Klein, L., et al., Coordinating occupant behavior for building energy and comfort management using multi-agent systems. Automation in Construction, 2012. 22, pp. 525–536.

[184] Anumba, C., et al., A multi-agent system for distributed collaborative design. Logistics Information Management, 2001. 14(5/6), pp. 355–367.

[185] Marcolino, L.S., et al., Agents vote for the environment: Designing energy-efficient architecture, Texas, 2015.

[186] Fabi, V., et al., A methodology for modelling energy-related human behaviour: Application to window opening behaviour in residential buildings. Building Simulation, 2013. 6(4), pp. 415–427.

[187] Abbasi, Y.D., et al. *Human Adversaries in Opportunistic Crime Security Games: Evaluating Competing Bounded Rationality Models.* in *Proceedings of the Third Annual Conference on Advances in Cognitive Systems ACS.* 2015.

[188] Tambe, M., Towards flexible teamwork. Journal of Artificial Intelligence Research, 1997, pp. 83–124.

[189] Mullen, T. and M.P. Wellman, *Some issues in the design of market-oriented agents*, in *Intelligent Agents II Agent Theories, Architectures, and Languages.* 1996, Springer. pp. 283–298.

[190] Jordan, J.S., The exponential convergence of Bayesian learning in normal form games. Games and Economic Behavior, 1992. 4(2), pp. 202–217.

[191] Kavulya, G., D.J. Gerber, and B. Becerik-Gerber. *'Designing in' complex system interaction: Multi-agent based systems for early design decision making.* in *ISARC 2011.* 2011. Seoul, Korea.

[192] Klein, L., et al., Towards optimization of building energy and occupant comfort using multi-agent simulation, 2011,

[193] Bullinger, H.-J., et al., Towards user centred design (UCD) in architecture based on immersive virtual environments. Computers in Industry, 2010. 61(4), pp. 372–379.

[194] Menges, A., Morphospaces of Robotic Fabrication, 2013.

[195] Keating, S.J., et al., Toward site-specific and self-sufficient robotic fabrication on architectural scales. Science Robotics, 2017. 2(5), p. eaam8986.

[196] Gerber, D.J., E. Pantazis, and A. Wang, A multi-agent approach for performance based architecture: Design exploring geometry, user, and environmental agencies in façades. Automation in Construction, 2017. 76, pp. 45–58.

[197] Associates, M.a. *RhinoCommon*. 2019 [cited 2019 03/07/20219]; Rhino Common API]. Available from: https://developer.rhino3d.com/guides/rhinocommon/.

[198] (DOE), U.S.D.o.E. *Energy Plus Documentation*. [cited 2019 01/03/2019]; Available from: https://energyplus.net/documentation.

[199] Preisinger, C., *Karamba3D API Documentation*. 2020; Available from: https://www.karamba3d.com/help/2-2-0/html/b2fe4d67-e7e2-4f96-bc84-ecd423bde1a7.htm.

[200] Pantazis, E. and D.J. Gerber, Design Exploring Complexity in Architectural Shells – Interactive form finding of reciprocal frames through a multi-agent system, Oulu, Finland, 2016, 1 (455–464).

[201] Turrin, M., P. von Buelow, and R. Stouffs, Design explorations of performance driven geometry in architectural design using parametric modeling and genetic algorithms. Advanced Engineering Informatics, 2011. 25(4), pp. 656–675.

[202] Cutler, B., et al., Interactive selection of optimal fenestration materials for schematic architectural daylighting design. Automation in Construction, 2008. 17(7), pp. 809–823.

[203] Shea, K., A. Sedgwick, and G. Antonuntto, *Multicriteria optimization of paneled building envelopes using ant colony optimization*, in *Intelligent Computing in Engineering and Architecture*. 2006, Springer. pp. 627–636.

[204] Felkner, J., E. Chatzi, and T. Kotnik. *Interactive particle swarm optimization for the architectural design of truss structures*. in *Computational Intelligence for Engineering Solutions (CIES), 2013 IEEE Symposium on*. 2013. IEEE.

[205] De Meyer, K., Explorations in stochastic diffusion search: Soft-and hardware implementations of biologically inspired spiking neuron stochastic diffusion networks, 2000.

[206] Nasuto, S. and M. Bishop, Convergence analysis of stochastic diffusion search. Parallel Algorithms And Application, 1999. 14(2), pp. 89–107.

[207] Rivières, J.D. and J. Wiegand, Eclipse: A platform for integrating development tools. IBM Systems Journal, 2004. 43(2), pp. 371–383.

[208] Reas, C. and B. Fry, *Processing: A programming Handbook for Visual Designers and Artists*. 2007, Boston: MIT Press.

[209] Robert McNeel and Associates. *Modeling Tools for Designers*. 2019 [cited February 2019; Available from: www.rhino3d.com.

[210] Preisinger, C. and M. Heimrath, Karamba – A toolkit for parametric structural design. Structural Engineering International, 2014. 24(2), pp. 217–221.

[211] Johannes Braumann, S.B.-C. *Parametric Robot Control: Integrated CAD/CAM for Architectural Design*. in *ACAADIA 2011: Integration through Computation*. 2011.

[212] Shea, K., R. Aish, and M. Gourtovaia, Towards integrated performance-driven generative design tools. Automation in Construction, 2005. 14(2), pp. 253–264.

[213] Radford, A.D. and J.S. Gero, Tradeoff diagrams for the integrated design of the physical environment in buildings. Building and Environment, 1980. 15(1), pp. 3–15.

[214] Andréen, D., et al., Emergent Structures Assembled by Large Swarms of Simple Robots, Ann Arbor, MI, 2016, 54–59

[215] Kalman, R.E., A new approach to linear filtering and prediction problems. Journal of Basic Engineering, 1960. 82(1), pp. 35–45.

[216] Bradski, G. and A. Kaehler, *Learning OpenCV: Computer vision with the OpenCV library*. 2008: O'Reilly Media, Inc.

[217] Bechthold, M., et al. *Integrated environmental design and robotic fabrication workflow for ceramic shading systems*. in *28th International Association for Automation and Robotics in Construction (ISARC)*. 2011. Seoul, Korea.

[218] Reinhart, C.F., J. Mardaljevic, and Z. Rogers, Dynamic daylight performance metrics for sustainable building design. Leukos, 2006. 3(1), pp. 7–31.

[219] Ander, G.D., *Daylighting performance and design*. 2003: John Wiley & Sons.

[220] Phillips, D., *Daylighting: natural light in architecture*. 2004, Burlington, MA: Elsevier, Architectural Press.

[221] Pawlofsky, T., et al., *Brickolage*, in *Stereotomy: Stone Architecture and New Research*, G. Fallacara and C. D'Amato, Editors. 2012, Presses des Ponts: Paris, France.

[222] Heydarian, A., et al., Immersive virtual environments versus physical built environments: A benchmarking study for building design and user-built environment explorations. Automation in Construction, 2015. 54(0), pp. 116–126.

[223] Heydarian, A., et al., Immersive virtual environments, understanding the impact of design features and occupant choice upon lighting for building performance. Building and Environment, 2015. 89, pp. 217–228.

[224] Clevenger, C.M. and J. Haymaker. *Frameworks and metrics for assessing the guidance of design processes*. in *ICED'09*. 2009. Stanford, CA, USA.

[225] Reas, C., *Processing: a programming handbook for visual designers and artists*. Vol. 6812. 2007, Cambridge, MA: MIT Press.

[226] *Shell Structures for Architecture: Form Finding and Optimization*, ed. S. Adriaenssens, et al. 2014: Taylor & Francis – Routledge.

[227] Pottmann, H., et al., Architectural geometry. Computers & Graphics, 2015. 47, pp. 145–164.

[228] Chilton, J. and H. Isler, *Heinz Isler*. The Engineer's Contribution to Contemporary Architecture. 2000, London: Thomas Telford Publishing 168.

[229] Kilian, A., Design innovation through constraint modeling. International Journal of Architectural Computing, 2006. 4(1), pp. 87–105.

[230] Piker, D., Kangaroo: form finding with computational physics. Architectural Design, 2013. 83(2), pp. 136–137.

[231] Block, P. and J. Ochsendorf, Thrust network analysis: A new methodology for three-dimensional equilibrium. International Association for Shell and Spatial Structures, 2007. 155, p. 167.

[232] Van Mele, T., et al., Geometry-based understanding of structures. Journal of the International Association of Shell and Spatial Structures, 2012. 53(4), pp. 285–295.

[233] Pizzigoni, A., Leonardo & The Reciprocal Structures. 2010.

[234] Larsen, O.P., *Reciprocal frame architecture*. 2008: Routledge.

[235] Kohlhammer, T. and T. Kotnik, Systemic behaviour of plane reciprocal frame structures. Structural Engineering International, 2011. 21(1), pp. 80–86.

[236] Thomas Kohlhammer, T.K., Systemic Behaviour of Plane Reciprocal Frame Structures. Structural Engineering International, 2010, pp. 80–86.

[237] Pugnale, A., et al. *The principle of structural reciprocity: history, properties and design issues*. in *The 35th Annual Symposium of the IABSE 2011, the 52nd Annual Symposium of the IASS 2011 and incorporating the 6th International Conference on Space Structures*. 2011.

[238] Nabaei, S.S. and Y. Weinand, Geometrical description and structural analysis of a modular timber structure. International Journal of Space Structures, 2011. 26(4), pp. 321–330.

[239] Tveit, P., Considerations for Design of Network Arches. Journal of Structural Engineering, 1987. 113(10), pp. 2189–2207.

[240] Weinand, Y., Innovative Timber Constructions, London, Great-Britain, 2011, CD.

[241] Chilton, J., Heinz Isler's Infinite Spectrum: Form-Finding in Design. Architectural Design, 2010. 80(4), pp. 64–71.

[242] Isler, H. *New Shapes for Shells*. in *International Association for Shell and Spatial Structures*. 1959. Madrid: North-Holland.

[243] Chilton, J. and C.-C. Chuang, Rooted in nature: aesthetics, geometry and structure in the shells of Heinz Isler. Nexus Network Journal, 2017. 19(3), pp. 763–785.

[244] Ostermeyer, Y., et al., Building inventory and refurbishment scenario database development for Switzerland. Journal of Industrial Ecology, 2018. 22(4), pp. 629–642.

[245] Harding, J., Dimensionality Reduction for Parametric Design Exploration, Zurich, Switzerland, 2016, 1, 275–287.

[246] Kohonen, T., The self-organizing map. Proceedings of the IEEE, 1990. 78(9), pp. 1464–1480. doi: 10.1109/5.58325.

[247] Schwinn, T. and A. Menges, Fabrication Agency: Landesgartenschau Exhibition Hall. Architectural Design, 2015. 85(5), pp. 92–99.

Index

https://doi.org/10.1515/9783110797435-008

www.ingramcontent.com/pod-product-compliance
Lightning Source LLC
Chambersburg PA
CBHW061357210326

41598CB00035B/6009